AF342184

IUTAM SYMPOSIUM ON FLOW CONTROL AND MEMS

IUTAM BOOKSERIES
Volume 7

Series Editors

G.L.M. Gladwell, *University of Waterloo, Waterloo, Ontario, Canada*
R. Moreau, *INPG, Grenoble, France*

Editorial Board

J. Engelbrecht, *Institute of Cybernetics*, Tallinn, Estonia
L.B. Freund, *Brown University, Providence, USA*
A. Kluwick, *Technische Universität, Vienna, Austria*
H.K. Moffatt, *University of Cambridge, Cambridge, UK*
N. Olhoff *Aalborg University, Aalborg, Denmark*
K. Tsutomu, *IIDS, Tokyo, Japan*
D. van Campen, *Technical University Eindhoven, Eindhoven, The Netherlands*
Z. Zheng, *Chinese Academy of Sciences, Beijing, China*

Aims and Scope of the Series

The IUTAM Bookseries publishes the proceedings of IUTAM symposia under the auspices of the IUTAM Board.

For a list of books published in this series, see final pages.

IUTAM Symposium on Flow Control and MEMS

Proceedings of the IUTAM Symposium
held at the Royal Geographical Society,
19–22 September 2006, hosted by
Imperial College, London, England

Edited by

J.F. MORRISON

*Department of Aeronautics, Imperial College,
London, England*

D.M. BIRCH

*Department of Aeronautics, Imperial College,
London, England*

and

P. LAVOIE

*Department of Aeronautics, Imperial College,
London, England*

 Springer

A C.I.P. Catalogue record for this book is available from the Library of Congress.

ISBN 978-1-4020-6857-7 (HB)
ISBN 978-1-4020-6858-4 (e-book)

Published by Springer,
P.O. Box 17, 3300 AA Dordrecht, The Netherlands.

www.springer.com

Printed on acid-free paper

All Rights Reserved
© 2008 Springer
No part of this work may be reproduced, stored in a retrieval system, or transmitted
in any form or by any means, electronic, mechanical, photocopying, microfilming, recording
or otherwise, without written permission from the Publisher, with the exception
of any material supplied specifically for the purpose of being entered
and executed on a computer system, for exclusive use by the purchaser of the work.

Table of Contents

MEMS DEVICES

SYNTHETIC JETS

SEPARATION CONTROL

DRAG REDUCTION AND MIXING

PASSIVE CONTROL

Introduction

Recent advances in technology for the fabrication, in bulk, of small sensors and actuators have enabled the use of these devices for flow control. It is probably true to say that our understanding of many aspects of fluid flow is sufficiently mature for there now to be ways in which it may be exploited for the technologically important area of active flow control. In this application, the use of MEMS (Micro-Electro-Mechanical Systems, or microstructures) is still in its infancy. Such devices are especially useful in turbulent flows found in engineering where much important information resides in 'small' eddies near surfaces. Similarly, the application of modern control theory to the distributed control of fluid flow has been exploited by only a few researchers in fluid mechanics. The design of a robust, distributed controller that is applicable to even a single specific flow-control problem is still some way off. Closed-loop control demonstration experiments are still lacking and there is a particular need for, and sharing of, proof-of-concept simulations and experiments.

The principal aim of the Symposium was to bring together many of the world's experts in fluid mechanics, control theory and microfabrication to discover the synergy that can lead to real advances and perhaps find ways in which collaborative projects may proceed. Industrial participants could expect to have direct access to world-leading practitioners across these disciplines and to be brought right up to date with the latest developments. Correspondingly, academic workers could expect to be exposed to 'real-world' problems. One session was devoted to applications of open- and closed-loop control to problems in both internal and external aerodynamics. The purpose of this session is to identify potential solutions to industry-specific problems. A further session was devoted to presentations of results from the *2nd European Forum on Flow Control*, April–June 2006, held at the University of Poitiers.

The meeting attracted approximately 120 participants from UK (50), France (28), Germany (6), Italy (1), USA (20), Australia (3), Israel (2), Canada (1), Switzerland (1), Spain (1), Sweden (1), China & Hong Kong (3), India (1) and Korea (1). Of these, approximately 12 participants came from the aeronautical and automotive industries.

Ten keynote talks were given on a variety of topics stemming from active flow control experiments and simulations to fundamental design issues concerning MEMS. They were:

- Miki Amitay (RPI, USA) "Synthetic jets and their applications for fluid/thermal systems".
- Thomas Bewley (UCSD, USA) "Multiscale retrograde identification, estimation and forecasting of chaotic nonlinear systems".
- Kenneth Breuer (Brown University, USA) "Models for adaptive feedforward control of turbulence".
- Haecheon Choi (Seoul National University, Korea), "Active and passive controls for form drag reduction".
- Mike Gaster (Queen Mary, London, UK) "Active control of laminar boundary layers disturbances".
- Mark Glauser (Syracuse University, USA) "Low-dimensional tools for closed-loop flow control in high Reynolds number turbulent flows".
- Dan Henningson (KTH, Sweden) "Model reduction and control of a cavity-driven separated boundary layer".
- John Kim (UCLA, USA) "Physics and control of turbulent boundary layers".
- Philippe Pernod (IEMN-LEMAC, France), "MEMS for flow control: Technological facilities and MMMS alternatives".
- Mark Spearing (University of Southampton, UK) "High power density MEMS: Materials and structures requirements".

A total of 32 oral presentations were given, together with 28 poster presentations. This volume provides written papers of nearly all oral and poster presentations.

In consultation with the Scientific Committee, and more generally with the Symposium at large, responses to two questions were sought:

- *Achievements to date – where are we in effective flow control?*; and
- *What are the remaining most important challenges?*

Responses were grouped into five groupings: sensors, actuators, flow definition, drag reduction and separation control.

- Sensors. Only one paper was offered concerning wall sensors (wall shear). This is perhaps surprising given their importance. Thermal sensors for measuring wall shear stress remain popular, despite being nonlinear and the inherent limitations to frequency response set by heat loss to the substrate. By comparison, techniques for the measurement of wall pressure are at a better stage of development, sensors are more robust and nearly linear. Typically the rms wall pressure is 10–20 times the rms wall shear stress. Development is still required for both and some key questions arose, such as accuracy and noise. For example, with filtering, how much freedom does a robust controller offer?
- Actuators. Many achievements to date such as Zero-Net-Mass-Flux (ZNMF) jets are built on silicon (bimorph, piezoelectric, small deflection, high frequency) and the semi-conductor industry. There was a strong focus on ZNMF jets, but are these necessarily the best for all control problems? There are potentially many different other types of actuator, and many innovations in new materials (e.g. polymers, C nanotubes – with/without doping, composites). Pernod introduced Magneto-Mechanical Microsystems (MMMs): here there some issues regarding

instabilities and/or non-continuum effects. In summary, it seems that the fluids community needs a better appreciation of what is available, and there are outstanding issues regarding the provision of cost-effective MEMS with a quick turn around.

- Flow Definition. We have a good understanding on how to apply modern control theory to fluids mechanics, and linear control theory seems promising. Key questions are:

  1. what is the minimum information required for flow control – density and location of sensors?
  2. Merits of blackbox vs. 'intelligent' control?
  3. How should a cost function be best defined?
  4. Need for better model reduction: smaller state-space models (controllability, observability are key); incorporation of better, and/or distributions of, sensors/actuators.

- Drag Reduction. We understand the fluid mechanics fairly well – but largely at low Reynolds number ('bottom-up'). It is not all clear that the fundamental processes at high Reynolds number are intrinsically the same ('top-down') – what are the implications for flow control? Bewley stressed the importance of overlapping/decentralised controllers (fast~local, slow~non-local) and practical problems require issues of realizability to be addressed.
- Separation Control. This is probably the goal that is closest to application in a real system. Different types of actuator (or even variations on the same basic design) may all achieve separation delay even though the actuator may induce different flow physics. This may enable a more straightforward design (fewer parameters) and permit a greater emphasis on other considerations (e.g. robustness). For closed-loop control, optimum design requires coupled actuator-algorithm design from the start: e.g. shear-layer response time depends on actuator speed.

In terms of applications, much of the focus and investment is on the aeronautical sector, while, in fact, both marine and automotive sectors offer vast energy savings. John Kim pointed out that worldwide ocean shipping consumes 2.1 million barrels of oil *per annum* whereas the airline industry only uses 1.5 million. It is therefore somewhat ironic that several effective methods are known for reducing skin-friction drag in water flows but few work in air. However, it suggests that more investment should be targeted towards drag reduction of ships and road vehicles.

However, the ACARE 2020 targets have largely been adopted by the European airline industry. The challenge of achieving 50% reduction in fuel burn implies a wing/fuselage drag reduction of about 20%, an improvement in engine efficiency of about 20%, with the remainder coming from improved traffic management. In theory, arrays of microjets, dimples, pimples or other actuators combined with suitable sensors and control systems could produce substantial reductions in drag. Whether this is possible or not remains an open question. An estimate for the number of sublayer streaks present at any one time on the fuselage of an Airbus A340-300 in cruise is 10^9, and shows the scale of the problem for active control. Clearly, advances in the application of model reduction techniques to wall turbulence are an essential prerequisite before any sophisticated control technique involving cost functions and adjoint equations can be used. Fundamental differences between the behaviour of boundary layers at operational Reynolds numbers and the low Reynolds numbers at which control schemes have showed some success have yet to be addressed.

It is likely that only open-loop methods for turbulent skin-friction reduction that do not require a control system are likely to be feasible for practical application by 2020. The only such methods currently known are spanwise oscillations, randomized roughness and riblets.

It is clear that the industry/academe divide remains: the horizons needed by the aeronautical industries are far too short for what is expected. However, in Europe, environmental issues constitute a significant driver for research funding. But there is a need to encourage mechanisms for discipline crossover/hopping. Moreover, the fluids community needs to engage with MEMS and control people.

Jonathan Morrison
Jean-Paul Bonnet

Acknowledgements

The Scientific Committee is indebted to IUTAM for the provision of financial support to some of the invitees in order that they might attend the Symposium. Additional sponsorship was also provided by ERCOFTAC, Airbus, QinetiQ, CCLRC, the Wing Technologies Centre, Department of Aeronautics, Imperial College and the Turbulence Platform Grant, Department of Aeronautics, Imperial College.

Scientific Committee

Peter Bearman (Imperial College)
Thomas Bewley (UCSD USA)
Jean-Paul Bonnet (Poitiers) Co-chair
Kenneth Breuer (Brown University, USA)
Peter Carpenter (University of Warwick, UK)
Kwing-So Choi (University of Nottingham, UK)
Hans Fernholz (Technical University of Berlin, Germany)
Mark Glauser (Syracuse University, USA)
John Kim (University of California, USA)
Michael Leschziner (Imperial College)
David Limebeer (Imperial College)
Beverley McKeon (CalTech, USA)
Jonathan Morrison (Imperial College) Co-chair
Andrew Pollard (Queen's University, Canada)
Sedat Tardu (L.E.G.I, France)

Local Organization

Jonathan Morrison
Tayo Nong

MEMS DEVICES

High Power Density MEMS: Materials and Structures Requirements

S. Mark Spearing
*School of Engineering Sciences, University of Southampton, Highfield,
Southampton SO17 1BJ, U.K.; E-mail: spearing@soton.ac.uk*

Abstract The materials and structures issues associated with high power density Microelectromechanical Systems (MEMS) are addressed. An overview of projects conducted at MIT and the key physical challenges affecting their performance goals is provided. These goals are translated into requirements for the materials and structural design. Particular emphasis is provided on the approaches taken to designing with brittle materials at low temperatures and designing with ductile creeping materials at high temperature. A novel approach to structural reinforcement using SiC in combination with Si is introduced.

Key words: Microelectromechanical systems, materials, structural design.

1 Introduction

Since 1994 several projects at the Massachusetts Institute of Technology have been conducted to design and build high power density microelectromechanical systems (MEMS), including a micro-gas turbine engine [1, 2], micro-rocket [3, 4], micro-hydraulic transducer [5] and a micro-solid oxide fuel cell [6]. These are graphically illustrated in Figure 1. All of these projects have had the goal of using microfabrication technology to create devices capable of producing useful power (1–10 W) from packaged volumes on the order of 1 cm^3. The overarching idea behind these projects is that, if such devices could be realized with reasonable efficiencies, and with power densities approaching those of large scale prime movers, and if they could be microfabricated such that many devices are produced in parallel, then the economic scaling enjoyed by other microfabricated products would apply and the unit cost would be very low. This in turn would open up opportunities in fields as diverse as: portable electrical power, vehicle propulsion, local cooling and distributed flow control. At this stage no fully functioning operating device has been produced which meets the overall design objectives. However considerable progress has been

J.F. Morrison et al. (eds.), IUTAM Symposium on Flow Control and MEMS, 3–13.
© 2008 *Springer. Printed in the Netherlands.*

Fig. 1 Pictures of the various microsystems and major subcomponents developed at MIT since 1994. Clockwise from top left: (a) micro-gas turbine [1], (b) micro-hydraulic transducer [5], (c) micro-solid oxide fuel cell [6] and (d) micro-rocket [3].

made, and the field is at a point where the main engineering challenges are understood and are being addressed. There is potentially commonality with requirements for microfabricated devices for flow control systems.

In order to understand the scope of the challenges to realizing working high power density MEMS, it is important to understand the primary physical requirements associated with achieving power densities in excess of 1 W/cm^3 from such devices. The following items particularly apply to the micro gas turbine engine. However the implications are common to most high power density microsystems: (1) high peak cycle temperatures (1600–2500 K), since the efficiency and power density scale with the temperature difference across the cycle. (2) High fluidic speeds ($\sim$500 m/s) since the rotor power density is proportional to the square of the tip Mach number. (3) Low friction bearings. (4) Reasonable component efficiencies in order to close the thermodynamic cycle and result in an acceptable overall system efficiency. Requirement (1) translates into a requirement for high temperature materials, and the design of an effective combustor [7, 8]. Requirement (2) translates into a requirement for an effective fluidic design [9] and a requirement for high strength materials. The stress levels in a rotor scale with the material density and the square of the rim speed, which means that the achievable power density is proportional to the allowable stress level. This equivalence of power and stress is fundamental to power producing machines and is probably the most challenging aspect of realizing high power density MEMS. Requirement (3) has the consequence that high power density MEMS devices typically employ gas bearings [10, 11] in preference to rolling elements with liquid lubricants, which would more commonly

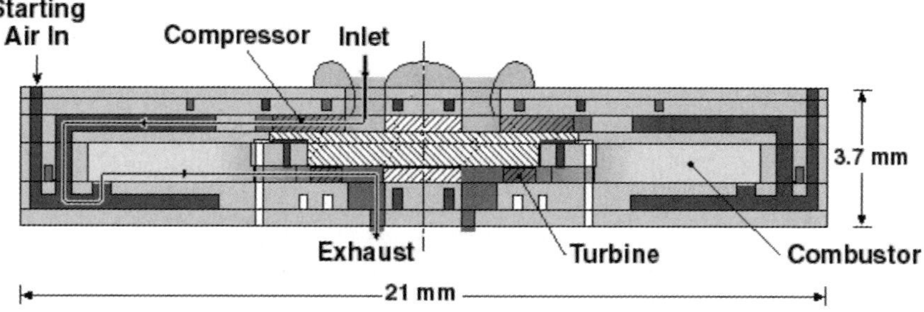

Fig. 2 Cross-section of the baseline microengine design.

be used in macroscale devices. Considerable work has been conducted on developing the design tools necessary to achieve bearings with the necessary stability and load carrying capacity. Requirement (4) as well as the demands of (3) implies a requirement for close tolerances (better than 1 μm) on blade tip clearances and bearing gaps. In addition to the requirements for efficiency of the mechanical components, there is a significant challenge to develop efficient electrical machines, which has resulted in the exploration of electrostatic induction generators [12] and motors as well as more conventional electromagnetic machines [13]. At the time of writing, the electromagnetic machines appear to be more promising, although their microfabrication is considerably more complex. Figure 2 shows a cross-section through the current baseline micro-gas turbine engine, which represents the simplest possible design that can achieve thermodynamic "break-even". The device shown in Figure 2 is sized to have the following performance metrics: a thrust of 0.1 N, a fuel consumption of 16 g/hr of hydrogen, a weight of 2 g, a turbine inlet temp of 1600 K, a rotor speed of 1.2×10^6 RPM and an exhaust gas temp of 1240 K. The initial demonstration device is entirely fabricated using silicon. This results in relatively poor performance for the engine, due to the temperature limitations of silicon. However, it exploits relatively mature microfabrication technology. The key process steps are the extensive use of high precision deep reactive ion etching, and wafer bonding to create the three dimensional structures. The bond lines can clearly be seen in Figure 2. Figure 3 shows a cross-section through a fabricated device. Figure 1a shows a die with the required fluidic connections. Packaging is a very important consideration for such devices and is covered in other references, e.g. [14].

Notwithstanding the significant challenges associated with other aspects of the mechanical and electrical design of high power density MEMS, it is apparent that the interrelated topics of materials, materials processing and structural design represent a common theme that directly affects each of the four principal physical requirements. The remainder of this paper will address the progress that has been made in achieving the necessary design requirements for the materials and structural design of high power density microsystems.

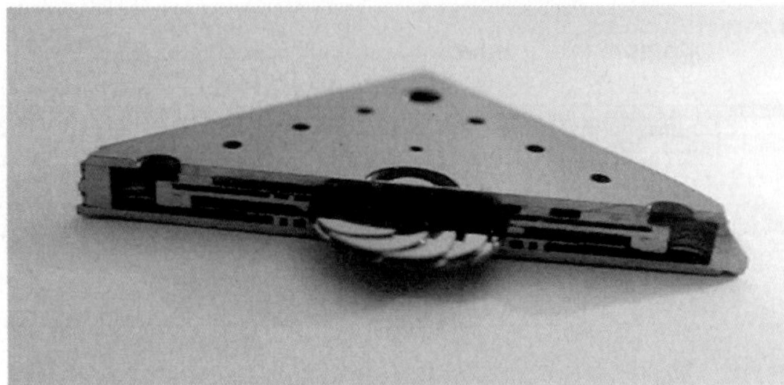

Fig. 3 Cross-section of a fabricated microengine turbocharger. The rotor diameter is 8 mm.

2 Low Temperature Structural Design

Material selection for the microengine was driven largely by the need to create a working device within strict time constraints. In order to achieve this, a conscious decision was taken to minimize programmatic risk associated with introducing new materials and developing new processes. As a result, the first generation, demonstration engine has been largely made out of Si. Nevertheless the material indices for microfabricated Si compare quite well with those for the metal alloys used in macroscale gas turbine power plant. The relatively low density of silicon and its high strength and stiffness make it an attractive material for rotating structures. Even its relatively poor temperature capability can be accommodated with careful design.

2.1 Mechanical Testing

Silicon is a brittle material and design for room temperature operation must account for the stochastic aspects of its strength. Axisymmetric flexural specimens were preferred for this purpose, in which the specimen is supported around its edge and loaded centrally [15]. A modified micro-hardness indenter was found to be very useful for this purpose. Specimens consisted of a square die, 10 mm × 10 mm, which were placed over a hole of diameter 5–8 mm. A load was applied centrally and the load at fracture recorded. Statistically significant data were obtained and fitted using a Weibull probability density function; a reference strength of 4.6 GPa and a Weibull modulus of 3.3 was obtained.

A concern for the design was the introduction of flaws by the deep reactive ion etching process. High surface roughness was observed at the intersections between horizontal surfaces and vertical etched walls. Since these locations also represent stress concentrations, locally reduced strength was of particular concern. To investigate this issue, a novel test specimen was developed: the radius-hub flexure specimen, which consists of a central hub defined by deep etching. The nominal stress

Table 1 Strength data from radius hub flexure tests.

	STS DRIE	DRIE + Wet Isotropic Etch	DRIE + SF6 Plasma Etch
Specimen Size	20	16	18
Polishing Thickness	NA	1.8 mm	2.7 mm
Reference Strength, σ_0	1.45 GPa	3 GPa	4 GPa
Weibull Modulus, m	7.5	3.9–5.7	3–6

concentration at the interface between the hub wall and the horizontal surface is calculated based on the nominal radius of the fillet and used to define a nominal local strength. Data from such tests is shown in Table 1. The effect of the locally increased roughness is pronounced. The reference strength obtained from STS DRIE specimens drops from 4.6 to 1.45 GPa. However, use of a secondary, "smoothing" etch is shown to be very effective in recovering much of this strength loss. An isotropic SF_6 plasma etch recovers the local strength to 4 GPa. Such a strength recovering etch is essential if microengine rotating parts are to be able to achieve the high speeds required for power production.

2.2 Structural Design

The strength data obtained using planar and radiused hub flexure specimens can be used to generate allowables for structural design. Finite element models were used to obtain stress distributions in the rotor. Figure 4 shows stress contours from a global model and the local detail of a turbine blade trailing edge. The stochastic properties of Si were introduced into the design using the probabilistic design code CARES [16] which applies the experimentally-determined Weibull probability density function at the integration points of the finite element model. This process allows for an assessment of the structural failure probability of the whole rotor. Results from such analysis can be presented graphically to allow design trades to be conducted. Figure 5 shows an example in which the effects of strength and blade height on the failure probability are compared. As a practical matter, true probabilistic design is rather unwieldy, so design was usually conducted to an allowable stress (typically 1 GPa) and then the failure probability was calculated to check the estimated reliability. It should also be noted that given the very low toughness of silicon the failure probability are heavily influenced by the introduction of flaws during processing and handling. This demands conservative design and care during fabrication.

3 Design for High Temperature Strength

Given the requirement for high temperature operation of the turbine spool, design for high temperature strength is of paramount importance. Silicon has not been used

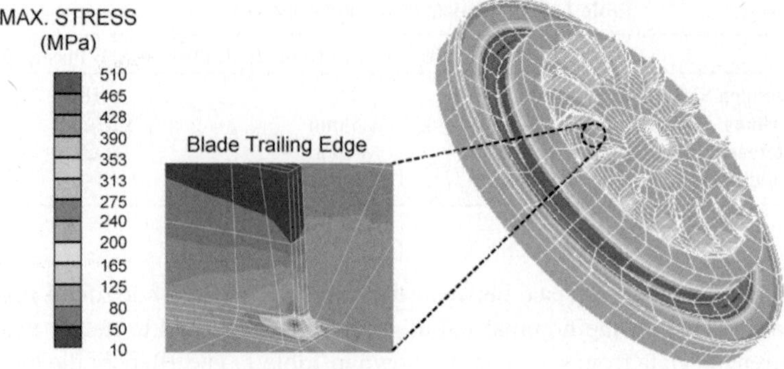

Fig. 4 Turbine stress contours obtained by a global/local finite element analysis.

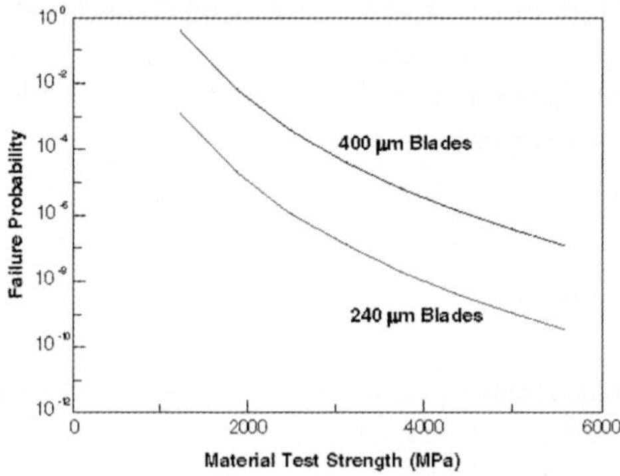

Fig. 5 Failure probabilities as a function of blade height and material test strength.

previously for high temperature, high stress applications, and so a constitutive model has been developed.

3.1 Material Model

The material model was based on an existing isotropic plasticity model. A complete description of the model is provided elsewhere [17].

The model is calibrated by uniaxial creep experiments using data sets such as those shown in Figure 6. Data was obtained from uniaxial compression tests conducted at two temperatures (600 and 800°C) and four stress levels at each temperature.

Once calibrated the model was validated by using it to predict the response of the monotonic loading of four point bend tests. Data for specimens tested at three temperatures and two loading rates are shown in Figure 7. The model provides a

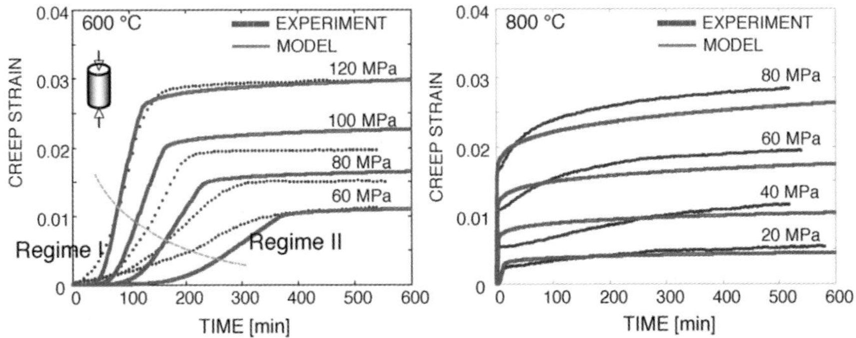

Fig. 6 Uniaxial creep calibration of the creep-plasticity model for silicon.

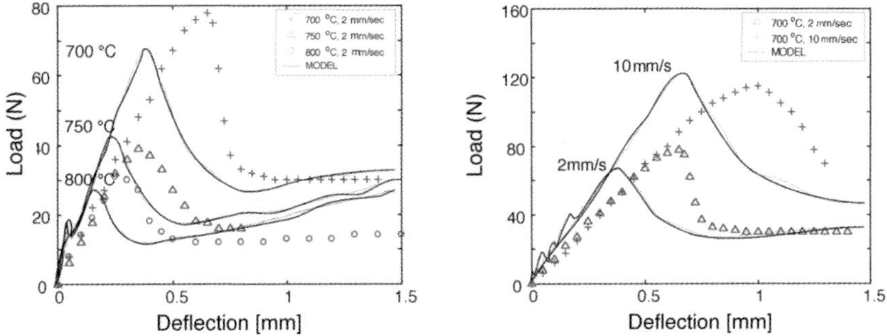

Fig. 7 Model validation in 4 point bending under constant displacement rate loading.

reasonable description of the data, thus confirming its validity for use in the structural design process.

4 Use of Silicon Carbide in Hybrid Structures

Given the temperature restrictions of silicon it is apparent that it is not an ideal material with which to construct high performance microturbomachinery. To this end more refractory materials have been assessed for use.

4.1 Mechanical Considerations in SiC/Si Hybrid Structures

The tolerances on the design of the bearings, in terms of defining the rotating gap and the mass balance of the turbine/compressor spool place severe limitations on the material options available. They stretch the limits of lithographic patterning, and greatly exceed those achievable by conventional machining, moulding or casting processes. This implies the continued use of lithographic patterning and etching or deposition processes. However, the candidate materials, chiefly refractory ceramics are not generally amenable to such processes. Etch rates are low, and the degree of anisotropy and control that can be achieved are very limited. With this background, a hybrid approach has been pursued that allows the continued use of silicon micro-fabrication, but provides structural reinforcement using silicon carbide. A detailed description of this approach can be found in reference [17]. The silicon carbide has a much higher stiffness (460 GPa Young's modulus vs. 165 GPa for silicon) and retains its strength to temperatures in excess of 1600 K. This means that the silicon carbide core of the spool carries the majority of the centrifugal load of the spool. Finite element analysis indicates that operation up to 1150K is achievable with the use of 30% SiC through the spool thickness. This has a significant effect on overall engine efficiency, as it can cut the heat flux from the turbine to the compressor to 5% of the all silicon case.

4.2 Process Considerations in SiC/Si Hybrid Structures

The use of silicon carbide reinforcement can be achieved with relative simple pro-cesses that are compatible with silicon microfabrication. These are illustrated in Figure 8. There are four key steps:

1. A circular pit is etched in a silicon wafer.
2. The pit is filled with chemical vapour-deposited silicon carbide.
3. The overburden of excess SiC is removed by grinding and polishing.
4. A second silicon wafer is bonded over the Si/SiC surface of the first wafer.

The bonded wafer pair moves through the rest of the process flow as though it were a monolithic silicon wafer. All the key dimensions (blades, bearings) are defined by silicon etching, which is well understood and can be controlled to the necessary tolerances.

4.3 Mechanical Test Results

The concept of Si/SiC structures has been partially verified by testing Si/SiC beams in four point bending. Results are shown in Figure 9. The specimens were created simply by depositing SiC on Si and cutting them to size with a die saw. The spe-cimens were then loaded in four point bending. Figure 9 shows that strength is re-tained at temperatures as high as 900°C (1173 K). FE predictions are superimposed on the data, further validating the Si Constitutive law referred to in Section 3. Hav-ing validated the concept and the model, the model has been used to guide trades

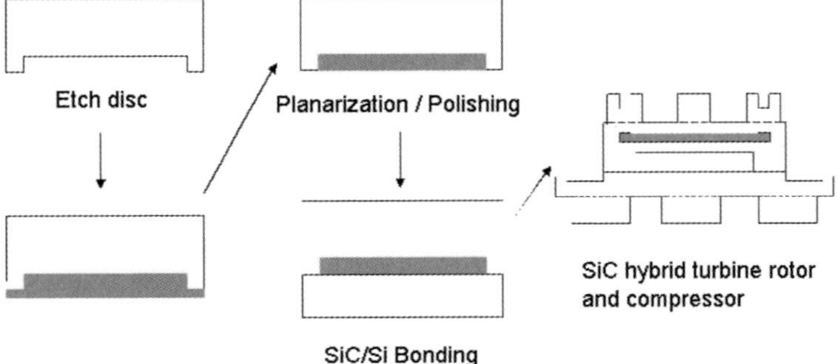

Fig. 8 Schematic of process steps required to make a Si/SiC hybrid spool.

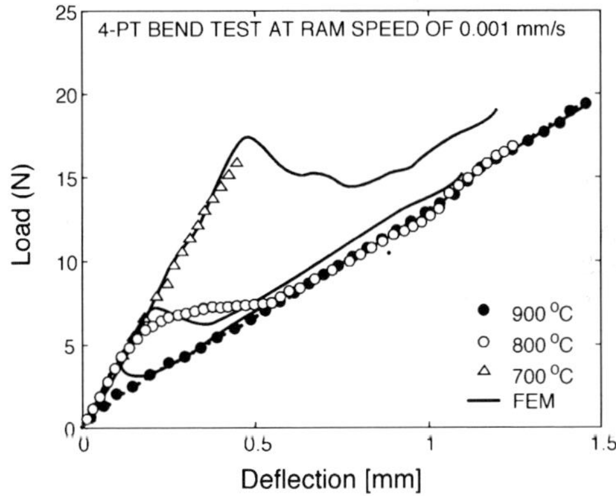

Fig. 9 Model-experiment comparison of Si/SiC hybrid beams in 4pt bending.

in the design of actual microengine rotors and also to make programmatic decisions regarding the pursuit of the various material and fabrication options available.

5 Summary

This paper has provided an overview of material characterization and structural design issues associated with the MIT microengine concept. Key principles are: (1) The use of Weibull statistics and micro-mechanical testing to provide design guidance for low temperature structures. (2) The development of a high temperature constitutive model for Si, and its validation against material test data. (3) The use of Si/SiC hybrid structures in order to increase the allowable operating temperatures while minimizing the deviation from Si processes.

Acknowledgments

Funding for the Power MEMS projects described herein has come from various sources under various grants. The support of the US Army Research Office and DARPA are gratefully acknowledged. The work described herein has been enabled by interactions with very many collaborators at MIT and elsewhere. These are gratefully acknowledged. The author holds a Royal Society-Wolfson Research Merit Award.

References

1. A. H. Epstein, Millimeter-scale, micro-electro-mechanical systems gas turbine engines, *J. Eng. Gas Turbines and Power, Trans. ASME* **126**(2), 2004, 205–226.
2. K-S. Chen, S. M. Spearing and N. N. Nemeth, Structural design of a silicon micro-turbo generator, *AIAA Journal* **39**(4), 2001, 720–728.
3. A. P. London, A. H. Epstein and J. L. Kerrebrock, High-pressure bipropellant microrocket engine, *J. Propulsion and Power* **17**(4), 2001, 780–787.
4. A. P. London, A. A. Ayón, A. H. Epstein, S. M. Spearing, T. Harrison, Y. Peles and J. L. Kerrebrock, Microfabrication of a high pressure bipropellant rocket engine, *Sensors and Actuators, Part A, Physical* **92**, 2001, 351–357.
5. D. C. Roberts, H-Q. Li, J. L. Steyn, O. Yaglioglu, S. M. Spearing, M. A. Schmidt and N. W. Hagood, A piezoelectric microvalve for compact high frequency high differential pressure micropumping systems, *J. MEMS* **12**(1), 2003, 81–92.
6. C. D. Baertsch, K.F. Jensen, J. L. Hertz, H. L. Tuller, V. T. Srikar T. Vengallatore, S. M. Spearing and M. A. Schmidt, Fabrication and structural characterization of self-supporting electrolyte membranes for a micro solid-oxide fuel cell, *J. Materials Research* **19**(9), 2004, 2604–2615.
7. C. M. Spadaccini, A. Mehra, J. Lee, X. Zhang, S. Lukachko and I. A. Waitz, High power density silicon combustion systems for micro gas turbine engines, *J. Eng. Gas Turbine and Power, Trans. ASME* **125**(3), 2003, 709–719.
8. C. M. Spadaccini, J. Peck and I. A. Waitz, Catalytic combustion systems for microscale gas turbine engines, *J. Eng. Gas Turbine and Power, Trans. ASME* **129**(1), 2007, 49–60.
9. L. G. Frechette, S. A. Jacobson, K. S. Breuer, F. F. Ehrich, R. Ghodssi, R. Khanna, C. W. Wong, X. Zhang, M. A. Schmidt and A. H. Epstein, High-speed microfabricated silicon turbomachinery and fluid film bearings, *J. Microelectromechanical Systems* **14**(1), 2005, 141–152.
10. C. W. Wong, X. Zhang, S. A. Jacobson and A. H. Epstein, A self-acting gas thrust bearing for high-speed microrotors, *J. Microelectromechanical Systems* **13**(2), 2004, 158–164.
11. L. X. Liu, C. J. Teo, A. H. Epstein and Z. S. Spakovszky, Hydrostatic gas journal bearings for micro-turbomachinery, *J. Vibrations and Acoustics, Trans. ASME* **127**(2), 2005, 157–164.
12. S. E. Nagle, C. Livermore, L. G. Frechette, R. Ghodssi and J. H. Lang, An electric induction micromotor, *J. Microelectromechanical Systems* **14**(5), 2005, 1127–1143.
13. D. P. Arnold, S. Das, J. W. Park, I. Zana, J. H. Lang and M. G. Allen, Microfabricated high-speed axial-flux multiwatt permanent-magnet generators – Part II: Design, fabrication, and testing, *J. Microelectromechanical Systems* **15**(5), 2006, 1351–1363.
14. Y. Peles, V. T. Srikar, T. A. Harrison, A. Mracek and S. M. Spearing, Fluidic packaging of microengine and microrocket devices for high pressure and high temperature operation, *J. Microelectromechanical Systems* **13**(1), 2004, 31–40.
15. K-S. Chen, A. Ayon and S. M. Spearing, Controlling and testing the fracture strength of silicon at the mesoscale, *J. Am. Ceram. Soc.* **83**(6), 2000, 1476–1484.

16. K-S. Chen, S. M. Spearing and N. N. Nemeth, Structural design of a silicon micro-turbo generator, *AIAA Journal* **39**(4), 2001, 720–728.
17. H. S. Moon, D. Choi and S. M. Spearing, Development of Si/SiC hybrid structures for elevated temperature micro-turbomachinery, *J. MEMS* **13**(4), 2004, 676–687.

MEMS for Flow Control: Technological Facilities and MMMS Alternatives

Philippe Pernod, Vladimir Preobrazhensky, Alain Merlen, Olivier Ducloux, Abdelkrim Talbi, Leticia Gimeno and Nicolas Tiercelin
Joint European Laboratory LEMAC: Institute of Electronics, Microelectronics and Nanotechnology (IEMN, CNRS), Laboratory of Mechanics of Lille (LML, CNRS), Wave Research Center (WRC, GPI, RAS), Ecole Centrale de Lille, BP 48, 59651 Villeneuve d'Ascq Cédex, France; E-mail: philippe.pernod@iemn.univ-lille1.fr

Abstract An introduction of MEMS in general is made. Then the paper provides an overview of essential MEMS devices already elaborated for different problems of flow control in aeronautics. In the last part, attention is focused on solutions based on Micro-Magneto-Mechanical Systems (MMMS).

Key words: MEMS, Micro-Magneto-Mechanical Systems (MMMS), flow control, microvalves, microjets.

1 Introduction

Micro-Electro-Mechanical Systems (MEMS) refer to micrometric-millimetric devices, integrating electronics with mechanical components and fabricated using integrated circuit batch-processing techniques. They usually combine sensors, actuators, and processing electronics, providing a high functionality, highperformance, low-cost integrated microsystem. They can sense and actuate mechanically on the micro scale, individually, or in arrays to act on the macro scale. Since their first appearance in the late 80s at Berkeley University (USA), MEMS, also known as micromachines, have made tremendous progress these last years. Numerous technological solutions for integration of phenomena of different physical nature, and original designs have been explored, with applications in various fields: integrated optics, accelerometers for airbags, pressure and biomedical sensors, labs on chips, mass data storage, new displays, micro-robots, etc.

Aerodynamic flow control is another application where MEMS may play a crucial role [1]. Their small size, low cost, low-energy consumption, possibility of integration of the sensing, actuating and processing technologies provide length- and time-scales matching the flows to be controlled. Nevertheless, if relevant solutions are already available for sensors, and processing, there are, up to now, very few

J.F. Morrison et al. (eds.), IUTAM Symposium on Flow Control and MEMS, 15–24.
© 2008 *Springer. Printed in the Netherlands.*

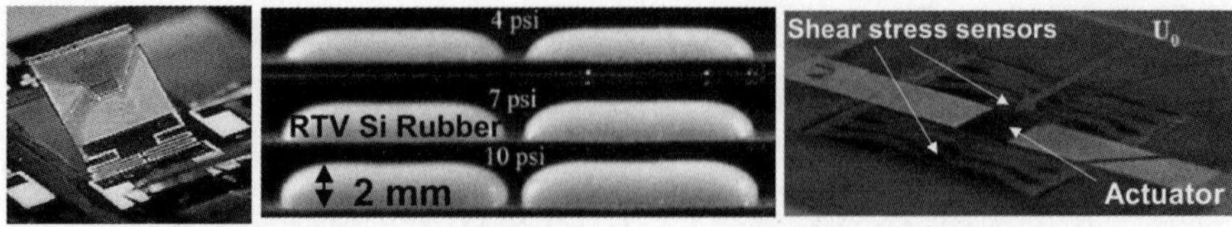

Fig. 1 Microballoon-Flap actuators. From C.M. Ho et al. [2]. (a) Electromagnetic-flap. (b) Microballoon actuators. (c) Microballoon-flap actuator between two shear stress imagers.

MEMS actuators fulfilling the requirements. The main reason is that most of the devices do not provide sufficient force, torque, displacement, or are not fast enough.

This paper presents some recent achievements in MEMS actuators for flow control. Then, it focuses on Micro-Magneto-Mechanical Systems (MMMS), our specific solution for fast dynamic actuators with high forces, torques and displacements. Demonstrators and future applications for flow control are presented.

2 Recent Achievements in MEMS Actuators for Flow Control

2.1 Microballoon-Flap Actuators [2]

Since the 1990s, Chih-Ming Ho's research group from University of California has been involved in the elaboration of MEMS-based technology for drag reduction in turbulent boundary layers and aerodynamic control of delta wing using the vortex-shift concept. The group developed MEMS shear stress sensors arranged in arrays using both rigid and flexible substrates. Several types of actuators were also successfully demonstrated. The first generation involved magnetic and electromagnetic flap actuators, the latter using patterned copper coils for control (Figure 1a). Amplitudes >100 μm were demonstrated at $f = 1.3$ kHz resonant frequency. However the flaps were stripped off by the free stream in wind tunnel experiments for speeds higher than 30 m/s. The next generation was the microballoon actuator (Figure 1b). It consists of a layer of silicon rubber spun on RTV with a pneumatic manifold underneath for pressurized air actuation. The resulting actuator is capable of extremely large forces (>100 mN), high actuation length (>1 mm), but low step response. These actuators have been used with success for manoeuvring delta-wing aircraft, but were found to be ineffective for turbulent boundary layer control. The last actuator developed is a combination of the two previous ones, the flap-type actuator being pneumatically actuated by a silicon membrane underneath. The resulting performance is a 10 mN actuator with 10 ms square wave response and 100 μm actuation height. This actuator has been used together with two shear stress imaging arrays in a two-dimensional turbulent wind tunnel. At Reynolds numbers Re $\sim 10k$, a time-averaged shear stress reduction of $\sim$4% downstream of the actuator has been achieved through constant flap actuation at 100 μm and 50 Hz.

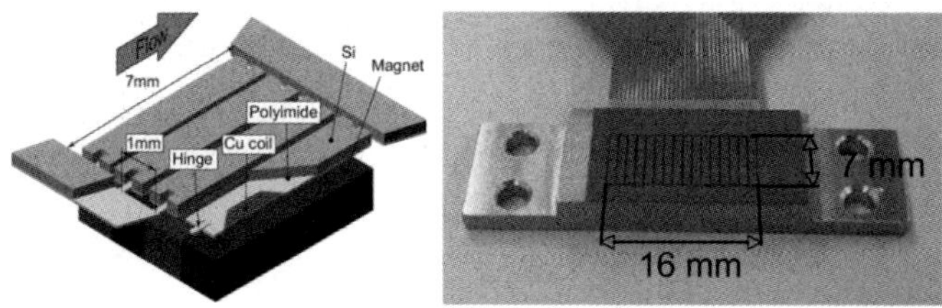

Fig. 2 Array of seesaw type magnetic actuators (fourth generation). From Kasagi et al. [3]. (a) Schematic view. (b) Photograph of an array of 16 flaps aligned in the spanwise direction.

2.2 Seesaw Type Magnetic Actuator Array [3]

Kasagi's group from University of Tokyo has developed four feedback control systems for wall turbulence combining sensors, actuators and a VLSI controller. The fourth generation is made of an array of 18 hot-film wall shear stress sensors aligned in the spanwise direction, the VLSI controller (not shown here), and an array of 16 flaps aligned in the spanwise direction (Figure 2). Silicon beams (7 mm $\times$ 1 mm $\times$ 300 μm) are suspended over permanent magnets by a pair of polyimide hinges (300 $\times$ 100 $\times$ 15 μm^3), with a spanwise spacing of 1 mm, which enables applications to Reynolds numbers Re $\sim$ 600. Copper coils (10 $\times$ 10 μm^2, 10 turns) fabricated on the backside of the beams are used to tilt the beam around their axis aligned in the streamwise direction. A tip displacement of 90 m, and a maximum tilt angle of 12°, are obtained for an energy consumption of 43 mW, compatible with VLSI. The resonant frequency is estimated at 500 Hz. Tests in a wind tunnel were made with the second generation of the system and demonstrated a 7% skin friction reduction in a turbulent channel flow. The fourth generation is expected to be tested in the near future.

2.3 MEKA-5 MicroElectroKinetic Actuator [4]

Dham's research group from the University of Michigan, proposed a microactuator without moving parts for drag reduction in turbulent boundary layers by manipulation of streamwise sublayer vortical structures. Electrokinetic pumping under a time varying applied electric field creates volume displacements of the sublayer vortices. Five prototypes have been developed leading to the MEKA-5 full-scale hydronautical array presented here (Figure 3).

It is composed of 25,600 individual microactuators with a 350 μm pitch, arranged in 40 $\times$ 40 cells, each composed of 4 $\times$ 4 actuators, for a total array dimension of 7 $\times$ 7 cm^2.

Fabrication involves laser drilling of electrokinetic channels in porous plastic substrates, and metallization processes for the electrodes and leadouts. The driv-

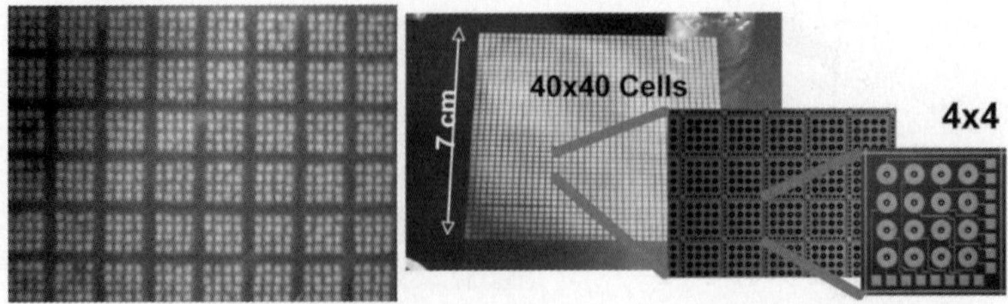

Fig. 3 MEKA-5 Electrokinetic array of 25 600 actuators. From Dahm et al. [4]. (a) Porous polymer flexible mylar. (b) Array of electrodes: 40×40 cells of 4×4 electrodes.

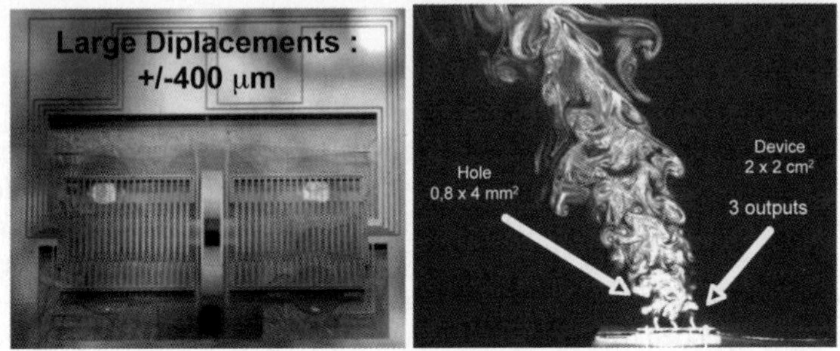

Fig. 4 Electrostatic microvalve for microjets. From FLOWDIT [5]. (a) Photograph of the electrostatic actuator. (b) View of the microjets for three valves.

ing voltage is about 20 V. Tests were performed successfully up to 20 kHz, and a theoretical bandwidth of up to 1 MHz is stated.

2.4 Electrostatic Microvalve for Microjets [5]

An electrostatic microvalve designed for flow control in automobile applications has been developed by the company FLOWDIT. Figure 4a shows a prototype with a large output section (600×1800 μm). An evolution containing 3 micro-jets on a single system has been developed (Figure 4b). Pulsed microjets with a mean velocity of 25 m/s at ~90 Hz have been obtained through holes of 0.8×4 mm^2 section for 3 mBars differential pressure and a 500 μW power consumption. These devices have been integrated in the roof of the last Citroën concept-car "C-SportLounge" for demonstration.

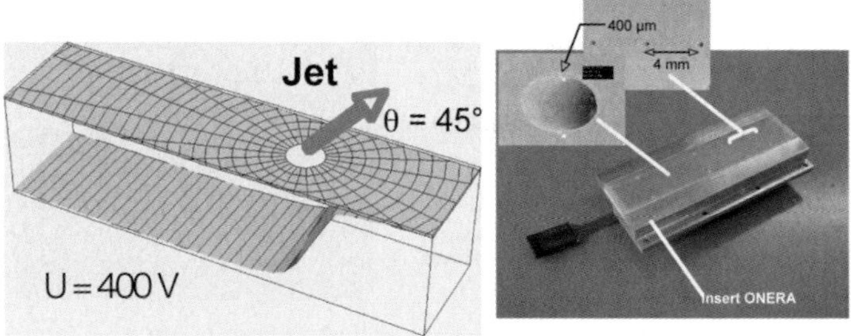

Fig. 5 "ZIP" Electrostatic microvalve for microjets. From J.R. Frutos et al. [6]. (a) Sketch of the valve. (b) Array of 15 packaged actuators.

2.5 *"ZIP" Electrostatic Microvalve for Microjets [6]*

The "ZIP" microvalve from University of Besançon (Figure 5) consists of a flexible S-shape film attracted alternatively by the upper and bottom electrodes by electrostatic forces in order to generate pulsed microjets through the upper hole. An array of 15 actuators, with 4 mm spacing holes, was fabricated for an experiment of reattachment of a boundary layer near an aircraft flap trailing edge (Aeromems European project). Jet velocities of about 100 m/s in the continuous mode were obtained for a 45° skew orifice of 400 µm diameter. The device is able to control pressure differences, ΔP, up to 27 kPa with a 400 V voltage. In pulsed mode, the maximum usable ΔP is 6 kPa, the maximum velocity is 40 m/s and typical frequency is about 200 Hz.

2.6 *Piezoelectric Microvalve for Microjets [7]*

A pulsed air-jet microvalve, based on a vibrating titanium cantilever (3 mm × 1 mm × 12.5 µm) covered on both sides by 30 µm PZT films has been presented by BAE systems. Peak jet velocities in the range of 100–200 m/s through 200 µm diameter holes are reported for a 90 V driving voltage and a nominal power consumption of 50 mW.

3 MMMS Alternatives

Micro-Magneto-Mechanical Systems (MMMS) are micro-system devices based on coupled magnetic and mechanical phenomena. Developed more recently than Micro-Electro-Mechanical Systems (MEMS), MMMS now provide efficient solutions in particular in the field of micro-actuators. A good review of the technological

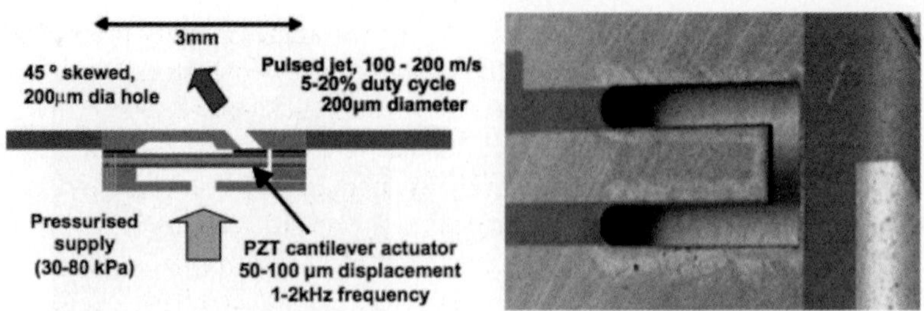

Fig. 6 Piezoelectric microvalve for microjets. From C. Warsop [7]. (a) Schematic layout of piezo-electric cantilever based microvalve. (b) Photography of the PZT-coated cantilever.

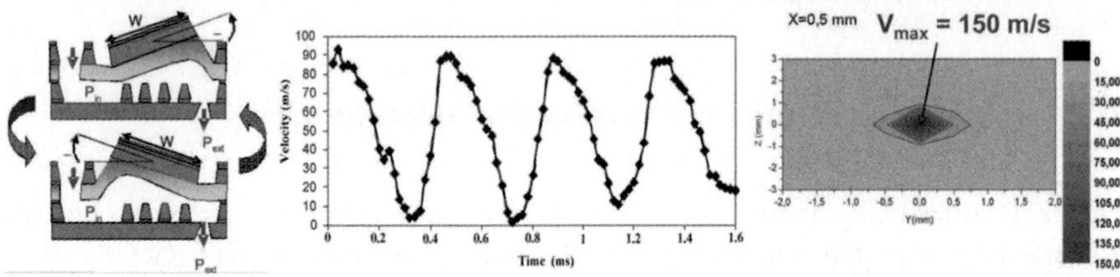

Fig. 7 Magnetostatic micro-valve based on a mechanical instability. (a) General concept in auto-oscillating mode. (b) Microjet velocity vs time at the hole output (hot-wire measurements). (c) Cross section of the velocity profile at hole output for 90° jet orientation.

specificities and already existing devices in various fields of applications is proposed in [8]. The specific approach of our joint European Laboratory LEMAC is to find, or artificially induce, magnetical or mechanical instabilities in the micro-devices in order to enhance their performances or enlarge their functionalities [9–12]. In the field of actuators for example, it is possible to enhance the sensitivity to the magnetic field of control, or to propose new strategies of control based on the specific properties of the system in the vicinity of the instability. The following paragraphs will present first a demonstrator of microvalve for separation control based on a mechanical instability using magnetostatic control, and second, the possiblities of more prospective solutions of actuation based on nanostructured magnetostrictive films with an induced magnetic instability.

3.1 Magnetostatic Micro-Valve Based on a Mechanical Instability

The microvalve, presented in Figure 7, is based on a polymer (PDMS) membrane opening and closing a microchannel made by silicon microfabrication technology. By an adequate design of the internal structure of the channel, of the membrane, and a well chosen inlet pressure, a mechanical instability can be induced for the position of the membrane [13]. This results in the pulsation of the air jet through the output hole (bottom side) without any actuation consumption.

Pulsed jets, with velocity modulations between approximately 0 and 100 m/s at 2.2 kHz have been obtained for a $10 \times 15 \times 1$ mm^3 demonstrator with a 1 mm diameter output hole oriented at $45°$ and a 0.5 bars differential pressure. We have shown that the oscillating frequency can be easily down-shifted to a few hundred hertz without any change in the design by loading the membane or applying on it a magnetostatic force using two mini-magnets. Upper frequencies are also possible to achieve by adapting the design. An electromagnetic control of the pulsation in a wide frequency range was demonstrated using a fixed mini-coil and a mini-magnet fixed on the membrane. When a normal orientation of the microjet was used, up to 150 m/s velocities were obtained for the same 0.5 bars differential pressure. Recent simulations show that this result can be improved to over 230 m/s by optimization of the internal design. If continuous jets are desired, the auto-oscillation can be suppressed by changing the parameters of the structure (internal design, stiffness of the membrane, applied magnetic force, value of the differential pressure) [13]. Further details on this microvalve are given in [14], and potential effects for flow control are considered in [15, 16].

3.2 MMMS Based on an Induced Magnetic Instability in Nanostructured Magnetostrictive Films

More prospective solutions of actuation based on an induced magnetic instability in nanostructured active films arc under consideration in the LEMAC. In some previous work, we demonstrated that a magnetic instability of Spin Reorientation Transition (SRT) type can be induced in magnetostrictive nanostructured films such as $N^*((\text{Tb}_x\text{Fe}_{1-x})_{4.5nm}/\text{Fe}_{6.5nm})$, $N^*((\text{Tb}_{0.4}\text{Fe}_{0.6})_{4.5nm}/(\text{Fe}_{0.6}\text{Co}_{0.4})_{6.5nm})$ and $N^*((\text{TbCo})_{4.5nm}/(\text{FeCo})_{6.5nm})$, where N, the number of bi-layers, is of order ten [17]. These films can easily be deposited by sputtering techniques on micro to centimetric structures such as cantilevers and membranes used for actuation of some devices. It was demonstrated that in the vicinity of SRT, the sensitivity to the driving magnetic field of the micro-actuator can be improved by one or two orders of magnitude, thus allowing the miniaturization of the coils necessary for the control of the actuator [12, 18]. In addition, elasticity of the magnetic material can be controlled by magnetic field or stress application, and the magnetoacoustic nonlinearity can be increased by four orders of magnitude. These peculiar properties allow one to propose new ways of controlling microactuators such as subharmonic excitation, or low frequency driving by nonlinear interaction of high frequency excitations [19, 20]. Some prototypes of microactuators based on micro-cantilevers [21] and membranes, and integrating electrodeposited microcoils have been developed [22]. A concept of microvalve for pulsed micro-jets is under consideration (Figure 8).

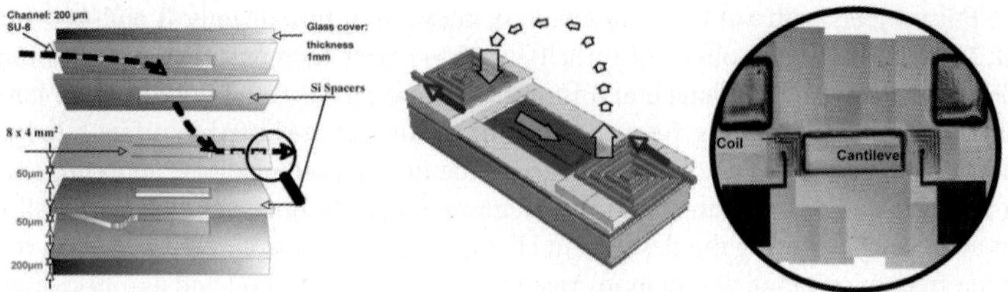

Fig. 8 Microvalve for pulsed micro-jets based on an actuation by nanostructured magnetostrictive films with induced SRT. (a) Concept. (b) Detail on the actuating part integrating a cantilever coated with the active film and the control microcoils. (c) Photograph of the fabricated actuating part.

4 Conclusion

With the recent progress made in micro-actuators, and the already available sensors and processing components, MEMS are now a mature approach to provide relevant integrated solutions for aerodynamic flow control. Various means of actuation (pneumatic, electromagnetic, electrokinetic, electrostatic, SRT-magnetostrictive), and various designs have recently been used to achieve first results in aircraft manoeuvring demonstration, drag reduction, and flow separation control. The available demonstrators can be divided in two main groups : micro-actuators creating wall surface deformations, and micro-valves for microjets. In the first category, the microballoon technique is interesting for the very large provided displacement (up to millimeters), and large forces ($>$100 mN), but is limited in the frequency response to a few tens of Hertz. The seesaw-type magnetic configuration allows frequencies in the range of a few hundred Hertz, with off-plane displacements of 100 μm. In both cases, the systems are millimetres in size and pitch. The electrokinetic actuator is an interesting transition to the second group, since only arrays of electrodes without moving parts are used to induce motion in the flow. A large number of actuators ($>$25000) are highly integrated (pitch $<$500 μm), and low voltage of about 20 V is required. Operation frequencies can achieve the kHz range (20 kHz demonstrated) and even supposedly MHz. But maximum velocities with the technique remains a question. Microjets are attractive solutions since the microvalves can be protected from the external flow. The available devices were mainly designed for flow control separation. Electrostatic solutions demonstrated up to now with jet velocities up to 40 m/s in dynamic mode with frequencies of 100 and 200 Hz and about 100 m/s in quasi-static operation. These parameters are already fitting the needs for automobile applications, and not far from the ones for aeronautic applications. Piezoelectric and MMMS solutions are already able to achieve velocities close to 200 m/s with frequencies of 500 Hz (for piezo), and even 2 kHz for MMMS. Performances are improving very quickly in each technology, and new concepts regularly appear.

Most of the presented devices are now under test in wind tunnel experiments for specific applications. The results of these tests will certainly provide in the near future information on the adaptation of the different proposed devices. Nevertheless, in any case, it is already clear that, in addition to further improvements of the performances of the devices, future research will have to focus on improvement of integration, fatigue life, mechanical limits, adaptation to harsh environment, reproducibility, cost reduction, and a larger number of produced elements.

Acknowledgments

The general MMMS research work summarized in this paper is carried out in part within the European Interreg IIIa program. MMMS applications to flow control are also made within the cooperative program GDR CNRS No. 2502, and benefits from the support or collaboration of the ADVACT European program, ONERA, Dassault Aviation, and MBDA.

References

1. Löfdahl, L. and Gad-el-Hak, M., MEMS applications in turbulence and flow control, *Progress in Aerospace Sciences* **35**, 1999.
2. Huang, A., Lew, J., Xu, Y., Tai, Y.C. and Ho, C.M., Microsensors and actuators for macrofluidic control, *IEEE Sensors J.* **4**(4), 2004.
3. Yamagami, T., Suzuki, Y. and Kasagi, N., Development of feedback control system or wall turbulence using MEMS devices, in *Proceedings 6th Symposium on Smart Control of Turbulence*, Tokyo, Japan, 2005.
4. Diez-Garias, F.J. and Dahm, J.A., *Electrokinetic Microactuator Arrays for Active Sublayer Control of Turbulent Boundary Layers*, Final Report, DARPA Contract, 2001.
5. Harambat, F., Lasserre, J.J., Beaudoin, J.F. and Edouard, C., Characterization of the flow induced by high amplitude clearance pulsed micro-jets, *European Drag Reduction and Flow Control Meeting*, Ischia, Italy, 2006.
6. Frutos, J.R., Vernier, D., Bastien, F., de Labachelerie, M. and Bailly, Y., An electrostatically actuated valve for turbulent boundary layer control, *IEEE Sensors*, 2005.
7. Warsop, C., *Aeromems II Advanced Flow Control Using MEMS. Results and Lessons Learned*, 5th Community Aeronautical Days 2006, Vienna, Austria, 2006.
8. Cugat, O. et al., *Micro-actionneurs électromagnétiques*, Lavoisier, Hermès Science Publications, 2002.
9. Pernod, P. and Preobrazhensky, V., High harmonic generation and turbulence of magnetoelastic excitations in hematite single crystal, *J. Appl. Phys.* **81**(8), 1997, 5709–5711.
10. Pernod, P. and Preobrazhensky, V., Dynamic control of elasticity by means of ultrasound excitation in antiferromagnet, *J. Magn. Magn. Mat.* **184**, 1998, 173–178.
11. Pernod, P. and Preobrazhensky, V., Nonlinear conversion of magnetoelastic modes in a magnetostrictive plate, *J. Magn. Magn. Mat.* **185**, 1998, 127–130.
12. Tiercelin, N., Preobrazhensky, V., Pernod, P., Masson, S., Ben Youssef, J. and Le Gall, H., New nanostructured active materials with giant magnetostriction and new driving modes of microactuators, *Mécanique & Industries* **4**, 2003, 169–174 [in French].
13. Ducloux, O., Talbi, A., Gimeno, L. Viard, R., Pernod, P. and Preobrazhensky, V., Self-oscillation mode due to fluid-structure interaction in a micromechanical valve, *Appl. Phys. Lett.*, 2007, accepted for publication.
14. Ducloux, O., Deblock, Y., Talbi, A., Pernod, P., Preobrazhensky, V.and Merlen, A., Magnetically actuated microvalves for active flow control, in *Flow Control and MEMS*, Proceedings

of the IUTAM Symposium held at the Royal Geographical Society, 19–22 September 2006, J. Morrison et al. (Eds.), Springer, Dordrecht, 2008.

15. Hiller, S.J., Ries, T. and Kürner, M., Flow control in turbomachinery using microjets, in *Flow Control and MEMS*, Proceedings of the IUTAM Symposium held at the Royal Geographical Society, 19–22 September 2006, J. Morrison et al. (Eds.), Springer, Dordrecht, 2008.

16. Garnier, E., Pruvost, M., Ducloux, O., Talbi, A., Gimeno, L., Pernod, P., Merlen A. and Preobrazhensky, V., ONERA/IEMN contribution within the ADVACT program: Actuators evaluation, in *Flow Control and MEMS*, Proceedings of the IUTAM Symposium held at the Royal Geographical Society, 19–22 September 2006, J. Morrison et al. (Eds.), Springer, Dordrecht, 2008.

17. Tiercelin, N., Ben Youssef, J., Preobrazhensky, V., Pernod, P. and Le Gall, H., Giant magnetostrictive supperlatices: From spin reorientation transition to MEMS. Static and dynamical properties, *J. Magn. Magn. Mat.* **249**(3), 2002.

18. Le Gall, H., Ben Youssef, J., Socha, F., Tiercelin, N., Preobrazhensky, V. and Pernod, P., Low field anisotropic magnetostriction of single domain exchange-coupled (TbFe/Fe) multilayers, *J. Appl. Phys.* **87**(9), 2000.

19. Tiercelin, N., Preobrazhensky, V., Pernod, P., Le Gall, H. and Ben Youssef, J., Nonlinear actuation of cantilevers using giant magnetostrictive thin films, *Ultrasonics* **38**, 2000, 64–66.

20. Tiercelin, N., Preobrazhensky, V., Pernod, P., Le Gall, H. and Ben Youssef, J., Sub-harmonic excitation of a planar magneto-mechanical system by means of giant magnetostrictive thin films, *J. Magn. Magn. Mat.* **210**, 2000, 302–308.

21. Tiercelin, N., Pernod, P., Preobrazhensky, V., Le Gall, H., Ben Youssef, J., Mounaix, P. and Lippens, D., Giant magnetostriction thin films for multi-cantilever structures driving, *Sens. & Actuators* **81**, 2000, 162–165.

22. Ducloux, O., Tiercelin, N., Deblock, Y., Pernod, P., Preobrazhensky, V. and Merlen, A., Cantilever resonance induced in situ for active flow control, in *Proceedings of the IEEE UFFC Ultrasonics Symposium*, 18–21 September 2005.

MEMS-Based Electrodynamic Synthetic Jet Actuators for Flow Control Applications

Janhavi S. Agashe, Mark Sheplak, David P. Arnold and Louis Cattafesta
*Interdisciplinary Microsystems Group, University of Florida, Gainesville,
FL 32611, U.S.A.: E-mail: {janhavia, sheplak, darnold, cattafes}@ufl.edu*

Abstract. Synthetic jet actuators are used in various applications, such as separation control, mixing enhancement, and thermal management. Each of these requires the design to be optimized to meet specific performance requirements. This paper presents a design and scaling analysis of an electrodynamic synthetic jet actuator using a lumped element modeling approach. Various performance parameters, such as the resonant frequency, output volumetric flow rate and velocity, and jet formation, are studied as a function of device scale and current density. The viability and potential advantages of microscale synthetic jets are discussed.

Key words: Synthetic jets, electrodynamic actuation, MEMS, flow control, scaling.

1 Introduction

Synthetic jet actuators have been used in many flow control applications, such as separation control, mixing enhancement, and thermal management [1–5]. A typical synthetic jet, also called a zero-net mass-flux actuator, consists of a vibrating diaphragm to drive oscillatory flow through a small orifice or slot. Although there is no source, a mean jet flow is established near the orifice as a result of the entrained fluid.

1.1 MEMS (Micro) Synthetic Jets

Microscale synthetic jets are attractive due to their small size, the potential reduction in manufacturing costs, and lower power consumption. Microscale synthetic jets using electrostatic, piezoelectric, and electrodynamic transduction schemes have been fabricated and tested [6-10]. However, previous efforts have primarily focused on

J.F. Morrison et al. (eds.), IUTAM Symposium on Flow Control and MEMS, 25–32.
© 2008 *Springer. Printed in the Netherlands.*

the microfabrication aspects of realizing small cavities, orifices, etc. The performance of these devices – output volume velocity, resonant frequency, power consumption, etc. – has widely differed depending on the transduction scheme and the
geometry chosen. Thus, a fundamental understanding of how device scale affects
these performance parameters is required for design optimization.

1.2 Modeling

There are numerous studies that have concentrated on the modeling, design, visualization, and/or measurements of synthetic jets. Several analytical and computational
approaches have focused on explaining the fundamental operational characteristics
and scaling of a synthetic jet [11–13]. A simple, reliable, and efficient modeling approach is essential for the design, optimization, and scaling analysis of an appropriate device for a particular application. Lumped element modeling (LEM) provides
such a method for a coupled electromechanical-acoustic system, such as a synthetic
jet. This analysis method has been previously applied to synthetic jets [3, 14]. This
paper complements the prior work by exploring the performance of microscale electrodynamic synthetic jet actuators.

2 Electrodynamic Synthetic Jets

2.1 Electrodynamic Transduction

In an electrodynamic synthetic jet, the driving system consists of a magnet and a
current carrying coil. The current flow in the coil interacts with the magnetic field,
producing a Lorentz force on the coil given by $F = BL_c i$, where B is the magnetic
flux density, L_c is the length of the coil, and i is the current through the coil. The
force produced is in the direction normal to both the current and the magnetic field,
as given by the right-hand rule.

2.2 Electrodynamic Flow Control Actuator

A schematic of the actuator is shown in Figure 1. A circular diaphragm of area A_D is
attached to an electrodynamic transducer comprising a coil and magnetic assembly.
The circular diaphragm forms the bottom portion of a cavity. The top of the cavity
is bound by a rigid plate containing a slot or orifice. The goal of the device is to
produce large volume displacements in order to force fluid in and out of the cavity.

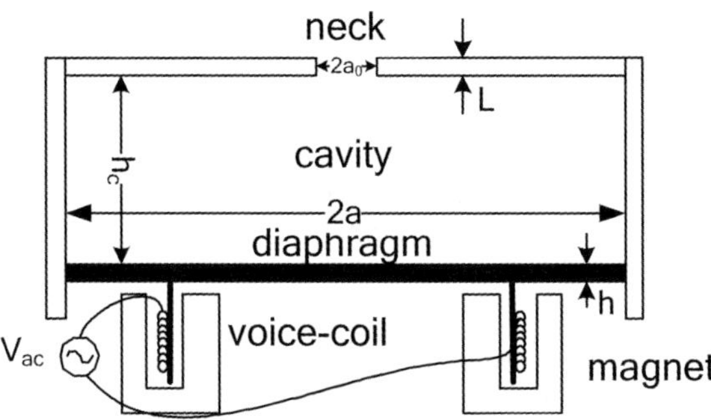

Fig. 1 Schematic of electrodynamic synthetic jet actuator.

2.3 Lumped Element Model

The fundamental assumption employed in LEM is that the characteristic length scales of the governing physical phenomena are much larger than the largest geometric dimension. In the case of a synthetic jet, the acoustic wavelength must be significantly larger than the device itself. If this assumption holds, then the temporal and spatial variations can be decoupled, and the governing partial differential equations for the distributed system can be "lumped" into a set of coupled ordinary differential equations. This approach provides a simple method to estimate the low-frequency dynamic response of a system for design and control-system implementation.

Figure 2 shows the complete lumped element model of the synthetic jet actuator. The coil is represented as a resistor R_{eC} and inductor L_{eC} in the series $(Z_{eC} = R_{eC} + j\omega L_{eC})$. The transduction factor T between the electrical and the acoustic domain is given by $T = BL_C/A_D$. The diaphragm is represented by an equivalent acoustic mass M_{aD}, acoustic compliance C_{aD}, and acoustic resistance R_{aD}. The cavity is represented with an acoustic compliance C_{aCav}. The neck is modeled as a combination of linear (R_{aN} and R_{aRad}) and nonlinear (R_{aO}) resistors and masses (M_{aN} and M_{aRad}). Note that the subscripts e and a denote quantities in the electrical and acoustic domains, respectively. The values of these parameters are determined from analytical expressions (see Table 1). The output volume velocity Q_{out} is calculated from the nonlinear lumped element model as a function of frequency using an iterative technique to account for the nonlinear neck resistance term.

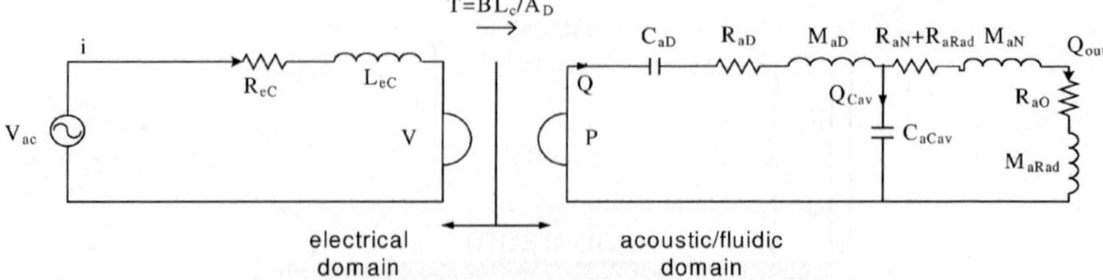

Fig. 2 Lumped element model of the electrodynamic synthetic jet actuator using a gyrator to couple the electrical and acoustic domains.

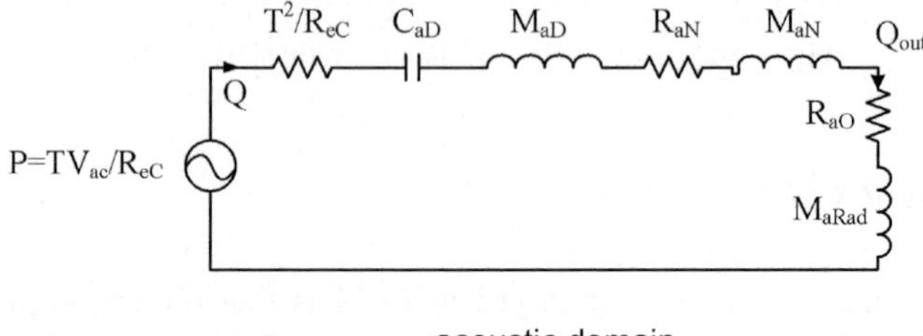

Fig. 3 Simplified lumped element model of the synthetic jet actuator at small scales.

3 Scaling Analysis

Using the lumped model, scaling of the device can be systematically studied to determine its influence on power requirements, resonant frequency, volume velocity, etc. A baseline device with the diaphragm radius a and thickness h of 25 mm and 0.5 mm, respectively, a cavity depth h_c of 6.6 mm, a ring-shaped magnet with a remnance of 1.2 T, and a 32 turn 31 AWG copper coil, and an orifice of radius a_0 and length L of 2.8 mm and 8 mm, respectively, was considered. The lumped parameters for this device were calculated, and then all linear dimensions of the device – for example, the diaphragm radius and thickness, the coil radius and length, the cavity radius and depth, and the orifice radius and height – were scaled by a scaling factor k.

The material properties of all the components are assumed to be constant. As a result, the magnetic flux density B associated with the coil also remains constant. Under these assumptions, Table 1 shows the variation of various lumped parameters with the scale factor k. Note that because damping decreases in microscale devices [15], a conservative scaling estimate using a constant damping ratio reveals the losses due to structural damping are negligible compared to other dissipation mechanisms. In particular, the neck resistance dominates at smaller scales. The scaling of the nonlinear neck resistance is unknown *a priori* due to its dependence on the output volume velocity. As shown in Table 1, the LEM of the synthetic jet can be further simplified by considering only the dominant terms at the smaller scales.

Table 1 Variation of lumped parameters with scale factor k.

Transduction factor	$T = \dfrac{BL_C}{A_D} \propto \dfrac{const \cdot k}{k^2} \sim \dfrac{1}{k}$	$R_{aD} = 2\,\zeta\,\sqrt{\dfrac{M_{aD}}{C_{aD}}} \propto \sqrt{\dfrac{k}{k^3}} = \dfrac{1}{k^2}$ (assumed constant)	Diaphragm Damping
Diaphragm compliance	$C_{aD} = \dfrac{9\pi a^6 (1-\upsilon^2)}{16 E h^3} \propto k^3$	$R_{aRad} = \dfrac{0.459 \rho_o c_o}{a_o^2} \propto \dfrac{1}{k^2}$	Neck radiation resistance
Cavity compliance	$C_{aCav} = \dfrac{\pi a^2 h_c}{\rho_o c_o^2} \propto k^3$	$R_{aN} = \dfrac{8\mu L}{\pi a_o^4} \propto \dfrac{1}{k^3}$	Neck linear resistance
Diaphragm mass	$M_{aD} = \dfrac{\rho_D h}{\pi a^2} \propto \dfrac{1}{k}$	$R_{aO} = \dfrac{0.5 \rho_o Q_{out}}{\pi^2 a_o^4} = \,?$	Neck nonlinear resistance
Neck radiation mass	$M_{aRad} = \dfrac{8\rho_o}{3\pi^2 a_o} \propto \dfrac{1}{k}$	$R_{eC} = \dfrac{\rho_e L_C}{A_C} \propto \dfrac{1}{k}$	Coil resistance
Neck mass	$M_{aN} = \dfrac{4\rho_o L}{3\pi a_o^2} \propto \dfrac{1}{k}$	$\omega L_{eC} \propto \omega \dfrac{1}{k} \ll R_{eC}$	Coil inductance

The corresponding simplified equivalent circuit model is shown in Figure 3, where the electrical elements have been transferred to the acoustic domain. The coil inductance is small and its impedance at low frequencies is negligible compared to R_{eC}, so L_{eC} is omitted from the analysis. Furthermore, the linear and nonlinear neck resistors that model major and minor losses in the orifice are retained. The cavity compliance is also omitted because it is very small for small scales. Thus at low frequencies, an incompressible flow assumption is appropriate.

The output of the synthetic jet is directly dependent on the current in the actuator coil. In the most conservative estimate, the current density in the coil can be assumed constant at all scales. However at smaller scales, the surface area to volume ratio is large, and the heat generated can be dissipated more effectively. Hence, microcoils can sustain higher current densities with current density scaling as $1/\sqrt{k}$ or even $1/k$ [16]. As a result, the performance of the synthetic jet was analyzed under these different assumptions.

3.1 Discussion of Various Performance Criteria

As can be seen from Figure 3, the device dynamics at small scales are controlled mainly by the diaphragm and the neck. The cavity does not play a significant role. The result is a second-order nonlinear system, and the resonant frequency of the device scales as $1/k$.

For constant current densities J, the centerline velocity obtained at resonance scales as $1/k^2$ for very small scales. However, with increased current densities, the

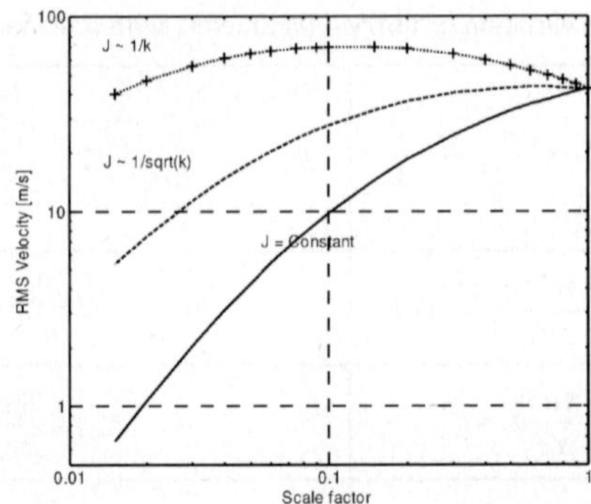

Fig. 4 Variation of peak rms centerline velocity for different current density vs. scale.

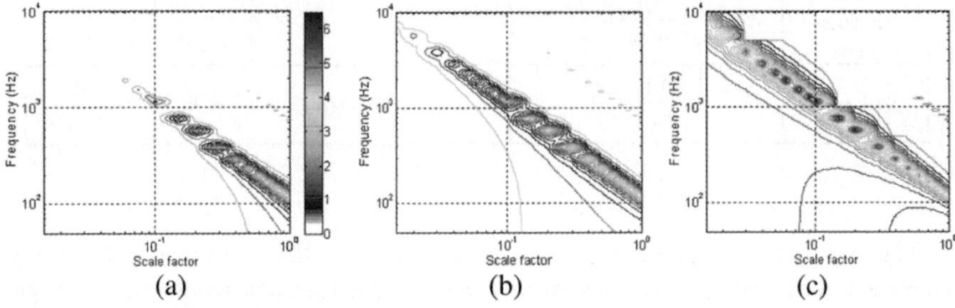

Fig. 5 Contours of jet formation criteria $1/St > 0.16$ for various current densities $J =$ (a) constant, (b) $1/\sqrt{k}$, (c) $1/k$. Same color scale for all.

scaling can be improved, as shown in Figure 4. The jet formation criteria can be analyzed by calculating the jet Strouhal number [17]. It can be seen from Figure 5 that, with increased current densities, jet formation can be ensured over a wide operating range even for smaller devices.

3.2 Fabrication Methods and Issues

The major challenge in miniaturizing the electrodynamic synthetic jet actuator is the fabrication of solenoidal (3-D) coils. Planar coils that are easier to fabricate but smaller in length are usually used in microscale devices [18]. The use of multiple planar coils may achieve better performance. Microscale magnets having high remnant magnetic flux density are also required to maintain large coupling between the electrical and acoustic domains. Various techniques such as electrodeposition and sputtering have been developed to incorporate permanent magnet materials in microscale devices. Moreover, at the microscale, clamped-clamped diaphragms made of polymers (e.g., parylene, polyimide, etc.) are easier to fabricate using spin or vapor deposition. Other geometries of the coil and the magnet may be used to achieve good performance.

4 Conclusions and Future Work

LEM has been used to perform a scaling analysis of an electrodynamic synthetic jet actuator. The analysis shows that the resonant frequency – which is determined by the driver – increases linearly with reduced device size. With higher current density possible in microcoils, the velocity output may be improved and jet formation can be ensured for a wider range of operating conditions. Future work will investigate the feasibility of manufacturing these microscale devices and verifying the scaling results.

References

1. Smith, B.L., Glezer, A., The Formation and Evolution of Synthetic Jets, *Physics of Fluids* **10**(9) (1998) 2281–2297.
2. Amitay, M., Smith, B.L., Glezer, A., Aerodynamic Flow Control Using Synthetic Jet Technology, in *Proceedings of 36th Aerospace Sciences Meeting and Exhibit*, AIAA 1998-0208 (1998).
3. McCormick, D., Boundary Layer Separation Control with Directed Synthetic Jets, in *Proceedings of 38th Aerospace Sciences Meeting and Exhibit*, AIAA 2000-0519 (2000).
4. Chen, Y., Liang, S., Aung, K., Glezer, A., Enhanced Mixing in a Simulated Combustor Using Synthetic Jet Actuators, in *Proceedings of 37th Aerospace Sciences Meeting and Exhibit*, AIAA 1999-0449 (1999).
5. Crook, A., Sadri, A.M., Wood, N.J., The Development and Implementation of Synthetic Jets for the Control of Separated Flow, in *Proceedings of 37th Aerospace Sciences Meeting and Exhibit*, AIAA 1999-3176 (1999).
6. Baysal, O., Koklu, M., Erbas, N., Design Optimization of Micro Synthetic Jet Actuator for Flow Separation Control, *Journal of Fluids Engineering* **128**(5) (2006) 1053–1062.
7. Coe, D.J., Allen, M.G., Smith, B.L., Glezer, A., Addressable Micromachined Jet Arrays, in *Proceedings of 8th International Conference on Solid-State Sensors and Actuators*, Vol. 2 (1995) 329–332.
8. Parviz, B.A., Najafi, K., Muller, M., Bernal, L.P., Washabaugh, P.D., Electrostatically Driven Synthetic Microjet Arrays as a Propulsion Method for Micro Flight Part I: Principles of Operation, Modeling, and Simulation, *Microsystem Technologies* **11**(11) (2005) 1214–1222.
9. Lee, C., Hong, G., Ha, Q., Mallinson, S., A Piezoelectrically Actuated Micro Synthetic Jet for Active Flow Control, *Sensors and Actuators: B, Chemical* **108**(1) (2003) 168–174.
10. Kercher, D.S., Jeong-Bong L., Brand, O., Allen, M.G., Glezer, A., Microjet Cooling Devices for Thermal Management Of Electronics, *IEEE Transactions on Components and Packaging Technologies* **26**(2) (2003) 359–366.
11. Rizzetta, D., Visbal, M., Stanek, M., Numerical Investigation of Synthetic-Jet Flowfields, *AIAA J.* **37**(8) (1999) 919–927.
12. Yamaleev, N., Carpenter, M., Ferguson, F., Reduced-Order Model for Efficient Simulation of Synthetic Jet Actuators, *AIAA J.* **43**(2) (2005) 357–369.
13. Tang, H., Zhong, S., Incompressible Flow Model of Synthetic Jet Actuators, *AIAA J.* **44**(4) (2006) 908–912.

14. Gallas, Q., Holman, R., Nishida, T., Carroll, B., Sheplak, M., Cattafesta, L., Lumped Element Modeling of Piezoelectric-Driven Synthetic Jet Actuators, *AIAA J.* **41**(2) (2003) 240–247.
15. Ekinci, K.L., Roukes, M.L., Nanoelectromechanical Systems, *Review of Scientific Instruments* **76**(6) (2005) 061101.
16. Cugat, O., Delamare, J., Reyne, G., MAGnetic Micro-Actuators & Systems MAGMAS, *IEEE Trans. Magn.* **39**(5) (2003) 3607–3612.
17. Holman, R., Utturkar, Y., Mittal, R., Smith, B.L., Cattafesta, L., Formation Criterion for Synthetic Jets, *AIAA J.* **43**(10) (2005) 2110–2116.
18. Lagorce, L., Brand, O., Allen, M.G., Magnetic Microactuators Based on Polymer Magnets, *J. Microelectromech. Syst.* **8**(1) (1999) 2–9.

Suction and Oscillatory Blowing Actuator

Gilad Arwatz, Ilan Fono and Avi Seifert
*School of Mechanical Engineering, Faculty of Engineering Tel-Aviv University,
Tel-Aviv 69978, Israel; E-mail: seifert@eng.tau.ac.il*

Abstract. Enhancing the ability to control flows in different configurations and flow conditions can lead to improved systems. Certain active flow control (AFC) actuators are efficient at low Mach numbers but the momentum and vorticity they provide limits the utilization to low speeds. At higher Mach numbers, robust, unsteady, efficient and practical fluidic actuators are a critical, largely missing, enabling technology in any AFC system. A new actuator concept, based on the combination of steady suction and oscillatory-blowing (SaOB) is presented. The actuator can achieve near-sonic speeds at about 1 kHz. It has no moving parts and therefore is expected to have superior efficiency and reliability. The operation principle of the SaOB actuator is presented along with two predictive computational models and their experimental validation.

Key words: Actuator, flow control, ejector, bi-stable, switching valve.

1 Introduction

The development of techniques for expanding the ability to control flows in a wide variety of configurations and flow conditions can lead to greatly improved systems. Probably the most common flow control application is the control of incompressible boundary layer separation which can augment the performance of flight vehicles in aspects such as increased lift and stall-margin, drag reduction, and noise and vibration attenuation. It can also enable more aggressive and/or shorter inlets and diffusers as well as thrust vectoring. It was shown that active flow control (AFC) can be effective in compressible flows [1, 2], but higher control authority is required in comparison to similar incompressible applications. Actuators for AFC can be divided into two categories, based on the fluidic output: Zero-Mass-Flux (ZMF or "synthetic jet") and Mass-Flux (MF, continuous or pulsed jets, steady or pulsed suction) actuators. Synthetic jets can be very efficient for active flow control [3–5], their main shortcoming being the magnitude of the momentum and vorticity flux

J.F. Morrison et al. (eds.), IUTAM Symposium on Flow Control and MEMS, 33–44.
© 2008 *Springer. Printed in the Netherlands.*

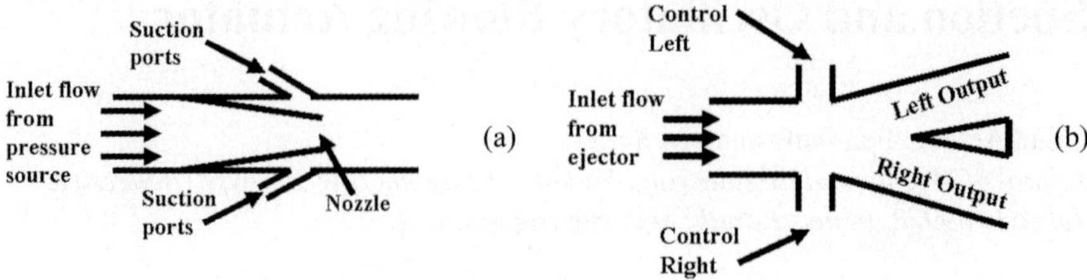

Fig. 1 A schematic rendering of the actuator: (a) ejector, (b) switching valve.

they can provide, currently limited by peak velocity of about $M = 0.3$. Therefore, robust, unsteady fluidic actuators are critical enabling technology components in any successful AFC system.

The new device combines steady-suction and oscillatory-blowing, both proven to be very effective AFC tools. The actuator is a combination of an ejector and bi-stable fluidic amplifier (Figure 1). The ejector (Figure 1a) is a simple fluidic device based on Bernoulli's law. When a jet stream is ejected into a bigger conduit it creates a low pressure region around it due to entrainment. If the cavity behind the jet is open to the free atmosphere or to a lower pressure environment, the pressure gradient will cause the external air to be sucked into the cavity around the internal jet [6, 7]. The common use of ejectors is to increase the flow-rate. However, if the fluid is drawn from an aerodynamic surface, a suction flow is created across the aerodynamic surface through slots or holes.

The bi-stable device is based on the principle of wall-attachment [8]: when a fluid jet is flowing in the proximity of a wall, low-pressure region is formed between the wall and the jet. The low pressure draws the jet towards the wall, deflecting it until it adheres to it. In the case of two near and symmetric walls (Figure 1b), the jet will randomly re-attach to either wall. If an appropriate pressure gradient is introduced between the control ports (Figure 1b) the jet will detach from one wall and re-attach to the opposite. By connecting the control ports by a tube the bi-stable fluidic amplifier can self-oscillate and serve as a switching valve. In this configuration the oscillation frequency is related to the tube length and diameter, the speed of sound in the tube, the resistance of the control ports and the flow-rate through the valve [9, 10].

2 Experimental Setup

In order to examine the actuator's performance, three scales of actuators were built. First, the feasibility of using the above concept as AFC actuator was tested with a large-scale setup. Based on this proof-of-concept, several medium-scale actuators were fabricated in order to determine the geometrical relations and evaluate performance. A small-scale actuator was fabricated using the successful geometry. The

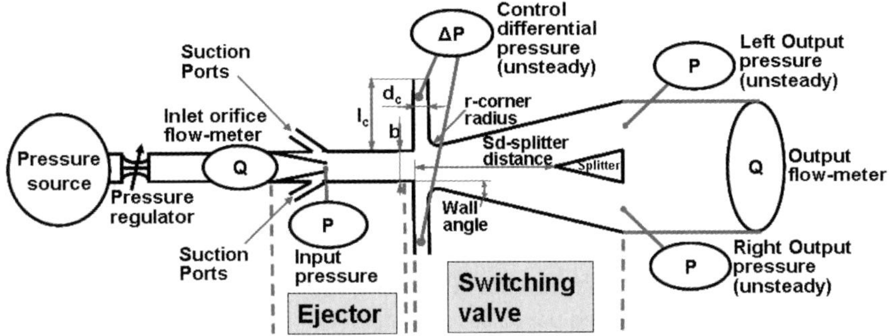

Fig. 2 Actuator (ejector and switching valve) geometry and experimental setup.

small-scale actuator is suitable for AFC applications with typical dimensions of 10–100 cm at near sonic Mach numbers. The actuators' scale was characterized by the switching valve inlet width (Figure 2b) $b = 8$ mm, 3 mm and 1.5 mm for the three device generations. The working fluid was shop-air. Steady and unsteady pressures, flow rates, velocity and frequencies were measured. More details can be found in [11]. Figure 2 shows a schematic description of the test experimental setup. Compressed shop-air was fed into the actuators' inlet. A computer controlled pressure regulator adjusted the required inlet pressure. An orifice flow-meter was placed at different locations, depending on the experiment requirements. Steady and unsteady pressures were measured at several locations, as shown in Figure 2. The unsteady pressures were low-pass filtered, amplified and together with all the other signals were acquired and digitized. The signals were ensemble-averaged in order to remove random noise and the turbulent fluctuations from the signals. The synchronization signal was the low-pass filtered differential pressure between the control ports (input).

3 Jet Deflection Model

3.1 Overview

Using the results of the preliminary experimental study, a simplified model for describing the operation of the switching valve was developed. It is based on analysis of the jet cross-flow motion between two inclined walls due to a cross-flow pressure gradient applied between the control ports. The model follows the analysis of Kirshner, and Katz [12], for the effect of control pressure gradient on a free-jet. The main addition to the model is the switching valve, requiring the addition of boundary conditions (walls and splitter) and the wall-attachment effect. The model assumptions are that the mass of the jet is concentrated at the jet centerline, the fluid is incompressible and inviscid, the jet axial velocity is constant (negligible spreading) and the cross-flow pressure gradient is uniform in the control region. The model is based on the simple idea of fluid elements flowing from the inlet nozzle to the valve outlet. When a fluid element reaches the control ports, it experiences a cross-flow acceleration due to the control pressure gradient. This acceleration integrates to change the vertical velocity, which causes the fluid element to deviate from the device centerline. An impermeability boundary condition is applied at the walls and the splitter. Since the jet velocity profile is fairly uniform in the control region, we assume that the interaction of a fluid element with other elements can be neglected. Therefore, we restrict the discussion to the motion of the jet axis, where we set $u = U_c$, with U_c being the jet velocity, assumed unchanged throughout.

We shall describe the dynamics of a fluid element at a transverse position Y_0 from the jet axis. A cross-flow pressure gradient $dp/dy = -g(x,t)$, initiated at $t = 0$, exerts a force on the fluid element accelerating it such that:

$$\frac{d^2y}{dt^2} = \frac{g(x,t)}{\rho}. \tag{1}$$

To numerically solve Equation (1), the basic equations of motion were used. Figure 3 shows the physical configuration of cross-flow pressure gradient applied between the two control ports. The pressure gradient acts on the jet only in the control region, i.e., from $x = 0$ to $x = X_1$. Three transport times can be defined: The axial transport time (the convection time from the switching valve inlet nozzle to the valve exit):

$$\tau = \int_0^t \frac{dx}{u_c},$$

the control transport time:

$$\tau_1 = \int_0^{x_1} \frac{dx}{u_c}$$

and the splitter transport time:

$$\tau_2 = \int_0^{S_d} \frac{dx}{u_c},$$

where all integrals start from the switching valve inlet nozzle.

The control pressure gradient can also be oscillatory, but of constant magnitude between $x = 0$ and zero elsewhere (next to Figure 3).

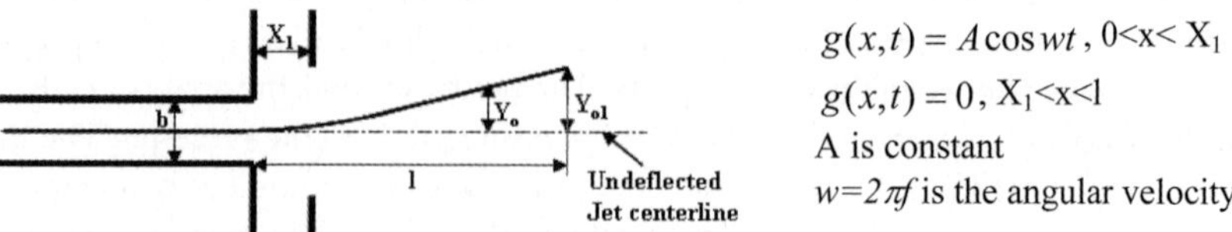

Fig. 3 The jet fluid elements (shown as a line) under the effect of a cross-flow pressure-gradient (on the left) and the applied pressure gradient (boundary conditions).

3.2 Minimum Switching Pressure

By using the model assumptions and the simplified equations of motion, it is possible to find an expression for the minimum switching pressure, defined as the smallest pressure gradient that will flip the jet from one side to the other. When a fluid element flows along one side of the inlet nozzle and reaches the control region, a pressure gradient will cause the fluid element to accelerate in the cross-flow direction, deflecting the fluid element from its original path. After a fluid element exits the control region it will keep moving in a straight line (the analysis ignores the wall-attachment effect, assuming that the wall-attachment pressure is negligible compared to the control pressure). If a fluid element path brings it to the opposite side of the splitter, the control pressure is large enough to cause full switching. The minimum switching pressure $\Delta P_{\min}$ is that which deflects the fluid element by the splitter height $(b/2)$:

$$\frac{\Delta P_{\min}}{\rho U_c^2} > \frac{2b^2}{X_1 S_d - \frac{1}{2}X_1^2} .$$ (2)

A larger splitter distance (S_d) and a smaller inlet nozzle width (b), with other parameters fixed, will reduce the minimum switching pressure. By using the same approach, it is possible to compute the maximum pressure $\Delta P_{\max}$, or the largest pressure gradient that the valve is able to produce between the control ports when the jet is fully deflected. This pressure gradient is important for the generation of a self-oscillating condition.

$$\frac{\Delta P_{\max}}{\rho U_c^2} = \frac{b \tan \alpha}{X_1} .$$ (3)

A larger wall divergence angle (α) will require a larger pressure gradient to reattach. Decreasing the inlet nozzle width (b) or increasing the length of the control region (X_1), with other parameters fixed, will reduce the maximum pressure. An almost perfect match between the experimental findings and the simplified theoretical model for the minimum and maximum pressures is shown in Figure 7.

3.3 Jet Deflection Model Results

Figure 4 presents the model input and output time-histories. The side walls and splitter are represented by dashed grey lines. The bi-stable characteristics are clearly seen: the wall-attachment, the rapid switching time and the slope change when the jet approaches the wall (wall attachment effect). The splitter effect on the jet deflection is that the jet can not exit from the splitter and therefore the jet output has a jump at zero deflection.

An option of the model was to consider a measured control input signal. Therefore, it is possible to more realistically compare the real valve to the computer model. An almost perfect match between the model and the experimental results is shown (Figure 5). This example is interesting because the model was able to re-

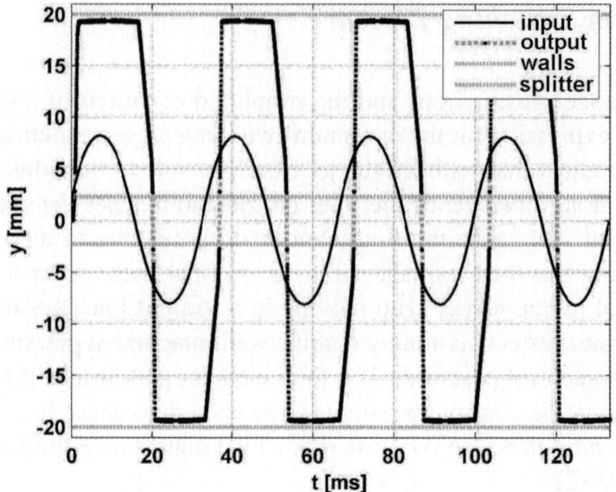

Fig. 4 Model input and output time histories for the large-scale valve at $P_{\text{in}} = 160$ kPa.

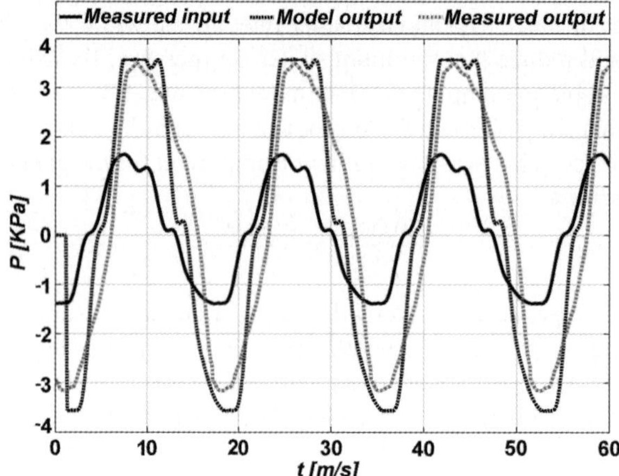

Fig. 5 A comparison between experiment and model outputs, for measured input signal. Large-scale valve, inlet pressure $= 80$ kPa, input flow-rate $= 1$ liter/sec, $f = 58$ Hz.

produce a condition when the control was too weak for complete wall-attachment (it can be seen that the jet did not reach the lower wall). Also, in Figure 5 a time-delay can be seen between the input and output related to the axial transport time (τ), while in the measured data it is about twice the axial transport time. This difference is related to the assumption that the jet's axial velocity is constant, and does not change along the valve due to friction, spreading or entrainment.

4 Self-Oscillations Frequency Model

As mentioned before, the valve can self-oscillate by connecting the two control ports with a tube (hereafter referred to as the "feedback tube"). To help understand this mechanism, note that in static operation (control ports sealed, jet stationary), a negative pressure (suction) proportional to the jet stagnation pressure and its turning, is created on the side where the jet is situated (port A). On the other port (port B) the pressure is near atmospheric, since the streamline there is nearly straight. It is possible to switch the jet by applying positive pressure on port A or negative pressure on port B. The pressure difference should be greater than a certain threshold, to overcome the tendency for wall-attachment (Coanda effect). When a tube connects the control ports, the valve starts to oscillate. The assumed self-oscillation mechanism is that the negative pressure created at port A propagates through the tube to port B. When the pressure difference between the ports passes the threshold, it causes the jet to switch. This mechanism repeats itself in the other direction and the jet self-oscillates. Several attempts were made to relate the frequency to the feedback tube volume [10]. Others tried to explain the mechanism using wave propagation and different combinations of convection and transition times [9, 13, 14]. Following the current experimental results, a simplified model that predicts the self-oscillation frequency was developed and will now be presented.

This model accounts for self-oscillation and for jet switching, as shown in Figure 6. The two associated time scales are the acoustic delay and the time constant of an electrical analog LR (flow inductance-resistance) circuit. The inductance, L, represents the flow inertia and the resistance, R, represents the viscous flow resistance of the feedback tube and the connections between the tube and the valve main flow path. The acoustic delay is

$$\tau\mathrm{ac} = (l_t + 2l_c)/a,$$

where l_t and l_c are the tube and the control port lengths, respectively, and a is the speed of sound in the tube. The time constant of an *LR* circuit is given by

$$\tau_{LR} = L/R.$$

According to electrical analogy of fluid systems [15] we can write (right-hand side of Figure 6):

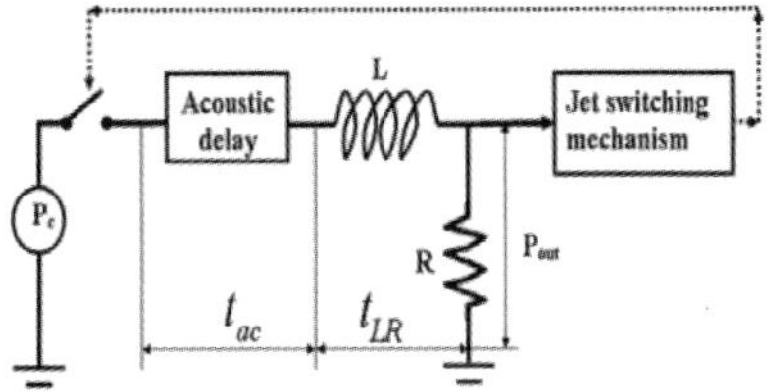

	Electric	Flow
	I- current [A]	Q- flow-rate [m³/s]
	V-Potential [Volt]	P-pressure [Pa]
	R-resistance: $R = \dfrac{V}{I}$ [ohm]	$R = \dfrac{P}{Q}$ [Pa*s/m³]=[kg/m⁴s]
	L-inductance: $V = L\dfrac{dI}{dt}$ [H]	$P = L\dfrac{dQ}{dt}$ [kg/m⁴]

Fig. 6 Oneway valve oscillations travel model and electrical analogy of fluid systems [15].

In order to fit the measured data, R (Equation (4)) was multiplied by the factor $\sqrt{l_t * d_t}$, and to get the proper dimensions, the formula was normalized by $\sqrt{l_c * d_c}$, when C in Equation (4) is an empirical constant.

$$R = C\frac{P_c}{Q}\sqrt{\frac{l_t * d_t}{l_c * d_c}} \ [\mathrm{Pa}/(\mathrm{m}^3)/\mathrm{s}]. \tag{4}$$

The magnitude of the control pressure (P_c) is created by the stationary jet at one side while the control ports are sealed. It is related to the inlet volume flow-rate by the relation

$$P_c = 32.1 * Q^2 \ [\mathrm{Pa}].$$

(for the small-scale valve, see Figure 7 and Equation (3)).

Dynamically, the flow-rate cannot instantly rise to its steady-state value (determined by the resistance) because of the inertia effect produced by the "inductance". The differential equation that describes the current increase from zero to its final value is:

$$P_c = L\frac{dQ}{dt} + QR \quad \left(E = L\frac{dI}{dt} + IR\right). \tag{5}$$

When the control ports are connected, the feedback tube allows this pressure pulse to propagate between the control ports and cause the jet to switch periodically. The minimum steady-state pressure that causes the jet to switch is about $0.3P_c$ (Figure 7 and Equation (3)) and the corresponding time-delay is $\tau_{LR} = 0.36\tau_{LR}$. Therefore, the oscillation frequency is:

$$f = \frac{1}{2(t_{ac} + 0.36\tau_{LR})} = \frac{1}{2\left(\dfrac{l_t + 2l_c}{a} + \dfrac{0.36}{C}\sqrt{\dfrac{l_c \cdot d_c}{l_t \cdot d_t}}\dfrac{Q\rho(l_t + l_c)}{AP_c}\right)}. \tag{6}$$

4.1 Self-Oscillations Frequency Model Results

Figure 8 presents a comparison between the self-oscillation model and the experimental results for the small-size valve. It is evident that the model fits the data very well over a wide range of feedback tube lengths, diameters and flow-rates.

5 Small-Scale Ejector

The purpose of the ejector in the new AFC actuator is to create suction to transfer the flow to the switching valve. It means that the ejector should operate against a backpressure (the ejector back-pressure is the valve inlet pressure). Since the ejector efficiency heavily depends on its capability to create low pressure, it is essential to

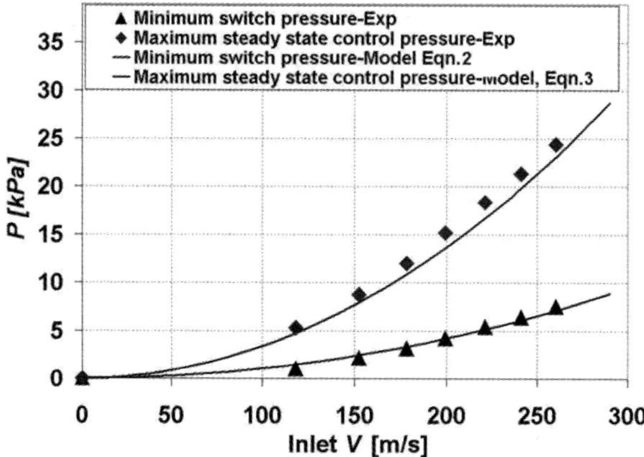

Fig. 7 The maximum (P_c) and the minimum switching pressure vs. inlet flow-rate.

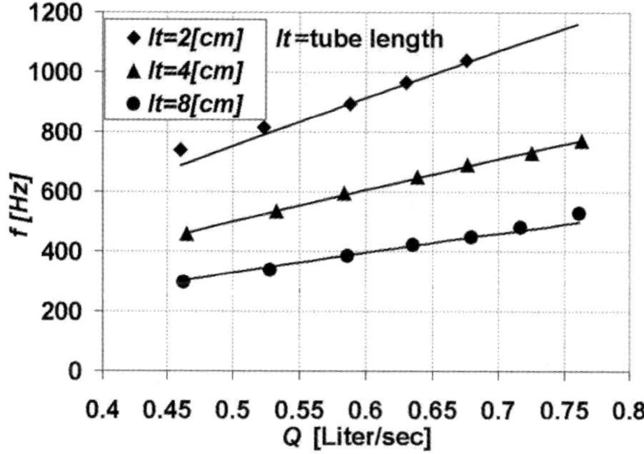

Fig. 8 Comparison between self-oscillation model and experimental results for small-scale valve for constant tube diameter and several tube lengths.

test and optimize the ejector to operate with an exit load (back-pressure). In order to increase the ejector's ability to create suction flow an increase in the jet energy is required, which may be achieved by increasing the jet velocity. Since the inlet nozzle is small and the inlet pressure is high, ejectors with converging nozzles will become choked, limiting the jet velocity to the speed of sound. To increase the jet velocity, a converging-diverging inlet nozzle was implemented. Assuming isentropic flow it is possible to predict the nozzle performance [16]. Figure 9 presents a comparison between several ejector inlet nozzles. It can be seen that the entrainment ratio Q_2/Q_1 (where Q_2 is the suction flow-rate and Q_1 is the inlet flow-rate) vs. the normalized back-pressure is superior for the converging-diverging inlet nozzle with a throat cross section of 1.01×1.2 mm^2. By examining the data shown in Figure 9 we can

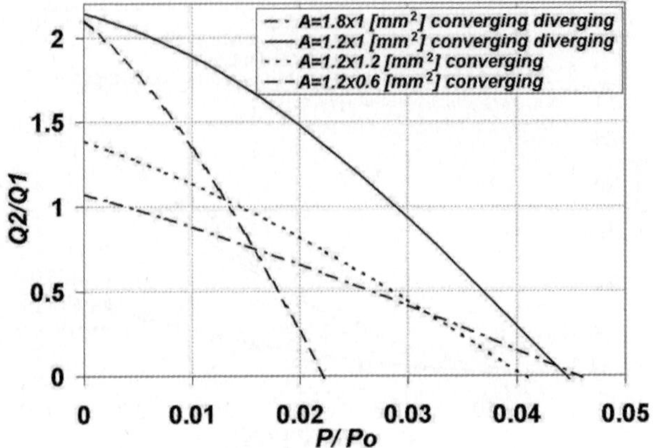

Fig. 9 A comparison between several small-scale ejectors inlet nozzles.

conclude that the supersonic flow improves the ejector performance, especially with back-pressure applied.

6 Fourth Generation Small-Scale Actuator Characteristics

Figure 10 presents data for the fourth generation valve connected to several small-scale ejectors. The figure presents the valve inlet flow-rate vs. inlet static pressure (solid black line) and the ejectors suction flow-rate vs. the valve inlet pressure. The superiority of the ejector with a converging-diverging inlet nozzle and with a throat cross section of 1.01×1.2 mm^2 can be clearly seen.

Seifert [17] compared different actuators according to the overall figure of merit (OFM), where data was available in the open literature. It was clear that the new actuator is significantly superior to electro-magnetic and plasma actuators and also as compared to piezo-fluidic actuators above output velocity of 60 m/s.

7 Conclusions

A new actuator concept based on the combination of steady-suction and oscillatory blowing was modeled, designed, fabricated and tested. The new actuator is based on the combination of an ejector and a bi-stable fluidic amplifier. The actuator can achieve near sonic speeds and high operating frequencies that are relevant to compressible flow control. Two theoretical models were developed: the *jet deflection model* and the *selfoscillations frequency model*. The former describes the switching valve operation based on a jet deflecting between two inclined walls due to a cross-flow pressure gradient, while the latter predicts the frequency as a function of

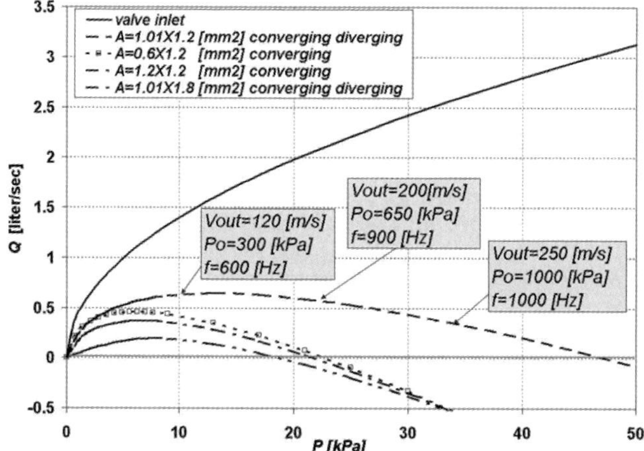

Fig. 10 Several possible operation points, showing different ejectors connected to the switching valve.

the feedback-tube length and diameter, valve inlet flow-rate and additional empirical relations. The shape of the ejector inlet nozzle and dimensions were found to be very important design parameters. A converging-diverging inlet nozzle improves the ejector performance, especially when back-pressure is applied. The ejector was shown to be capable of doubling the total mass flow-rate through the actuator. The entrained flow can be sucked into the device from the external or internal boundary layer to be controlled. Detailed static and dynamic tests showed that the actuator is capable of producing the necessary frequencies and amplitudes relevant for active separation control in compressible flows. The new actuator is light, compact, robust and has no moving parts; it therefore provides superior efficiency compared to existing actuators at similar flow conditions.

Acknowledgments

The authors would like to thank the following members of the technical staff: Eli Kronish, Mark Vasserman, Avraham Blas and Eli Nevo, for invaluable technical assistance. Special thanks to Mr. Shlomo Paster, the laboratory mechanics specialist, for the special efforts undertaken to make the experiment possible. The authors would also like to thank Assaf Nahum, Eli Ben-Hamou, Oksana Stalnov, Vitalei Palei, Yoni Yom-Tov and Yuri Borisinkov, members of the Meadow aerodynamic laboratory.

References

1. Seifert, A., Greenblatt, D. and Wygnanski, I., Active separation control: An overview of Reynolds and Mach numbers effects, *Aerospace Science and Technology* **8**, 2004, 569–582.
2. Seifert, A. and Pack, L.G., Effects of compressibility and excitation slot location on active separation control at high Reynolds numbers, *J. Aircraft* **40**(1), Jan./Feb. 2003, 110–119.
3. Seifert, A., Eliahu, S., Greenblatt, D. and Wygnanski, I., Use of piezoelectric actuators for airfoil separation control (TN), *AIAA Journal* **36**(8), 1998, 1535–1537.
4. Amitay, M., Smith, D.R., Kibens, V., Parekh, D.E. and Glezer, A., Aerodynamic flow control over an unconventional airfoil using synthetic jet actuators, *AIAA Journal* **39**(3), 2001, 361–370.
5. Margalit, S., Greenblatt, D., Seifert, A. and Wygnanski, I., Delta wing stall and roll control using segmented piezoelectric fluidic actuators, *AIAA J. Aircraft*, May/June 2004 (previously AIAA Paper 2002-3270).
6. Ouzzane, M. and Adieus, Z., Model development and numerical procedure for detailed ejector analysis and design, *Applied Thermal Engineering* **23**(18), 2003, 2337–2351.
7. Da-Wen, S. and Eames, I.W., Recent development in the design theories and applications of ejector – A review, *Journal of the Institute of Energy* **68**, 1995, 65–79.
8. Kirshner, J.M., *Fluid Amplifiers*, McGraw-Hill Book Company, 1966.
9. Viets, H., Flip flop jet nozzle, *AIAA Journal* **13**, 1975, 1375–1379.
10. Raman, G., Rice, E.J. and Cornelius, D., Evaluation of flip-flop jet nozzles for use as practical excitation devices, *ASME Journal of Fluids Engineering* **116**, 1994, 508–515.
11. Arwatz, G., Development and modeling of suction and oscillatory blowing actuator for flow control applications, Master Thesis, Tel-Aviv university, Israel, August 2006.
12. Kirshner, J.M. and Katz, S., *Design Theory of Fluidic Components*, Academic Press, New York, 1975.
13. Simoes, E.W., Furlan, R. and Pereira, M.T., Numerical analysis of a microfluidic oscillator flowmeter operating with gases or liquids, in *Technical Proceedings of the Fifth International Conference on Modeling and Simulation of Microsystems, MSM 2002*, San Juan, Puerto Rico.
14. Tippetts, J.R., Ng, H.K. and Royle, J.K., An oscillating bi-stable fluid amplifier for use as a flowmeter, *Fluidics Quarterly* **5**(1), 1973.
15. Fox, J.A., *Transients Flow in Pipes, Open Channels and Sewers*, Ellis Horwood Limited, 1989.
16. Streeter, V.L. and Wylie, E.B., *Fluid Mechanics*, First SI Metric Edition, McGraw Hill Ryerson, 1981.
17. Seifert, A., Closed-loop AFC system: Actuators, in *Active Flow Control 2006*, Papers Contributed to the Conference "Active Flow Control 2006", R. King (Ed.), NNFM, Vol. 95, Springer-Verlag, Berlin, 2007.

Numerical Investigation of a Micro-Valve Pulsed-Jet Actuator

Karen L. Kudar and Peter W. Carpenter
*School of Engineering, University of Warwick, Coventry CV4 7AL, U.K.;
E-mail: {karen.kudar, p.w.carpenter}@warwick.ac.uk*

Abstract. A micro-valve pulsed-jet vortex-generator driven by piezoelectric actuation was successfully modelled numerically to determine the feasibility and response characteristics of such a design. This includes: modelling the dynamic motion of a unimorph cantilever and the fluid-structure interaction occuring between the unimorph and the fluid flowing over such a structure; the unsteady developing channel flow that would occur through the outlet orifice was also modelled. The response time of the actuator was found to be governed by the micro-valve opening rather than the time taken to establish the jet. However, the resistance of the pulsed-jet actuator was shown to be governed by the outlet orifice.

Key words: Micro-scale valve, unimorph, fluid-structure interaction, pulsed-jet.

1 Introduction

The study described in the present paper was carried out as part of an EC research programme called AEROMEMS II that was managed by BAE Systems. One of the aims of the programme was to develop pulsed-jet actuators for the control of flow separation over the rear flaps of the wings of large aircraft. In order that the pulsed jet act effectively as a vortex generator, the velocity ratios between orifice exit and mainstream flow need to be of the order of one to two (see [1, 2]). Estimating a flow velocity of 50–100 m/s over the rear flap, we have used 100 m/s as the target valve exit velocity. Further details on the practical motivation for the design of the actuator are given in [3].

The initial simple valve design modelled here comprises a unimorph-cantilever that deflects to allow air to escape when required (see Figure 1a). In its resting position the valve is closed and a seal is formed. An electric field is then applied across the thickness of the piezoelectric material and, as it is bonded to an elastic material which cannot contract, a moment is generated that forces the unimorph

J.F. Morrison et al. (eds.), IUTAM Symposium on Flow Control and MEMS, 45–51.
© 2008 *Springer. Printed in the Netherlands.*

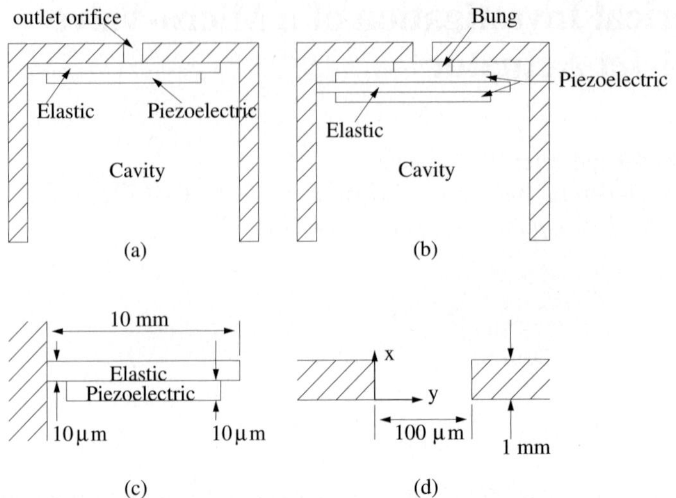

Fig. 1 Diagram of (a) the initial design of the valve using a unimorph-cantilever, (b) the improved design using a bimorph and with a gap over its top surface to ensure valve can open, (c) the typical dimensions of the unimorph and (d) the typical dimensions of the outlet orifice.

to bend. Once the unimorph bends a small gap is formed that allows air to escape from the cavity, which is at an elevated pressure, through the outlet orifice into the boundary layer on the surface above. Switching the electric field off, leaves the unimorph free to return to its resting position and the valve would be closed once again. The following study was carried out to establish the feasibility of such a design and to determine such characteristics as the response times and pressures required for the valve to function as required.

2 Numerical Methods

Owing to the comparatively thin nature of the unimorph we used a combination of the simplified piezoelectric constitutive equations and thin-plate theory to model its motion. The cavity pressure acting on the underside of the unimorph and the fluid flowing over its upper surface interact with the unimorph and affect its motion; to calculate the pressure generated by the fluid motion, lubrication theory was used.

The outlet orifice cross-section has a high aspect ratio and the non-linear developing flow characteristics will occupy most, if not all of the orifice length and cannot be neglected. The flow can be modelled initially as laminar boundary layers, growing in space and time, at each wall (in x direction, see Figure 1d). The boundary layers are assumed to have a parabolic velocity profile and a constant mean velocity at the centre of the channel; this is substituted into the boundary-layer, isothermal-energy and continuity equations which are then integrated between the wall ($y = 0$)

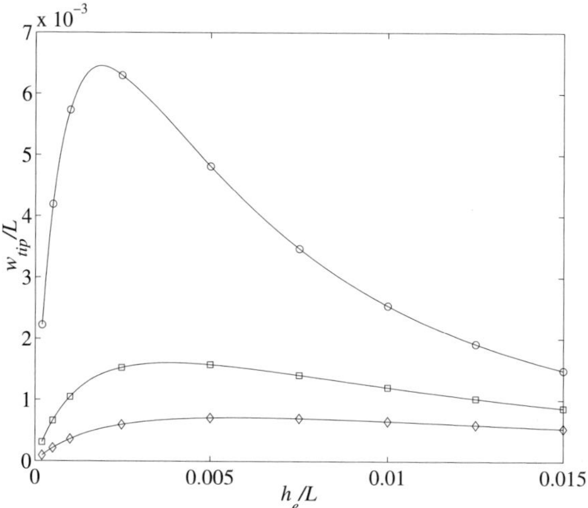

Fig. 2 The variation of non-dimensional tip deflection (w_{tip}/L) with non-dimensional thickness of the elastic layer (h_e/L) for a unimorph with piezoelectric layer thickness of $0.005L$ ($\circ$), $0.010L$ ($\square$) and $0.015L$ ($\diamond$) for fixed electric field strength.

and the centre of the channel ($y = h$) to find the variation in centreline velocity with space (x) and time.

3 Results

3.1 Unimorph with No Fluid-Structure Interaction

Figure 2 shows the tip deflection produced by a unimorph with various thicknesses of the elastic and piezoelectric layers. It can be seen that the thinner the piezoelectric layer the more the unimorph deflects. Also, there is an optimum thickness ratio (of elastic to piezoelectric) that produces the largest deflection of $h_e/h_p \approx 0.39$, where h_e and h_p are thicknesses of the elastic and piezoelectric layers, respectively. This compares well with experimental findings (see [6]). The dynamic motion of the unimorph can be seen in Figure 3. The motion of the unimorph is shown to open in a whiplash-type action; this is consistent with the contraction of the piezoelectric material travelling down the length of the unimorph in a wave-like motion. It should be noted that this could possibly be disadvantageous as the time taken to open is increased slightly by the unimorph deflecting upwards.

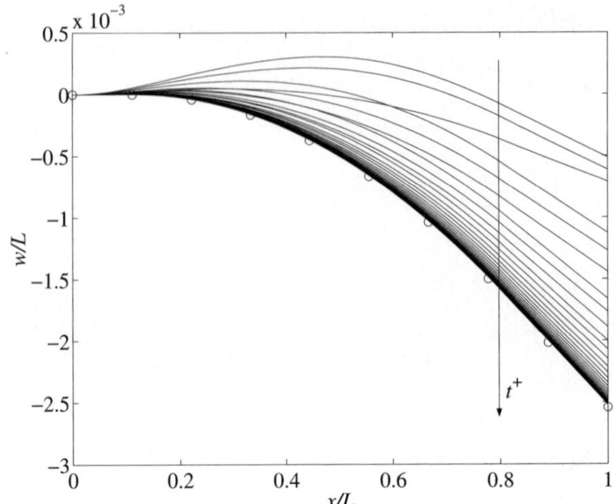

Fig. 3 Dynamic motion of the unimorph (–) converging towards the steady-state solution (○).

3.2 Unimorph with Fluid-Structure Interaction

The valve as it was initially designed (as depicted in Figure 1a) can neither open nor close successfully. Due to the positioning of the unimorph (flush against the ceiling of the cavity) there is a large pressure loading on the underside of the unimorph that means the valve cannot open. By ensuring that there is a gap between the top of the unimorph and roof of the cavity, as depicted in Figure 1b, an equalising pressure acts over most of the unimorph thereby allowing the valve to open; a bung will be required to ensure that the valve is closed when the unimorph is in its resting position. On closing, the squeeze-film effect (see [7] for details) that is generated by the flow between the unimorph and the cavity ceiling means that the valve cannot close. One possible solution (although not implemented here) would be to substitute a bimorph for the unimorph thereby providing a force to close the valve such that it can overcome the squeeze-film effect (see Figure 1b).

After incorporating the gap-bung design improvement into the model the valve can now open when there are relatively large cavity pressures. Figure 4 shows a plot of the unimorph tip deflection with time as the cavity pressure varies; the effects of the fluid-structure interaction are apparent. When the cavity pressure is at a sufficiently low level, the opening force of the unimorph is relatively strong and can overcome the influence of the pressure forces acting on the unimorph; the valve opens in a manner similar to that found with no fluid-structure interaction. For larger cavity pressures the combined action of the reduced pressure in the fluid flowing over the upper surface and the cavity pressure acting on the under-side of the unimorph prevents the valve from opening as far. It can also be seen that as the cavity pressures get larger the valve cannot open at all; hence a limiting cavity pressure still exists.

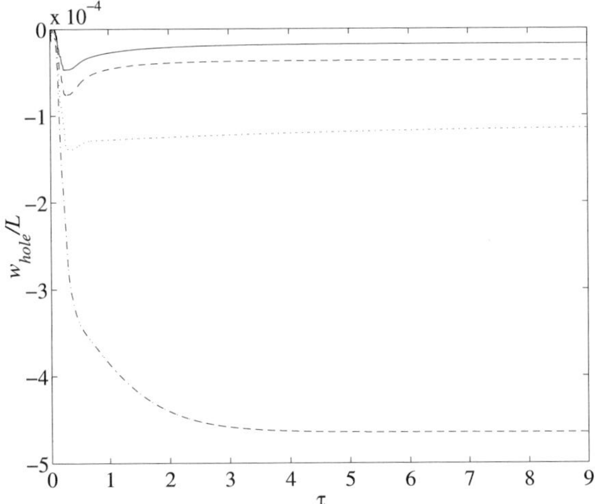

Fig. 4 Variation of the non-dimensional deflection of the unimorph, where the outlet orifice is positioned (w_{hole}/L) with non-dimensional time (τ), for different cavity pressures: $P_{\text{cav}} = 45$ kPa (–), 35 kPa (– – –), 25 kPa ($\cdots$), 15 kPa (–·–).

3.3 Outlet Orifice

The variation of the centreline velocity with time at the entrance and exit of the outlet orifice is plotted in Figure 5. In both cases approximately the same time is required to reach a fully-developed state. This implies that regardless of the length of the orifice it would take approximately the same time to reach a fully-developed state. Therefore, considering that it requires a smaller pressure drop to produce a jet flow speed of similar magnitude it would be beneficial to have a shorter length of orifice.

4 The Complete Valve

In their present form the unimorph and the outlet-orifice codes cannot be combined together. However, an estimate of the combined effect of the whole valve can be made by considering the two separate parts as a whole. For example, pressure drops across the micro-valve must be of the order of 1 kPa for the valve to be able to function and also to drive a jet of the required velocity. The pressure drop across the outlet orifice to drive a jet of the order of 100 m/s must be of the order of 10 kPa. Adding these pressure drops together in series gives an overall pressure drop of the order of 10 kPa. Thus the flow resistance of the micro-valve is comparatively minor compared with that of the exit orifice. Typical response times of the outlet orifice are of the order of 1 µs and the opening time of the unimorph is of the order of

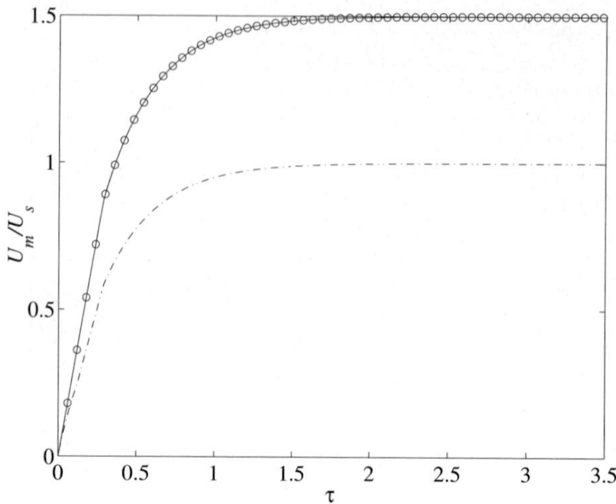

Fig. 5 Numerical results for the variation of the non-dimensional centreline velocity (U_m/U_s) with time (τ) at the orifice inlet (–·–) and at the orifice exit (–). (o) represents the unsteady, fully-developed flow solution.

100 µs. Therefore, the valve response time is wholly dependent on the response of the unimorph and is of the order of 100 µs.

Both required pressure and response times seem acceptable and for practical purposes are of realistic magnitude and on a superficial level appear to be unproblematic. Hence, theoretically, at least, the valve design considered here is feasible provided the design alterations, previously mentioned, are made.

Acknowledgments

The research reported herein was supported partly by the U.K. Engineering and Physical Sciences Research Council and partly by the AEROMEMS II project. The AEROMEMS II project is a collaboration between BAE Systems, Dassault, Airbus Deutschland GmbH, EADS-Military, Snecma, ONERA, DLR, LPMO, Manchester University, LML, Warwick University, UB, Cranfield University, NTUA, and Auxitrol. The project is funded by the European Union and the project partners.

References

1. Suzuki, T., Nagata, M., Shizawa, T., Honami, S.: Optimal injection condition of a single pulsed vortex generator jet to promote the cross-stream mixing. *Experimental Thermal and Fluid Science* **17** (1998) 139–146.

2. Magill, J.C., McManus, K.R.: Exploring the feasibility of pulsed jet separation control for aircraft configurations. *J. Aircraft* **38** (2001) 48–56.

3. Warsop, C., Hucker, M., Press, A.J., Dawson, P.: Pulsed air-jet actuators for flow separation. *Flow, Turbulence and Combustion* **78**(3/4) (2007) 255–281.

4. Lockerby, D.A., Carpenter, P.W.: Modeling and design of microjet actuators. *AIAA Journal* **42** (2004) 220–227.

5. Kudar, K.L.: Flow control using pulsed jets. PhD Thesis, University of Warwick, UK (2004).

6. Li, X., Shih, W.Y., Aksay, I.A., Shih, W.H.: Electromechanical behaviour of pzt-brass unimorphs. *J. Amer. Ceram. Soc.* **87** (1999) 1733–1740.

7. Sherman, F.S.: *Viscous Flow*. International edn., McGraw-Hill Publishing Company, Singapore (1990).

Characterization of MEMS Pulsed Micro-Jets with Large Nozzles

Jean-Luc Aider[1,3], Fabien Harambat[1], Jean-Jacques Lasserre[1],
Jean-Françcois Beaudouin[1] and Christophe Edouard[2]
[1]*PSA Peugeot-Citroën, Research and Innovation Department, route de Gisy,
78943 Vélizy-Villacoublay, France; E-mail: fabien.harambat@mpsa.com*
[2]*Flowdit, 18 Avenue Guy de Maupassant, Z.A. de l'Agavon, 13170 Les Pennes
Mirabeau, France; E-mail: contact@flowdit.com*
[3]*Laboratoire PMMH, UMR CNRS 7636, ESPCI, 10 rue Vauquelin,
75231 Paris Cedex 05, France; E-mail: aider@pmmh.espci.fr*

Abstract. In this paper, we present the first characterization of new pulsed micro-jets. The specificity of these new actuators is their large nozzle for MEMS device: the total exit cross-section on a single actuators can reach up to 3×3.28 mm$^2 \simeq$ 10 mm^2. Two different geometries are presented: one with a single slot and one with three slots on a single MEMS device. We measure the jet velocities as a function of the inlet pressure. Maximum jet velocity can reach up to 28 m/s with a pulsation frequency around 100 Hz. Other phenomena are discussed, like the deviation of the jet axis from the vertical and the wrong evaluation of the jet velocity close to the nozzle. We finally apply the three slots actuators to a backward-facing step flow and show a small reduction of the recirculation length when the pulsation frequency is equal to the shear layer frequency.

Key words: Flow control, MEMS actuators, pulsed micro-jets, backward-facing step flow.

1 Introduction

Shape optimization to reduce the aerodynamic forces made possible a significant reduction of the mean drag of vehicles in the 1970s. Now that we are facing a new oil crisis, it appears necessary to achieve a new significant reduction of the drag coefficient of future vehicles. Such a reduction will only be possible with a "flow optimization", i.e. a modification of the flow structures using flow control systems. This objective is ambitious in two ways: it implies a better understanding of the base flow and it also requires technological innovations making the use of the system realistic for industrial applications. First attempts of flow control with vortex generators on vehicles have been recently achieved [1, 2]. Nevertheless, the need for very small, low consumption and cheap mechanical systems is important. Only

J.F. Morrison et al. (eds.), IUTAM Symposium on Flow Control and MEMS, 53–58.
© 2008 *Springer. Printed in the Netherlands.*

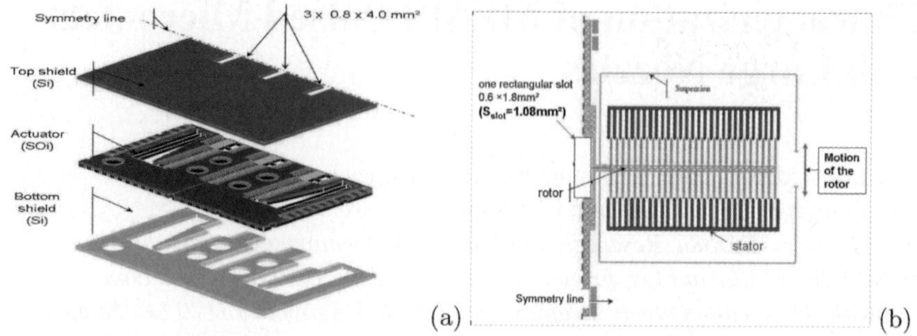

Fig. 1 Side view of the three layers constituting the MEMS actuator (a): one SOI layer between two SI layers. Upper view of the comb fingers structure etched in the SOI layer (b). The red parts are not moving while the blue parts are driven by electrostatic forces.

the MEMS actuators, and especially pulsed micro-jets, seem to be able, in the future, to fulfill these constraints [3].

2 The MEMS Micro-Jets

The MEMS actuators are full silicon systems manufactured in only three steps in clean rooms (photolithography, wet oxidation and double side etching). Each actuator is composed of one SOI (380-2-200 μm) substrate packaged between two Si (380 μm) substrates. All substrates are processed by DRIE (Deep Reactive Ion Etching or Plasma etching) etching and then glued. The system is then monolithic (full silicon) so that it is more robust and can endure thermal deformation (same thermal properties for all the system).

The top SI substrate contains the three etched exits, while the bottom SI substrate contains the etched holes for electric and pneumatic connections (Figure 1a). The springs and comb drives are etched into the SOI substrate. The deformation of four springs enables the comb drives motion which is created by electrostatic forces (Figure 1a). The high amplitude motion of the comb fingers (up to 0.8 mm which is very large in the MEMS world) is obtained thanks to a specific design of the comb fingers, patented by the two companies [4].

The advantage of electrostatic forces is also a low power consumption. In our case, the power consumption P is less than 500 μW for 200 V input. The motion of the comb fingers structures drives the shutter, which will close the jet nozzle with a typical frequency $f = 100$ Hz.

The MEMS actuators overall planar dimensions are 2×2 cm^2 for a 1.3 mm thickness (Figure 1a). The slot of the mono-jet actuator has a rectangular cross-section 0.6×1.8 mm $= 1.08$ mm^2, while the three-jets actuator has three rectangular slots whose blowing cross-section area is $S_j = 0.8 \times 4.1$ mm $= 3.28$ mm^2. The three jets actuators has then a very large total blowing cross-section: $S_{tot} = 3 \times S_j \approx$

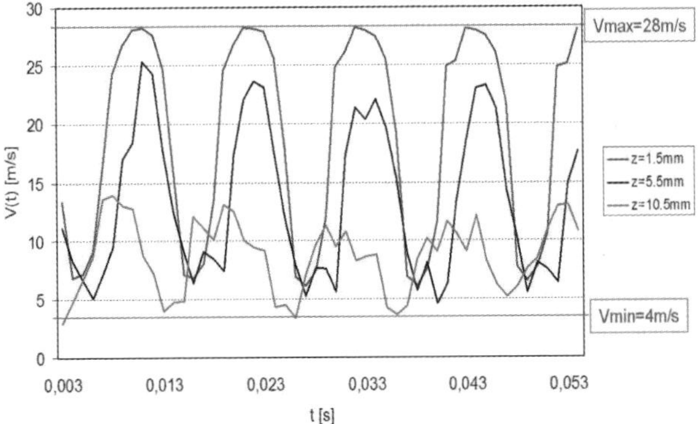

Fig. 2 Vertical velocity time series for three different heights above the nozzle: $z = 1.5$, 5.5 and 10.5 mm.

10 mm^2. These large blowing sections are necessary to have a significant effect on the large boundary layers encountered in automotive aerodynamics (typically $\delta \approx 3$ cm for $U_0 = 40$ m/s).

3 Characterization of the MEMS Actuators

3.1 The Single Slot Actuator

The hot-wire measurements are carried out with a single wire probe located above the slot, with the wire along the length of the slot (Figure 2) for an inlet pressure $P_{in} = 9$ mbar. The measurements close to the nozzle tends to induce a blockage on the jet so that the maximum instantaneous vertical velocity is probably slightly higher than 28 m/s obtained when the probe is located close to the nozzle ($z = 1.5$ mm). This will be confirmed in the following section.

Even if the time series shown in Figure 2 underestimates the real jet velocity, we see that the actuators work well with a maximum jet velocity $U_j \simeq 28$ m/s and the right frequency $f = 90$ Hz for $z = 1.5$ mm. We can also see the reduction of the jet velocity for increasing height, but we still measure $U_j \simeq 14$ m/s for $z = 10.5$ mm above the nozzle. We also observed a slight increase of the maximum jet velocity when we increased the pulsation frequency from 30 to 90 Hz.

3.2 The Three Slots Actuator

We first show in Figure 3 a vertical velocity profile as a function of the height above the central slot for a given inlet pressure ($P_{in} = 900$ Pa) and a given frequency ($F_j =$

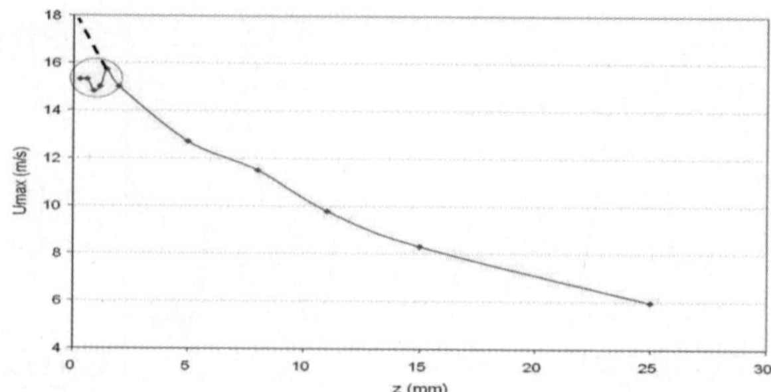

Fig. 3 Vertical evolution of the jet velocity measured by a hot-wire above the nozzle for an inlet pressure $P_{\mathrm{in}} = 900$ Pa and a pulsation frequency $F_j = 103$ Hz.

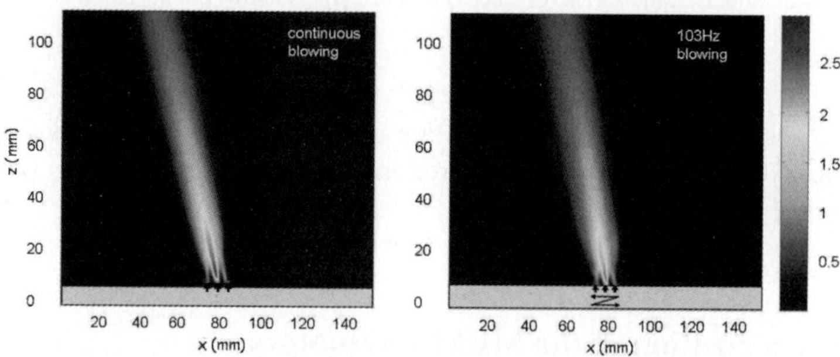

Fig. 4 Velocity fields obtained by PIV measurements above the three slots for the continuous jets (a) and the pulsed jets (b) showing a clear deviation of the jet axis compared to the vertical.

103 Hz). The same comments as in the previous section hold for this actuator: we probably underestimate the maximum jet velocity for all the reasons listed above. The discrepancy of the measurements close the jet nozzle can be observed clearly on the graph: we observe a non-physical drop of the jet velocity for $z < 2$ mm. The maximum jet velocity should be at least 18 m/s at the nozzle exit.

We also performed PIV measurements above the three slots to characterize the jets velocity field. Figure 4 shows a contour of velocity magnitude above the three slots for the continuous (a) and pulsed jet (b). In both cases, a clear deviation of the jet axis from the vertical axis can be observed ($\alpha \simeq 12.7°$ for the pulsed jet and $\alpha = 16.04°$ for the continuous jet). This behavior has been observed for different pressures and pulsation frequencies. It probably is related to the actuator geometry but no explanation has been found for the moment.

Figure 5 shows two types of mean velocity profiles measured at a given height $z = 0.5$ mm above the three slots. In Figure 5a we show mean velocity profiles taken along the central slot. $x = 0$ is the center of the slot, which is 4 mm long. Both the continuous and the pulsed jet exhibit a nearly flat profile with a slight reduction in the center of the profile. Figure 5b shows velocity profiles measured across the three slots. Once again, we cannot see clear difference between the pulsed and continuous

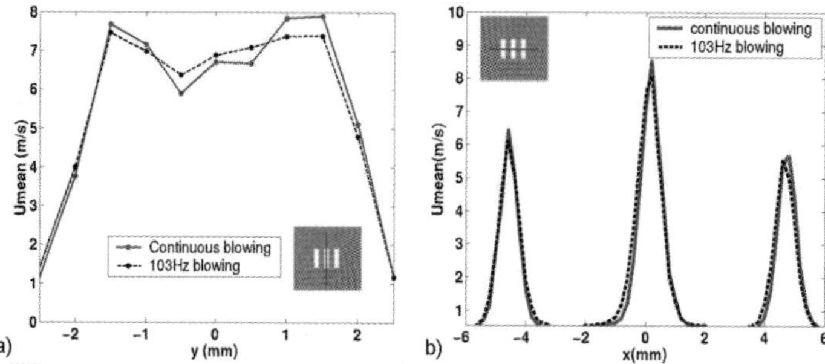

Fig. 5 Velocity profiles obtained by PIV measurements above the 3 slots for the continuous jets and the pulsed jets. The first profile (a) is taken along the central slot for $z = 0.5$ mm while the second profile (b) is taken across the three slots for $z = 0.5$ mm.

jets. The only noticeable feature is the higher maximum jet velocity obtained for the central slot: $U_j \simeq 8$ m/s for the central slot instead of $U_j \simeq 6$ m/s for the side slots. In fact, the only difference between the continuous and pulsed jets was observed in the RMS values of the vertical velocity, which are higher in the pulsed jet than in the continuous case ($U_j$rms $\simeq 2.5$ m/s for the pulsed jet instead of $U_j$rms $\simeq 2$ m/s for the continuous case).

4 Application to the Backward-Facing Step Flow

Finally, we carried out a preliminary flow control experiments on a backward-facing step flow. We integrated a single three-slot actuator at the edge of the step, with the slots parallel to the edge. The free-stream velocity is $U_0 = 15$ m/s so that the pulsation frequency matched the shear layer frequency $f = 91$ Hz. The step height was fixed $h = 6$ cm, giving a Reynolds number $Re = U_0 h / \nu = 64000$.

For a pressure inlet $P_{\text{in}} = 9$ mbar, corresponding to a mean jet velocity $u_j \simeq 8$ m/s, we measured a reduction of the mean recirculation length $\Delta X_r \simeq 20$ mm $= 0.34h$, for a mean recirculation length $X_r \simeq 5.5h$ (Figure 6). The reduction of the recirculation length was observed only when the pulsation frequency matched the shear layer frequency, confirming the role of the pulsation for an efficient control of a separated flow. Of course, the reduction of the recirculation length is not necessarily correlated to a drag reduction which could not be measured in this experiment.

5 Conclusion

We presented in this paper new MEMS pulsed micro-jet, whose specificity is the very large nozzle section. Thanks to technological innovations, it was possible to

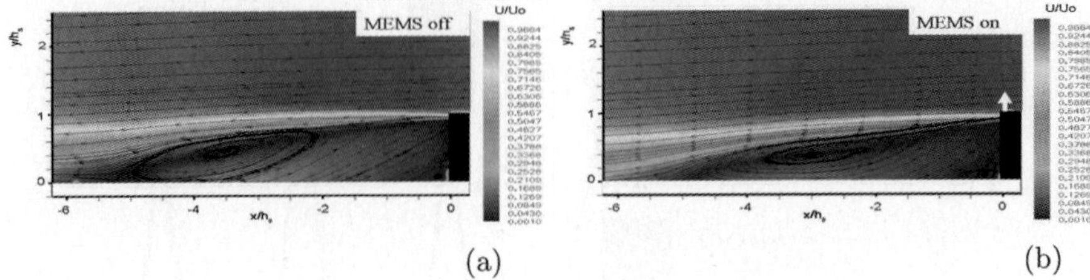

(a)   (b)

Fig. 6 Modification of the mean recirculation bubble downstream of a backward-facing step. The flow is coming from right to left and the step edge is located at $x/h = 0$. (a) The recirculation bubble for the reference flow, (b) the modification of the recirculation bubble when the pulsed micro-jets are turned on.

use electrostatic forces to open and close the large rectangular openings of these full silicon systems. With the single slot actuator, we could reach a jet velocity close to 30 m/s, which should be improved by recent technological improvements. The three-slots actuators has a very large total blowing cross-section (nearly 10 mm^2) so that the flow rate is very large for a MEMS jet. Even with this large flow rate, we could measure nearly 18 m/s for an pressure input $P_i = 9$ mbar. Larger pressure inputs are currently explored. We also underlined some problems, like velocity measurements close to the nozzle, and some unexplained deviation of the jet axis. We finally demonstrated the efficiency of a single three-slots actuators on a backward-facing step flow. When the frequency of the jet matched the shear layer frequency, we observed a clear reduction of the recirculation length. We expect the new generation of these systems to reach easily at least 40 m/s, which is expected for automotive applications.

References

1. Aider, J. L., Lasserre, J. J., Herbert, V.: Active vortex generators to reduce drag and lift of a vehicle. European Patent No. EP 04.29001 (2003).
2. Aider, J. L., Beaudoin, J. F., Wesfreid J. E.: Drag reduction of a 3D bluff body using vortex generators. In *Proceedings of 4th Conference on Bluff Body Wakes and Vortex Induced Vibrations*, Santorini, Greece, June 21–24 (2005).
3. Greenblatt, D., Wygnanski, I. J.: The control of flow separation by periodic excitation. *Prog. Aerospace Sci.* **36** (2000) 487–545.
4. Lasserre, J. J., Edouard, C., Giovannelli, G.: Optimization of electrostatic forces in MEMS devices. French Patent No. FR 06.02387 (2006).

Magnetically Actuated Microvalves for Active Flow Control

Olivier Ducloux, Yves Deblock, Abdelkrim Talbi, Leticia Gimeno, Nicolas Tiercelin, Philippe Pernod, Vladimir Preobrazhensky and Alain Merlen
Joint European Laboratory LEMAC: IEMN-DOAE – UMR CNRS 8520, LML – UMR 8146, Ecole Centrale de Lille, BP 48, 59652 Villeneuve d'Ascq Cedex, France; E-mail: olivier.ducloux@iemn.univ-lille1.fr

Abstract. This paper describes the design, fabrication and characterization of silicon based, high flow rate microvalves for the active control of separated air flows. The fabricated system provides pulsed microjets with an outlet speed reaching 150 m/s at an actuation frequency ranging from static actuation to 2.2 kHz, using either electromagnetic actuation or a self oscillating mode. After a brief introduction, the microvalve dimensioning and fabrication process are presented. The actuation techniques used are then described and discussed.

Key words: MEMS, microfluidics, electromagnetic actuation, flow control.

1 Introduction

Flow separation is generally accepted to consist in the detachment of a fluid from a solid surface. In the case of aircraft flight, this phenomenon generally happens on the front edge of the wings and is associated with loss of lift, drag increase, pressure recovery losses, etc. It has been proven that this separation can be managed by blowing oscillatory air jets in the boundary layer through submillimetric holes situated on the surface near the separation area [1, 2].

Two types of air blowing techniques exist: pulsed jets provide only air injection whereas synthetic jets provide successive air injection and sucking. The commonly used method to achieve such jets consists in using bulky mechanical solutions that imply high airborne load and important power consumption [3]. Moreover, most of these solutions are massive and are not suited to be placed in airfoils for aeronautical applications. Many experiments have also been conducted with the use of acoustic waves diffused through fences as jet-generators to prove the efficiency of microjets as a momentum provider for active flow control applications. Much work has recently been achieved in the development of microscale actuators for an easier integration and independent control of each microjet [4]. This paper describes the

J.F. Morrison et al. (eds.), IUTAM Symposium on Flow Control and MEMS, 59–65.
© 2008 *Springer. Printed in the Netherlands.*

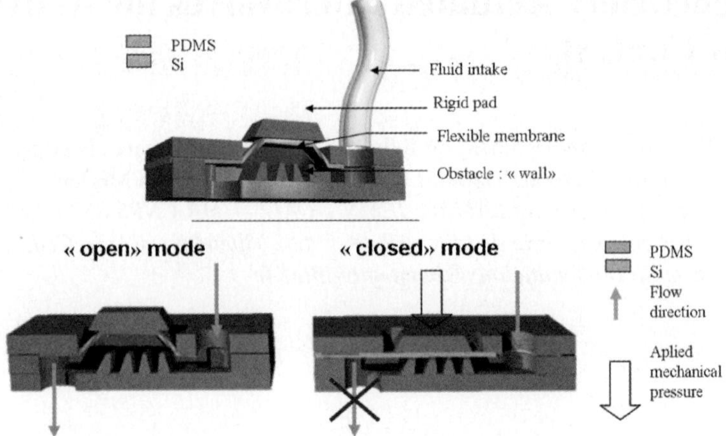

Fig. 1 General layout of the microvalve and actuation principle.

design, fabrication and characterization of silicon-based, high flow rate microvalves adapted to the creation of high speed and high frequency pulsed microjets.

2 General Layout

The microvalve solution developed in the LEMAC/IEMN-LML laboratory is based on a simple geometry: an internal silicon channel covered by a flexible membrane (polydimethylsiloxane, PDMS) (cf. Figure 1) and designed to be compatible with several actuation techniques, depending on the application. When fed with a pressurized source of air, the membrane blows up, providing the valve aperture and the creation of a high speed jet at the valve exhaust. A rigid silicon pad, processed on the flexible membrane, is pushed towards several silicon walls processed within the channel, preventing the gas from passing through the valve.

The system is designed in order to provide very high speed jets (150 m/s) at moderate inlet-outlet pressure difference (0.5 bar). The membrane and channel dimensions are chosen in order to minimize the pressure drop in the channel in open mode.

3 Fabrication Process

To achieve simultaneously large dimension objects with the compulsory high precision (wall thickness = 150 μm), silicon based microfabrication techniques were used on commonly spread, 380 μm thick, $\langle 100 \rangle$ crystalline silicon wafers. Moreover,

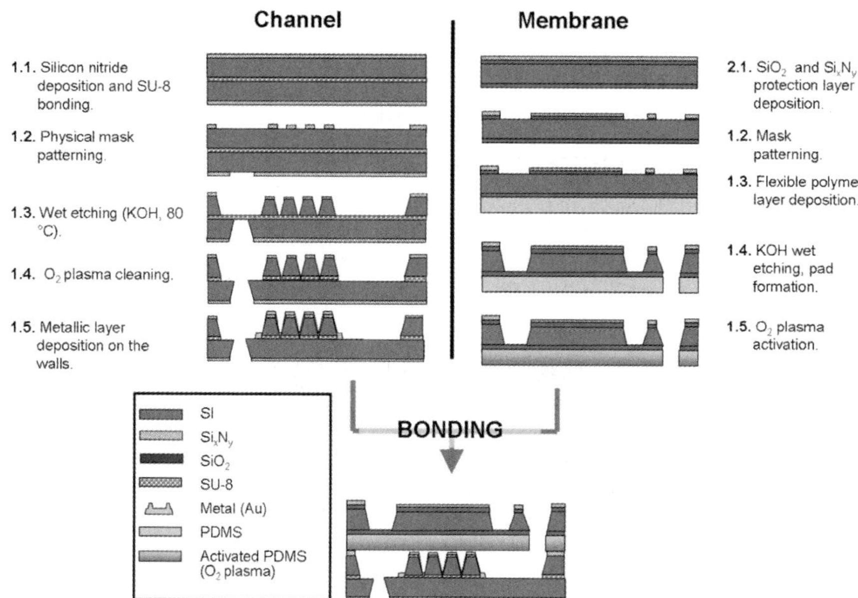

Fig. 2 Microvalve fabrication process, involving three layers of silicon independently processed and bonded together.

such fabrication techniques allow batch processing and cost reduction, as packs of 15 to 30 valves are to be used for wind tunnel experiments.

A specific fabrication process was successfully developed and used to achieve the large microchannel. The final process is developed in Figure 2.

Channel fabrication is achieved by wet etching (KOH, 80°C) two adhesively bonded wafers (SU-8 2002, 2 μm) through a silicon nitride mask (2000Å thickness) (cf. Figure 2, steps 1.1 to 1.3). The bonding SU-8 resist layer acts as a stopping layer for the etching step. The last step consists in RF sputtering a 2000Å Ti/Au layer on the wall pattern in order to prevent PDMS from sticking to the walls during the final bonding. On a third silicon wafer, the membrane and pad patterns are backside processed using ICP/RIE CF4/CHF3 dry etching in a 2000Å thick Si_3N_4 layer. A 60 μm thick PDMS layer is subsequently spin coated on the front side, then cured (step 2.3.). After KOH releasing of the fabricated membranes, both parts are plasma-activated [5] and contacted for final bonding (step 2.4). A final cleaving yields individual valves or ready-to-mount bars (cf. Figure 3). The fabricated microvalves are finally bonded to a vectoring plate (2 mm thickness) providing a good control of the microjet orientation for wind tunnel tests.

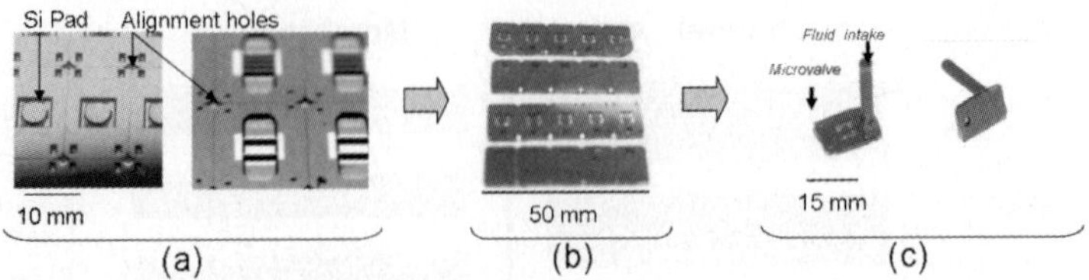

Fig. 3 Fabricated microvalves: (a) detailed fabricated wafers before final bonding, (b) cleaved bars and (c) individual prototypes.

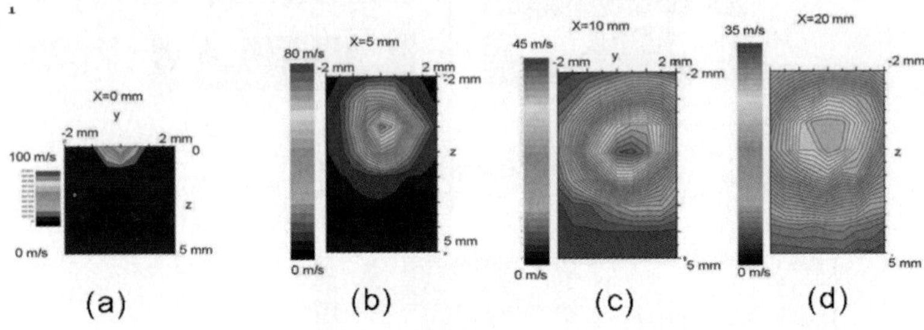

Fig. 4 Measured outlet speed, 30° vectoring plate, 1mm diameter outlet hole. (a), (b), (c), and (d) represent perpendicular cuts of the microjet at a distance respectively equal to 500 μm, 5 mm, 10 mm and 20 mm of the outlet hole.

4 Continuous Jet Mode Characterization

The fabricated microvalves were first of all characterized in the case of static/quasi-static mode actuation. The outlet speed was measured by hot wire anemometry using a commercial 1.25 mm length, 5 μm diameter 55P11 DANTEC hot wire anemometer. The measurements were achieved on a micro-displacement bench permitting the precise control of the anemometer's position and orientation. Independently, a shadowgraph system permitted the monitoring of the microjet shape, and leakage diagnosis on the measured microvalves. Figure 4 shows cartographies of the measured outlet speed obtained using a 30° inclined, 2 mm long, 1 mm diameter vectoring channel. This figure shows a high outlet speed value reaching 100 m/s in the centre of the jet near the outlet hole (a), which remains quite high (80 m/s) 5mm away form the outlet hole (b). Moreover, a low microjet aperture of approximately 20° can be observed as a consequence of the guidance effect obtained with the vectoring plate. The maximum outlet speed reached 150 m/s in the case of a vectoring channel perpendicular to the microvalve plane.

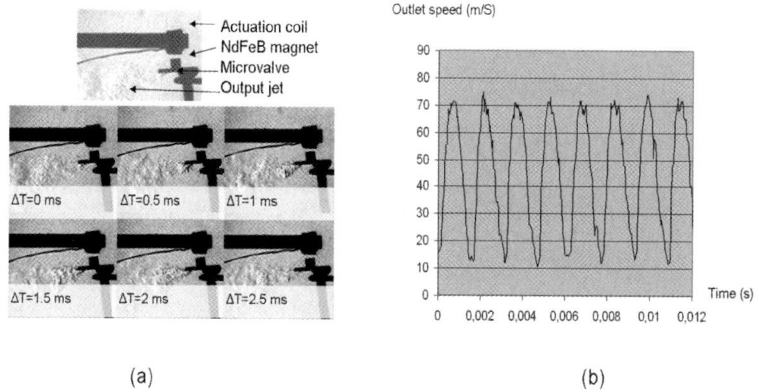

Fig. 5 (a) Stroboscopic shadowgraph of the exhaust jet, (b) exhaust speed vs. time (4 wall valve, 550 Hz actuation frequency, 0.5 A coil current, 70 m/s max. outlet speed). The high actuation current can be reduced by increasing the coil winding number.

5 Actuation

5.1 Low Frequency: Electromagnetic Actuation

Electromagnetic actuation means provide the compulsory high force and displacement needed for the aimed high flow rate, low frequency applications. A permanent NdFeB magnet (cylindrical, 3 mm diameter, 2 mm height), bonded to the pad and coupled with a 100 windings, 200 µm diameter wire coil are used to actuate the valve. When fed by the actuation current, the coil produces a magnetic field gradient on its proximity, generating a force on the permanent magnet directed normal to the membrane plane.

The exhaust speed is measured using hot wire anemometry. Measurements are made at the centre of the jet, 500 µ away from the outlet hole. As the measured value is an integration of the speed along the wire length, the values presented are slightly underestimated. Figure 5 shows the good impermeability in closed mode during dynamical actuation (550 Hz, 0.5 A).

5.2 High Frequency: Self-Oscillating Actuation

The presence of silicon walls under the flexible PDMS membrane, permit not only a good impermeability in closed mode, but also the presence of a pressure drop in opened mode under certain geometrical conditions (number of walls, membrane stiffness and displacement). As the system is pressurized, the induced dynamical pad displacement yields dynamical actuation of the microvalve.

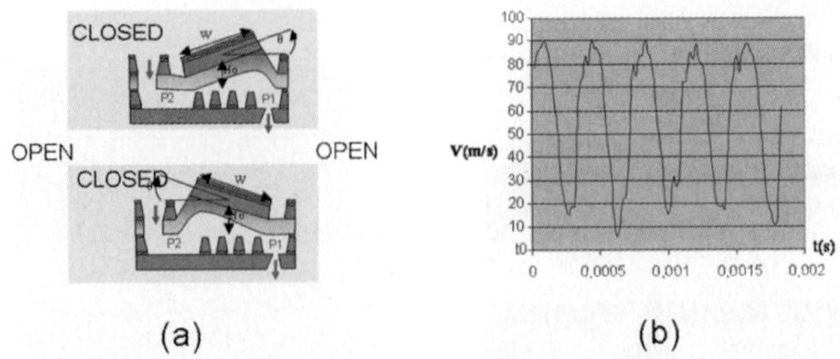

Fig. 6 (a) Self-oscillation actuation principle, (b) measured outlet speed, 4 wall valve, 1.5 bar inlet pressure, 2.4 kHz actuation frequency, 90 m/s outlet speed, 10 m/s leakage speed.

The experienced pad movement, described in Figure 6, corresponds to a mechanical instability of the global system (membrane + rigid pad) due to the strong coupling between the first flexion and torsion modes via the pressure drop in the fluid under the membrane [6]. The high actuation frequency obtained can be tuned in the range of 1 to 2.2 kHz by changing the mobile part inertia or by adding a spring-like interaction on the mobile part using two mini-magnets. Moreover, no external electrical energy is needed for actuation, which makes this actuation concept interesting for embedding in airfoils.

6 Conclusion

High flow rate, large frequency range microvalves were designed, fabricated and characterized. Microjet speeds reaching 150 m/s were measured at the outlet of the microsystem using a vectoring plate for the control of the microjet orientation. Moreover, specific actuation techniques were also setup for dynamical actuation of the microvalve: electromagnetic actuation, consisting in a coil-magnet coupled system, was used in the range 0 to 600 Hz and a self-oscillating actuation mode was successively used in the range of 1 to 2.2 kHz without external energy feeding.

Acknowledgements

This work was achieved within the cooperative program GDR CNRS No. 2502, and is supported by the ADVACT research program and collaborations with Dassault Aviation and MBDA.

References

1. Greenblatt, D., Wygnanski, I.J., The control of flow separation by periodic excitation, *Progress in Aerospace Sciences* **36**, 2000, 487–545.
2. Garnier, E., Pruvost, M., Ducloux, O., Talbi, A., Gimeno, L., Pernod, P., Merlen A., Preobrazhensky, V., ONERA/IEMN contribution within the ADVACT program: Actuators evaluation, in *Flow Control and MEMS*, Proceedings of the IUTAM Symposium held at the Royal Geographical Society, 19–22 September 2006, J. Morrison et al. (Eds.), Springer, Dordrecht, 2008.
3. Gilarranz, J.L., Rediniotis, O.K., Compact, high power synthetic jet actuators for flow separation control, AIAA-2001-0737, 2001.
4. Pernod, P., Preobrazhensky, V., Merlen, A., Ducloux, O., Talbi, A., Gimeno, L., Tiercelin, N., MEMS for flow control: Technological facilities and MMMS alternatives, in *Flow Control and MEMS*, Proceedings of the IUTAM Symposium held at the Royal Geographical Society, 19–22 September 2006, J. Morrison et al. (Eds.), Springer, Dordrecht, 2008.
5. Duffy, D.C., McDonald, J.C., Schuller, O.J.A., Whitesides, G.M., Rapid prototyping of microfluidics systems in Poly(dimethylsiloxane), *Analytical Chemistry* **70**(23), 1998, 4974–4984.
6. Ducloux O., Talbi, A., Gimeno, L., Viard, R., Pernod, P., Preobrazhensky, V., Merlen, A., Self-oscillation mode due to fluid-structure interaction in a micromechanical valve, *Appl. Phys. Lett.* **91**, 2007, 034101.

Micromachined Shear Stress Sensors for Flow Control Applications

Mark Sheplak, Louis Cattafesta and Ye Tian
*Interdisciplinary Microsystems Group, University of Florida, Gainesville,
FL 32611, U.S.A.; E-mail: sheplak@ufl.edu*

Abstract. This paper reviews existing microelectromechanical systems-based shear stress sensors in the context of their suitability for various flow control situations. The advantages and limitations of existing devices for use in flow control systems are discussed. Unresolved technical issues are summarized and recommendations provided for future sensor development.

Key words: Shear stress sensors, MEMS, flow control.

1 Introduction

The measurement of mean and fluctuating wall shear stress in a turbulent boundary layer finds applications both in industry and the scientific community. Fluctuating data can provide physical insight into complex flow phenomena, including turbulent viscous drag, transition to turbulence, flow separation, and shock-wave/boundary layer interactions. Furthermore, time-resolved, fluctuating shear stress data is a vector field that offers advantages over pressure sensing for separation detection flow state estimation for flow control [1–4].

Unfortunately, existing conventional measurement technologies cannot provide accurate fluctuating wall shear stress data [5]. To accurately measure time-resolved turbulent wall shear stress, the sensor must possess sufficient spatial resolution to avoid spatial averaging. In addition, the device must possess sufficient measurement bandwidth to avoid excessive low-pass filtering of the data. Both of these requirements must be met to maintain spectral fidelity of the turbulence. For example, at high Reynolds numbers, according to Kolmogorov scaling, the spatial length scales of interest can be O(100 μm) or less and the required bandwidth can be O(1 kHz) or more [5]. In addition, a flat frequency response function is desired for turbulence measurements so that the spectra and statistical moments are accurately estimated.

J.F. Morrison et al. (eds.), IUTAM Symposium on Flow Control and MEMS, 67–73.
© 2008 *Springer. Printed in the Netherlands.*

From a flow control perspective, the stringent spatial and temporal resolution requirements for quantitative turbulence measurements may be relaxed depending upon the flow physics and control strategy [4]. For example, in separation control applications, large-scale structures dominate the flow dynamics. Therefore, the shear stress sensors are not required to accurately sense the fine turbulent scales. While the measurement resolution may be less stringent, the measurements must resolve the desired low-frequency scales, and there may be additional requirements in terms of closely spaced sensors for array applications [4]. If correlation analysis is required for state-estimation, negligible phase lags between individual sensors over a wide frequency range are also desirable. For certain control strategies, such as disturbance rejection approaches for separation control, the sensing requirement may be only qualitative in nature, opening up opportunities for thin-film thermal sensors [6]. If quantitative data is needed, however, the characterization of the sensor/compensation system dynamic response is necessary to bound measurement uncertainty for time-resolved data.

The inherent small physical size, the ability to be placed in close proximity, and batch-fabrication characteristics of microfabricated transducers offer the potential to meet the demands of flow control applications. Realizing the potential performance advantages of microfabricated shear stress sensors via MEMS technology, a number of researchers have presented an assortment of micromachined devices at various stages of technical maturity. Overviews of the operational characteristics and performance of these devices has been presented in a review paper [5]. This paper examines unresolved technical issues that must be addressed before these devices can be considered reliable sensing tools for flow control.

1.1 Ideal Traits of a Flow Control Sensor

Whether it is used for feedback control or state estimation, an ideal sensor possesses several desirable traits. In addition to sufficient spatial and temporal resolution, it should be non-intrusive and should not actuate the flow. For example, it is well known that thermal-based wall shear stress sensors locally heat the flow, thus perturbing the velocity profile [5]. As will be shown below, the sensor system should also possess known constant gain and phase response as phase lags can lead to controller instabilities [7]. The sensor should also possess a linear response to the largest expected fluctuations. In many flow control applications, such as separation control, large fluctuation levels may be present, which can destabilize the controller [7–10]. The sensor should be immune to unwanted inputs such as electromagnetic interference. A more subtle example is direct feedthrough of an actuator signal to a sensor (e.g., via acoustic vs. hydrodynamic path), which can manifest itself as a notch in an experimentally obtained transfer function and lead to undesirable pole-zero cancellation and poor performance [8–10]. Because flow control is inherently distributed, the sensing scheme must be amenable to arrays. For such applications, gain and phase matching between the sensors is important to avoid sensing errors.

From a practical perspective, the sensing system should be sufficiently robust, consume minimal power, and be economically feasible from a cost perspective.

1.2 Flow Control System Issues

The gain and phase characteristics of a sensor or sensor array, along with any estimation algorithm, are part of the inherent dynamics of a flow control system that are typically ignored. For example, Figure 1a shows a block diagram of a standard, idealized feedback control in which the output signal y is fed back with unity gain. This implies that the magnitude of the frequency response of the sensor is flat and devoid of phase lags. Furthermore, the dynamics of the actuator are ignored, implying that the fluid dynamics dominate relative to those of the actuator. In this idealized case the transfer function of the closed-loop control system is

$$T = \frac{Y}{R} = \frac{PC}{1+PC}.$$

A more descriptive representation is shown in Figure 1b that includes the dynamical blocks of the sensor and actuator, in which case the closed-loop transfer function is

$$T = \frac{Y}{R} = \frac{PHCA}{1+PHCA}.$$

This equation explicitly shows how the dynamics of the actuator and sensor directly affect the closed-loop dynamics and, in turn, how this influences controller design. Actuator and sensor resonances, bandwidth, lags, nonlinearities, and saturation require, at a minimum, the design of a robust controller. In a more realistic situation, the lack of careful consideration of the overall control system design can doom a control system [7].

1.3 Advantages of Shear Stress Sensing for Flow Control

For realistic 3-D feedback flow control applications, wall measurements are more practical than velocity field measurements. Furthermore, when the control objective is often skin-friction or pressure drag reduction, a direct measurement of these quantities may be preferable [3]. The question arises about the advantages of pressure versus shear stress sensing for feedback flow control. As previously stated, shear stress is a vector property that offers the advantage of flow direction over pressure. Recent studies indicate that depending on the application, it may be advantageous to use skin friction alone or in combination with pressure for state estimation [1–4]. An additional advantage of skin-friction over pressure sensing is the inherent immunity to direct actuator feedthrough via acoustic excitation.

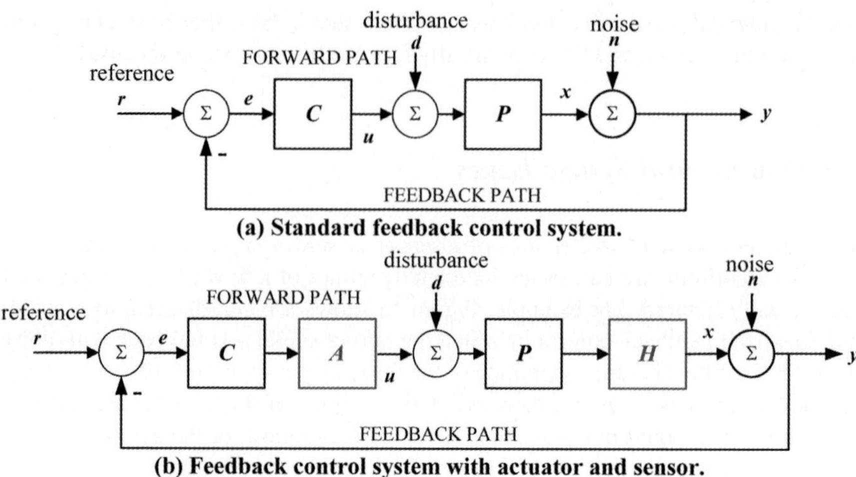

(a) Standard feedback control system.

(b) Feedback control system with actuator and sensor.

Fig. 1 Block diagrams of a feedback control system including various dynamical blocks: controller C, actuator A, fluid dynamic plant P, and sensor H.

From an implementation perspective, while pressure sensing alone has proven sufficient in feedback control of flow-induced cavity oscillations [11], the results for separation control are mixed. Figure 2 shows the unsteady power spectra results of a separation control experiment on a 6 inch chord NACA 0025 model using zero-net mass-flux actuators and a lift/drag balance at two different angles of attack [9, 10]. Figure 2a shows that at 12°, when effective open-loop control results in a large increase in lift/drag, a reduction in the fluctuating pressure levels on the surface results. In this case, unsteady pressure sensing used for feedback control is as effective as the balance. Figure 2b shows that at 20°, where effective open-loop control increases lift/drag, the unsteady pressure levels on the surface increase. Here, feedback control using just unsteady pressure sensing fails. These results suggest that unsteady pressure alone may be insufficient for separation control. Here, surface shear stress may be required, and ongoing research is exploring this possibility [3, 4].

2 Existing MEMS Sensors

Conventional measurement technologies have been shown insufficient for obtaining accurate mean and fluctuating 3-D wall shear stress data [5, 12–14]. This opportunity has motivated the development of micromachined sensors for various applications. MEMS shear stress sensors, like their conventional counterparts, are broadly classified as direct and indirect techniques [5, 12–14]. Direct sensors measure the shear force acting on the flow surface. Typically, this is achieved by employing a "floating element" balance. Indirect techniques require an empirical or theoretical

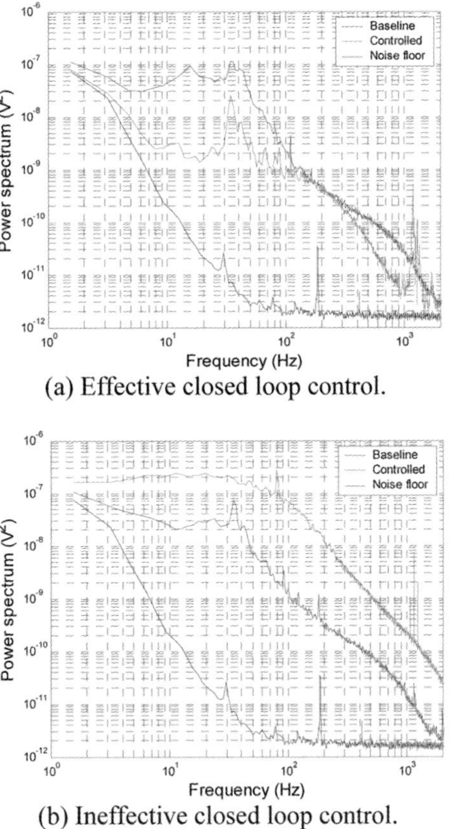

(a) Effective closed loop control.

(b) Ineffective closed loop control.

Fig. 2 Power spectra of unsteady pressure signals measured near trailing edge of a NACA 0025 airfoil separation control experiment showing the noise floor, baseline uncontrolled, and feedback controlled cases for $Re_c = 10^5$; (a) AoA $= 12°$ and (b) AoA $= 20°$ [9, 10].

correlation to relate the measured property to the wall shear stress. The MEMS community has produced a variety of different indirect transduction schemes, such as hot-film sensors, micro-optical systems to measure near-wall velocity gradients, and mechanical micro-fences. The respective advantages and disadvantages of these devices for flow control applications are summarized below. The interested reader can find the detailed reviews and associated references for all sensors discussed below in [5, 14].

2.1 Indirect MEMS Sensors

All indirect shear stress sensors require a correlation between a measured flow property (heat transfer, velocity profile in the sublayer, etc.) and the desired wall shear stress. Typically, the calibrations for these devices are only valid under very specific flow conditions. For example, laser-based optical MEMS (MOEMS) sensors measure the velocity gradient in the viscous sublayer of a boundary layer and relate that to the wall shear stress. Micro-fence sensors infer the wall shear stress by placing a small fence within the viscous sublayer of a turbulent boundary layer. The static pressure drop across the upstream and downstream side of the fence is then related to the wall shear stress via a calibration curve for a known velocity profile. The extension of these techniques to realistic complex 3-D flows is an unresolved challenge.

Thermal-based shear stress sensors possess several additional limitations when used for quantitative wall shear stress measurements. Specifically, the uncertainty of the dynamic responses of these thermal techniques – due to heat conduction to the wall, calibration difficulties, flow perturbation due to heating, and errors in response due to large fluctuations with respect to the mean ($\sim$40%) – have not been quantified [5]. There is considerable evidence that the uncertainty of thermal sensors can be quite large in gas flow applications. In particular, a recent computational study suggests that perturbations due to heat transfer from the sensor to the flow can alone result in mean shear stress errors of 5% or greater [15]. In addition, a single thermal sensor is unable to discern the direction of the wall shear stress, thus limiting their usefulness in the vicinity of separating and reattaching flows.

2.2 Direct MEMS Sensors

Direct sensors measure the integrated force produced by the wall shear stress on a flush-mounted, movable, "floating" element. The floating element is attached to either a displacement transducer or is part of a feedback force-rebalance configuration. Floating element techniques appear to be better suited for obtaining quantitative, time-resolved data provided that a stable, low-noise transduction scheme can be developed that is immune to both EMI and transverse motions. From a packaging perspective, the sensor system must possess backside electrical or optical interconnects to provide a truly flush- mounted device. The robustness of the sensors to debris must also be addressed by covering the sensor gaps. Finally, the transduction scheme should permit the ability to realize arrays O(10 s) to O(100 s) of sensors to map wall shear stress fields.

3 Conclusions

While MEMS shear stress sensors possess great promise, all existing devices are fairly immature and require further development to become reliable measurement tools for feedback control. While microfabrication technology is fairly well-established, sensor development is hampered by a dearth of research activity when compared to actuator-related research and theoretical, computational, and experimental flow control efforts. This is vividly illustrated by this paper being the lone sensor representation at this IUTAM Symposium on Flow Control and MEMS.

References

1. Rathnasingham, R., Breuer, K.S.: Active Control of Turbulent Boundary Layers. *J. Fluid Mech.* **495** (2003) 209–233.
2. Aamo, O.M., Krstic, M., Bewley, T.R.: Control of Mixing by Boundary Feedback in 2D Channel Flow. *Automatica* **39** (2003) 1597–1606.
3. Surana, A., Grunberg, O., Haller, G.: Exact Theory of Three-Dimensional Flow Separation. Part I. Steady Separation. *J. Fluid Mech.* **564** (2006) 57–103.
4. Alam, M.R., Liu, W., Haller, G.: Closed-Loop Separation Control: An Analytic Approach. *Phys. Fluids* **18** (2006) 043601.
5. Naughton, J.W., Sheplak, M.: Modern Developments in Shear Stress Measurement. *Prog. Aero. Sci.* **38** (2002) 515–570.
6. Liu C., Huang C.-B., Zhu Z., Jiang F., Tung S., Tai Y.-C., Ho C.-M.: A Micromachined Flow Shear-Stress Sensor Based on Thermal Transfer Principles. *J. MEMS* **8** (1999) 90–99.
7. Banaszuk, A, Mehta, P.G., Hagen, G,: The Role of Control in Design: From Fixing Problems to the Design of Dynamics. In *Proceedings of International Symposium on Advanced Control of Chemical Process*, Gramado, Brazil (2006).
8. Ogata, K.: *Modern Control Engineering*, 4th edition, Prentice Hall, Upper Saddle River, NJ (2001).
9. Tian, Y., Song, Q., Cattafesta, L.: Adaptive Feedback Control of Flow Separation. In *Proceedings in 3rd Flow Control Conference*, San Francisco, CA AIAA-2006-3016 (2006).
10. Tian, Y., Cattafesta, L., Mittal, R.: Adaptive Control of Separated Flow. In *Proceedings in 44th AIAA Aerospace Sciences Meeting and Exhibit*, AIAA 2006-1401, Reno, NV (2006).
11. Cattafesta, L., Alvi, F., Rowley, C., Williams, D.: Review of Active Control of Flow-Induced Cavity Oscillations. In *Proceedings of 33rd AIAA Fluid Dynamics Conference and Exhibit*, Orlando, FL (2003) AIAA Paper 2003-3567.
12. Haritonidis, J.H.: The Measurement of Wall Shear-Stress. In *Advances in Fluid Mechanics Measurements*, M. Gad-El-Hak (Ed.), Springer-Verlag, New York (1989) pp. 229–261.
13. Fernholtz, H.H., Janke, G., Schober, M., Wagner, P.M., Warnack, D.: New Developments and Applications of Skin-Friction Measuring Techniques. *Meas. Sci. Technol.* **7** (1996) 1396–1409.
14. Sheplak, M., Cattafesta, L., Nishida, T.: MEMS Shear Stress Sensors: Promise and Progress. In *Proceedings of 24th AIAA Aerodynamic Measurement Technology and Ground Testing Conference*, Portland, OR (2004) AIAA Paper AIAA 2004-2606.
15. Appukuttan, A., Shyy, W., Sheplak, M., Cattafesta, L.: Mixed Convection Induced by MEMS-Based Thermal Shear-Stress Sensors, *Numerical Heat Transfer A* **43**(3) (2003) 283–305.

SYNTHETIC JETS

Synthetic Jets and Their Applications for Fluid/Thermal Systems

Michael Amitay
*Mechanical, Aerospace and Nuclear Engineering Department, Rensselaer
Polytechnic Institute, 110 8th Street, Troy, NY 12180, U.S.A.;
E-mail: amitam@rpi.edu*

Abstract. The present paper discusses the formation and evolution of finite span
synthetic jets and their application for performance enhancement of fluid/thermal
systems. PIV measurements revealed that the synthetic jet field has a unique flow
pattern, where along its slit, the flow is two-dimensional near the orifice, while
farther downstream the vortex pair lines develop secondary counter-rotating 3-D
structures. Moreover, the streamwise and spanwise spacing between these structures
vary with stroke length and formation frequency. Next, a couple of examples of the
implementation of the synthetic jets to improve system performance are presented,
including a scaled Cessna 182 model, and spray cooling. Using synthetic-jet-based
flow control for flight control showed comparable effects to those of conventional
ailerons at moderate deflection angles. For heat transfer, synthetic jets were shown
to alter the global and detailed characteristics of a water spray and thus augment its
cooling performance.

Key words: Synthetic jets, active flow control, spray cooling enhancement.

1 Introduction

Over the last decade synthetic jets have been used for various flow control applications. A synthetic jet is synthesized at the edge of an orifice by a periodic motion of a
diaphragm mounted on one (or more) walls of a sealed cavity. When the diaphragm
moves towards the orifice, a vortex pair (for a 2-D slit) is formed at the edge of
the orifice and is advected by its own self-induced velocity such that when the diaphragm moves away from the orifice, the vortex pair is far enough and is not affected
by the fluid that is drawn into the cavity. Therefore, a synthetic jet has a zero-net-
mass-flux but it allows momentum transfer to the flow. With a proper design, the
diaphragm and the cavity are driven near resonance; therefore, only small electrical
power input is needed, making the synthetic jet a very efficient and attractive actu-

J.F. Morrison et al. (eds.), IUTAM Symposium on Flow Control and MEMS, 77–93.
© 2008 *Springer. Printed in the Netherlands.*

ator for various flow control applications [1–4]. Another advantage of the synthetic jet is that no plumbing or mechanical complexities are needed.

In recent years, plane and round synthetic jets have been investigated both experimentally (see [5–12] among many others) and numerically (e.g., [13–19] and more). The studies prior to 2002 are detailed in the review paper by Glezer and Amitay [20].

Synthetic jets have been used as versatile actuators for active flow control; in control of free jets (e.g., [21–23]), 2-D airfoils (e.g., [24]), small unmanned Aerial Vehicles (e.g., [25, 26]), and in electronic cooling (e.g., [27, 28]).

The present paper discusses the flow field associated with a finite span synthetic jet and presents two applications: (1) flight control using synthetic-jets-based flow control, and (2) active heat transfer enhancement of spray cooling.

2 Experimental Setup

The experiments were conducted at the Flow Control Research Lab at RPI. Various measurement techniques were used for each project and are described below.

2.1 Formation and Evolution of Finite Span Synthetic Jets

The mechanisms associated with an isolated finite span synthetic jet actuator were investigated in detail in a specially designed facility that was constructed to take advantage of advanced optical diagnostic techniques. Spatial and temporal measurements were obtained using Particle Image Velocimetry (PIV) in planes across and along the jet's orifice complemented with hot-wire anemometry. The PIV system utilized two 120 mJ Nd:YAG lasers and a 1376×1040 pixel resolution thermoelectrically cooled 12-bit CCD camera. The flow was seeded with smoke particles, $O(1 \ \mu m)$, produced using incense. The streamwise and cross-stream velocity components (U, V and U, W in the x-y and x-z planes, respectively) were computed from cross-correlation of pairs of successive images with 50% overlap between interrogation domains.

The calibration of the synthetic jet was obtained by placing the hot-wire sensor at the jet's orifice plane. The synthetic jet was issued from a slit having a width of $h = 0.5$ mm and a length of $L_z = 25.4$ mm, and was driven by a piezo-ceramic disk at a frequency of 300 Hz (the schematic of the synthetic jet is presented in Figure 1).

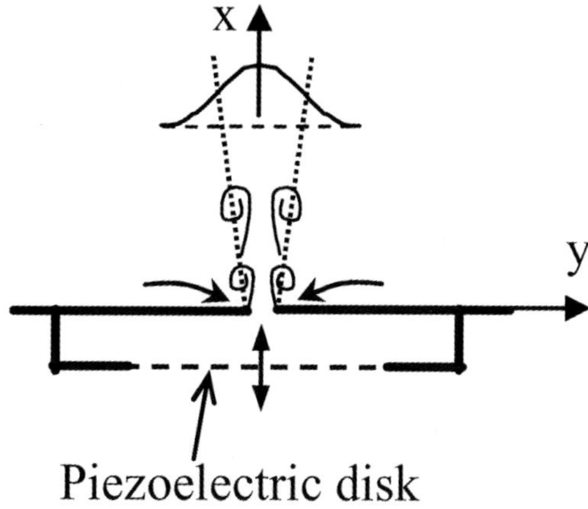

Fig. 1 Synthetic jet schematics.

2.2 *Flight Control Using Flow Control*

The experiments were conducted in a closed-return low speed wind tunnel, having a test section measuring 608×608 mm. The forces and moments acting on the model were measured, using a six component sting balance, about the quarter chord of the wing along the model centerline. The experiments were conducted at Reynolds numbers of 67,300 and 134,600, respectively, and angles of attack from $0°$ to $16°$.

A 1/24th Cessna model was constructed from stereolithography, where the main wing consists of a NACA 2412 section, and a NACA 0005 was used for both the horizontal and vertical tails. The main wing was designed such that different wingtips could be used (Figure 2) where aileron deflections from $0°$ to $18°$ in $3°$ increments could be achieved. Each aileron's streamwise length is 20% of the root chord on the outer span portions (25% of the span for each aileron) of the wing.

In addition to the aileron deflection wingtips, several wingtips with embedded synthetic jets were also designed, where the synthetic jets were mounted at $x_{sj}/\bar{c} = 0.25$ (near the separation point for high angles of attack). The synthetic jet performance was measured using the momentum coefficient, C_μ [1], which was varied between 1.5×10^{-4} and 8.7×10^{-3} (both wingtips actuated) and between 7.4×10^{-5} and 4.3×10^{-3} (when a single wingtip is actuated).

The synthetic jet-instrumented wingtips had an 18% thick Clark-Y airfoil section. Each wingtip consisted of three synthetic jets (7.5 mm apart) each having a 0.5 mm wide slit and extending 25.4 mm along the span. Each synthetic jet actuator was driven at a frequency of 750 Hz using a piezoceramic disk. In addition, a hot-film shear stress sensor was placed at 35% chord, downstream the synthetic jet exit.

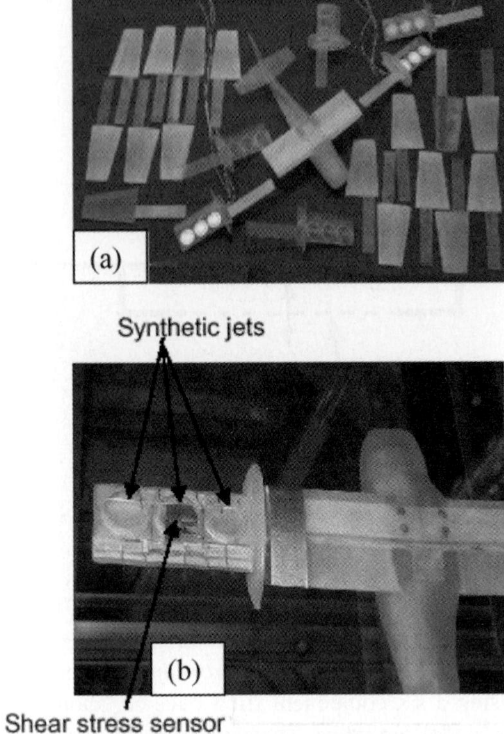

Fig. 2 Cessna wind tunnel model.

2.3 *Active Heat Transfer Enhancement of Spray Cooling*

Here, flow control of sprays was investigated using synthetic jets and was applied in spray cooling of a heated surface in the non-boiling regime. A special facility was designed (Figure 3), consisting of an enclosure that allowed non-intrusive optical measurement techniques and accommodated a spray nozzle, equipped with a flow control module, as well as an insulated heater for spray cooling measurements. The enclosure featured a 485 mm $\times$ 425 mm $\times$ 435 mm volume with an open top and a drain in the bottom. The spray nozzle was mounted above the enclosure, attached to a micrometer slide for a vertical positioning control of the spray.

In the present experiments, water was used as the working liquid, and the spray was created using an air-assisted siphoning atomizing nozzle from Delavan (model 30609-2), featuring an orifice diameter of 1.7 mm. The water was siphoned into the nozzle from a larger container that was mounted on a motorized, computer controlled traverse, allowing the water intake to be at a range of L_w of 0–49 cm above the level of the spray nozzle orifice. For the spray nozzle used here, the water flow rate through the spray nozzle depended on the water intake level elevation, L_w, and

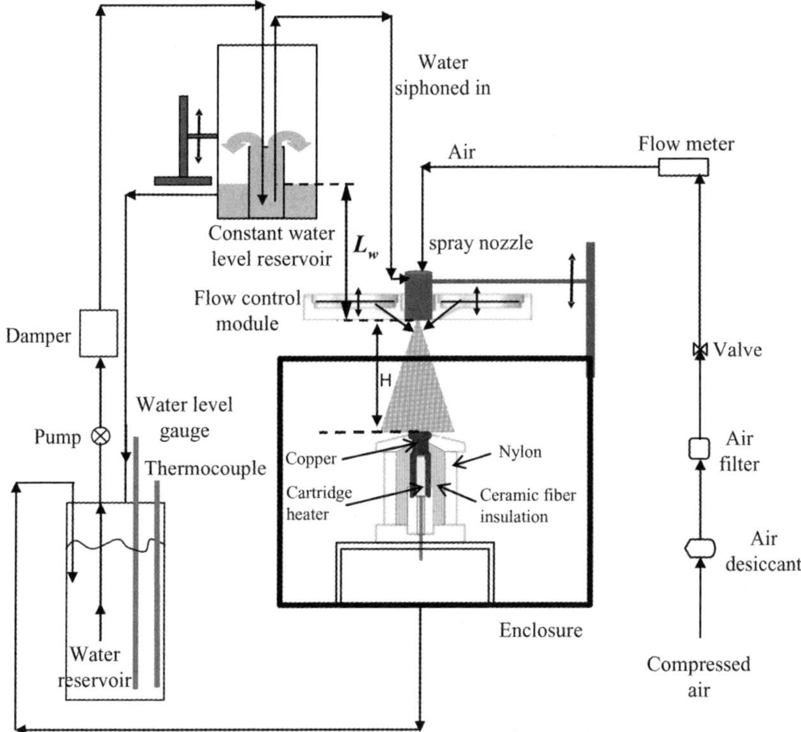

Fig. 3 Setup for the spray cooling experiments.

the air flow rate. The air flow to the nozzle was provided with a pressurized air line with a flow rate range from 5.5 l/min to 16.5 l/min.

A flow control module was fitted onto the spray nozzle such that the spray nozzle was in the center and four synthetic jet actuators, located around the circumference of the spray nozzle, are at $90°$ to each other. Each synthetic jet was driven at a frequency of 1000 Hz and issued from a slit orifice 5 mm in length and 0.5 mm in width at 7.3 mm from the spray orifice center and inclined at $30°$ with respect to the plane of the nozzle orifice to allow vectoring of the spray by providing streamwise and cross-stream velocity components to the spray.

Global spray measurements were carried out with PIV, which provided the velocity field for the spray droplets at five air flow rates, three water intake level to nozzle level distances, L_w, and three momentum coefficient values, C_μ, at each setting. For each case, a set of 2000 samples was acquired, and averaged velocity fields were computed. In these measurements, no flow tracers were added to the air flow; therefore, the velocities in this two-phase flow were only the velocities of the water droplets themselves, but not of the air present in the spray.

Detailed spray characteristics (droplets size, concentration, distribution, and velocity of individual droplets) were obtained using a combined Shadowgraphy and

Particle Tracking Velocimetry (PTV). The same dual-pulse laser and optical arm were used as for the PIV experiments, along with a special diffuser attachment for scattering the laser light. The CCD camera, with a long-range microscope attached and the light source were mounted on an optical table. Shadowgraphy measurements were acquired at 45 locations in the spray, such that the effect of active flow control could be determined for the entire three-dimensional spray.

For the spray cooling experiments, the heater module was placed below the spray nozzle. The module was embedded in an oxygen-free piece of copper, where the top surface was a flat circle with a diameter of 19 mm that was exposed to the spray. The copper piece had 15 T-type thermocouples (0.2 mm in diameter), embedded in the cylindrical top region to measure the temperature distribution on the heater. In order to minimize heat losses from the copper piece, it was placed inside a solid nylon cylinder insulation sleeve and ceramic fiber insulation.

3 Results

3.1 Synthetic Jets

As was shown by Cannelle and Amitay [12], the flow field of a finite synthetic jet exhibits three-dimensionalities, which were affected by the stroke length, the aspect ratio, and the Reynolds number. Representative results from their work are presented in this section.

The effect of the stroke length on the normalized spanwise and cross-stream mean vorticity fields is presented in Figures 4a–4c and 4d–4f, respectively (the vorticity is normalized by the slit width, h, and the average orifice velocity, U_o). Figures 4a and 4d correspond to $L_o/h = 16.6$ and $\mathrm{Re}_{U_o} = 85$, Figures 4b and 4e represent the case where $L_o/h = 25.9$ and $\mathrm{Re}_{U_o} = 133$ while Figures 4c and 4f show the case where $L_o/h = 49.4$ and $\mathrm{Re}_{U_o} = 254$. In all three cases, the formation frequency was 300 Hz. Note that contours representing negative vorticity are shown with dashed lines. At the lowest stroke length used in the experiments ($L_o/h = 16.6$, Figure 4a), the downstream development of the synthetic jet can be divided into two regimes (similar to the work by Chen et al. [9]) where near the jet exit ($0 < x/h < 10$) the vorticity is concentrated on both sides of the orifice (with opposite sense) and the jet moves downstream almost vertically (i.e., no spreading). Farther downstream, at $x/h > 10$, the jet widens and the spanwise vorticity is diffused in both x and y directions. When the stroke length is increased to $L_o/h = 25.9$ (Figure 4b) the downstream extent of the near field region is reduced to $x/h = 8$. Farther downstream the jet is slightly wider than at the lower stroke lengths. When L_o/h is increased to 49.4 (Figure 4c), the near region is even shorter ($x/h < 5$), while the jet is wider and has a higher peak vorticity farther downstream.

The effect of the stroke length on the cross-stream vorticity field (in a plane along the jet's orifice, x-z plane) is presented in Figures 4d–4f. At $L_o/h = 16.6$ (Figure 4d)

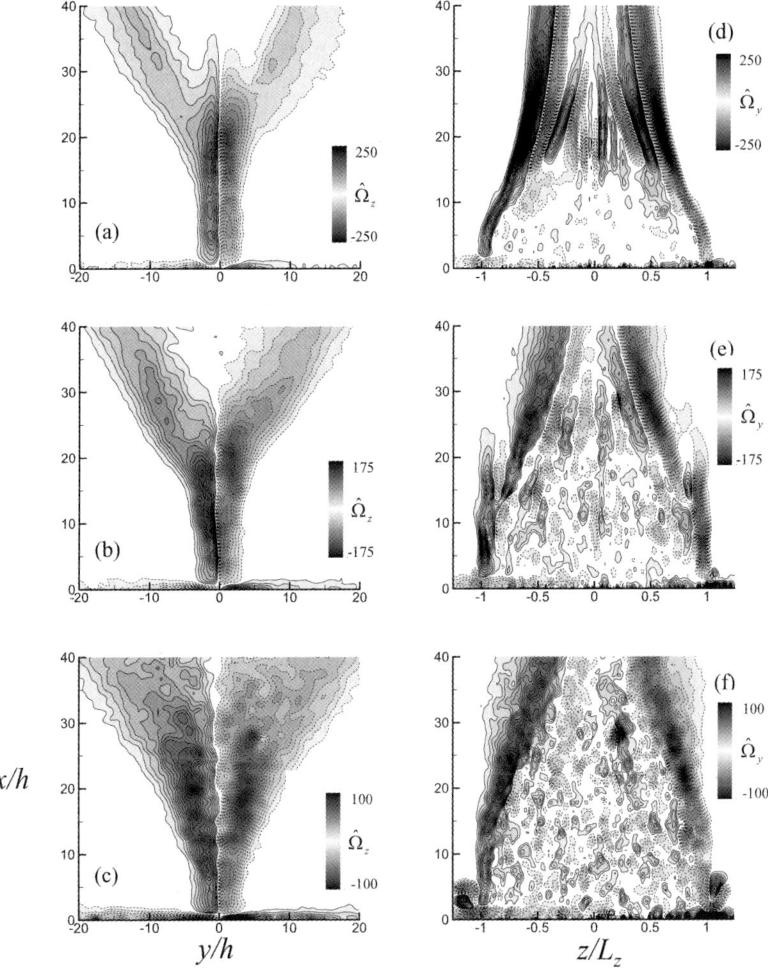

Fig. 4 Mean spanwise (a–c) and cross-stream (d–f) vorticity fields for a formation frequency of 300 Hz with $L_o/h = 16.6$ and $\mathrm{Re}_{U_o} = 85$ (a and d), $L_o/h = 25.9$ and $\mathrm{Re}_{U_o} = 133$ (b and e) and $L_o/h = 49.4$ and $\mathrm{Re}_{U_o} = 254$ (c and f). Dashed lines represent negative vorticity.

the flow field is uniform across most of the span ($-0.95 < z/L_z < 0.95$, where L_z is half of the length of the slit) for $0 < x/h < 10$; thus, the vorticity is practically zero. However, there is vorticity concentration along each side of the slit (with opposite sense) at $z/L_z = \pm 1$ due to the induced velocity towards the orifice edges. Farther downstream, the vorticity field exhibits a unique streaky-like behavior where streamwise streaks of opposite cross-stream vorticity dominate the flow field, which correspond to spanwise instability waves. These vortical streaks reach their maximum magnitude at $x/h = 18$, which corresponds to the location where increased spreading of the synthetic jet in the x-y plane occurs (see Figure 4a). Moreover, the

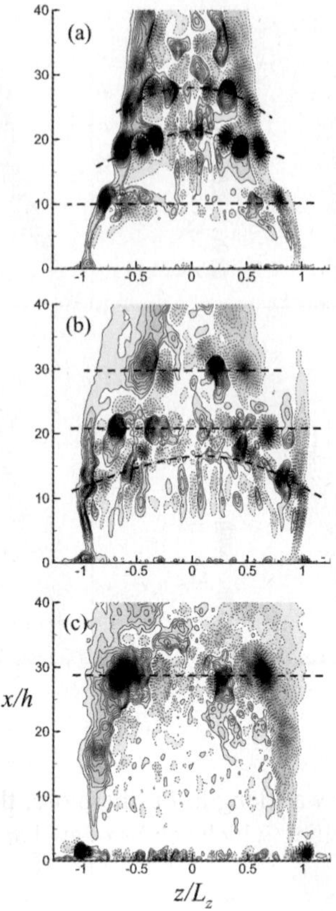

Fig. 5 Phase-locked cross-stream vorticity fields at a formation frequency of 300 Hz and (a) $L_o/h = 16.6$, (b) $L_o/h = 25.9$ and (c) $L_o/h = 49.4$. $\phi = 120°$. Dashed lines represent negative vorticity.

sudden widening of the jet at $x/h = 10$ in the x-y plane is accompanied by an abrupt change in the rate of the jet narrowing in the x-z plane. When the stroke length is increased to $L_o/h = 25.9$ (Figure 4e) the flow field is less organized; however, the counter-rotating structures (having a similar normalized spanwise wavelength) are clearly visible for $x/h > 20$. As L_o/h is further increased to 49.4 (Figure 4f), the vorticity field consists of non-organized, random, small structures for $0 < x/h < 20$, and the streamwise counter-rotating vortices can still be observed for $x/h > 20$. However, their spanwise wavelength is larger and their coherence is reduced.

To further understand the formation of the secondary counter-rotating stream-wise structures, the phase-locked cross-stream vorticity fields in a plane along the jet's slit (x-z plane) were calculated from the phase-locked velocity fields and are presented in Figures 5a–5c at a phase of $\phi = 60°$ (during the blowing portion of the cycle) at $L_o/h = 16.6$, 25.9 and 49.4. At the low L_o/h the cross-stream vorticity fields (Figure 5a) exhibit a unique pattern where the vortex line that was formed in the previous cycle (located at $x/h \approx 10$) seems to be two-dimensional. However, it consists of secondary counter-rotating vortices near the edges ($0.5 < |z/L_z| < 1$). The previous vortex pair line (at $x/h \approx 18$) develops three-dimensionalities where the center moves faster than the sides (horseshoe structure, marked by the dashed lines). Similar formation of a horseshoe-like pattern was also observed by Mumford [29] and Antonia et al. [30] in the fully-developed region of a plane jet. Moreover, this entire vortex line consists of secondary counter-rotating 3-D spanwise ("roller") structures. These structures form the streamwise vorticity streaks in the mean cross-stream vorticity field that are shown in Figure 5d. In addition, the vorticity field exhibits streamwise streaks ("ribs") between the new vortex line and the previous one (clearly visible at $12 < x/h < 17$). These structures (also observed by Smith and Glezer [5] using flow visualization) were formed during the suction portion of the previous cycle. Similar flow patterns were observed by Bernal and Roshko [31], and Nygaard and Glezer [32] in a mixing layer.

As L_o/h is increased (Figure 5b), similar behavior is observed where the secondary spanwise vortices are larger, and both the streamwise and spanwise wavelengths increase compared to the corresponding wavelengths for the lower L_o/h. As the downstream distance increases these structures increase in size as they coalesce. Moreover, at this L_o/h the counter-rotating streamwise ribs between the new vortex line and the previous one are clearly visible. Even at the highest L_o/h used in the present experiments (Figure 5c) a system of counter rotating vortices is visible (at $x/h = 28$) where these secondary vortices are much larger with a larger non-dimensional spanwise wavelength. Furthermore, these structures also consist of small-scale non-uniformities, which can be related to the transition to a turbulent jet.

3.2 Flight Control Using Flow Control

The implementation of synthetic jets for separation control and for roll control on a scaled Cessna 182 model is shown in this section. Figure 6 shows the change in the roll coefficient (with respect to the corresponding baseline values) with the momentum coefficient of the synthetic jets. At $\alpha = 0°$ a very small roll moment is obtained and it is similar for all momentum coefficients tested. When the angle of attack is increased to $6°$, proportional control is obtained, where as C_μ increases, ΔC_r decreases. At higher angles of attack, where the baseline flow is separated over the wingtips, the effect of the momentum coefficient is different. At $\alpha = 10°$, for low C_μ, increase in C_μ yields a higher rolling moment while for $C_\mu > 1.73 \times 10^{-3}$, as C_μ increases ΔC_r decreases. At angles of attack past the stall angle, low momentum coefficient is not sufficient to reattach the flow over the wingtip and thus low rolling moment is achieved. However, high momentum, coefficient jets can reattach the flow and create a significant rolling moment. It is noteworthy that the magnitudes of the rolling moments obtained with synthetic jets is similar to those obtained with ailerons at deflection angles up to $12°$.

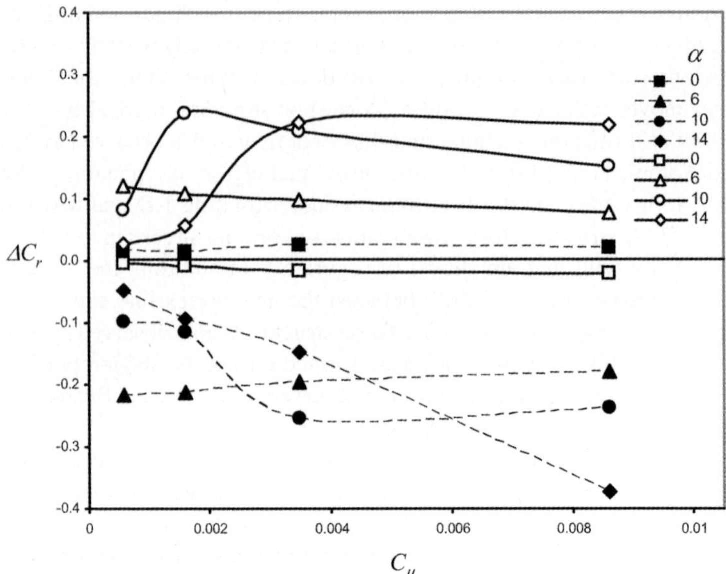

Fig. 6 The change of roll moment vs. C_μ at different angles of attack. Dashed and solid lines correspond to the starboard and port synthetic jets activated, respectively.

Next, a simple closed-loop control scheme is used to suppress separation over the wingtips using a dynamic shear stress sensor to detect the separation and synthetic jet actuators to reattach the flow. The separation was observed qualitatively, using tuft flow visualization, and detected quantitatively, using the shear stress sensor. When the flow separates, the RMS levels of the shear stress increases dramatically (not shown) and this was used as a trigger for the closed-loop control to activate the synthetic jets.

Figure 7 present the time trace of the shear stress sensor output as the angle of attack increases monotonically. Note that in this figure the time axis also represents increase in the angle of attack. As time progresses (i.e., as the angle of attack increases) the shear stress output voltage increases (corresponding to a decrease in the shear stress) as well as the RMS level. When the angle of attack was $\sim 8°$, the flow was on the verge of separation, and when the RMS threshold reached the predetermined value the synthetic jets were activated *before the wingtips ever stalled*. This suggests that through the selection of a simple close-loop control, for an appropriate shear stress RMS threshold, synthetic jets cannot only recover a wing from stall, but avoid it altogether.

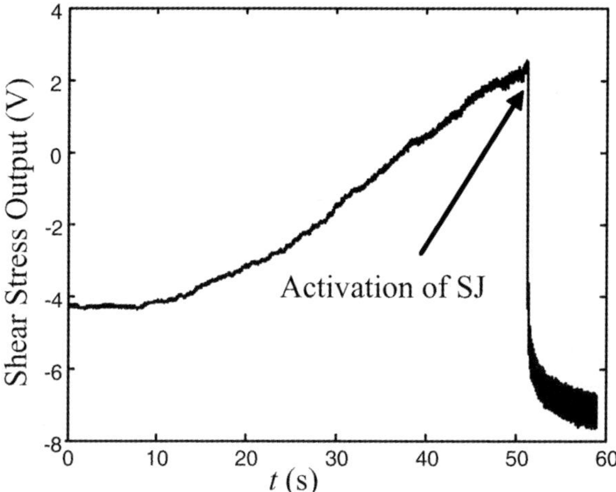

Fig. 7 The change of the shear stress sensor output with time (as the angle of attack increases).

3.3 Active Heat Transfer Enhancement of Spray Cooling

This section presents the effect of synthetic jet actuators on the global and detailed performance of a full cone spray, which then is used as a means to enhance the performance of spray cooling heat transfer.

As was shown by Pavlova et al. [33, 34], when the synthetic jet was activated, the spray was vectored away from the synthetic jet, and the vectoring angle became larger with the increase of momentum coefficient. Therefore, using two opposing synthetic jets spray flapping can be achieved, which might be used to enhance spray cooling. The synthetic jets were driven with a special ramped function in order to achieve gradual sweeping of the spray from one side to another (see the schematic in the bottom of Figure 8). Figure 8 presents phase-averaged velocity vector fields for the flapping experiments at a flow rate ratio of $Q_a/Q_w = 166$. Here, the full sweeping cycle occurred over $\tau_p = 1$ s (similar results were obtained for cycles of 0.25, 0.5 and 2 s). In each vector field, the relative strength of the synthetic jets on the left and right of the spray is represented by the length of the arrows.

At $t/\tau_p = 0$ (Figure 8a) both synthetic jets have the same strength; thus, resulting in a symmetric (about the x-axis) spray. At $t/\tau_p = 0.05$ (Figure 8b), the right synthetic jet, SJ_2, is stronger than the left jet, SJ_1, and the spray is vectored to the left with an angle of $\sim 4°$. As time progresses to $t/\tau_p = 0.1$ (Figure 8c), the left synthetic jet is off, and the right synthetic jet is at its maximum strength, resulting in further vectoring of the spray to the left with an angle of $\sim 10°$. At mid-cycle, the spray is returned to its original position, as the left synthetic jet is ramped up, while the right synthetic jet is ramped down, resulting in control jets of equal strength (Figure 8d), and thus a symmetric spray. At $t/\tau_p = 0.55$ (Figure 8e), the left synthetic jet is stronger than the right, resulting in a spray that is vectored to the right with an

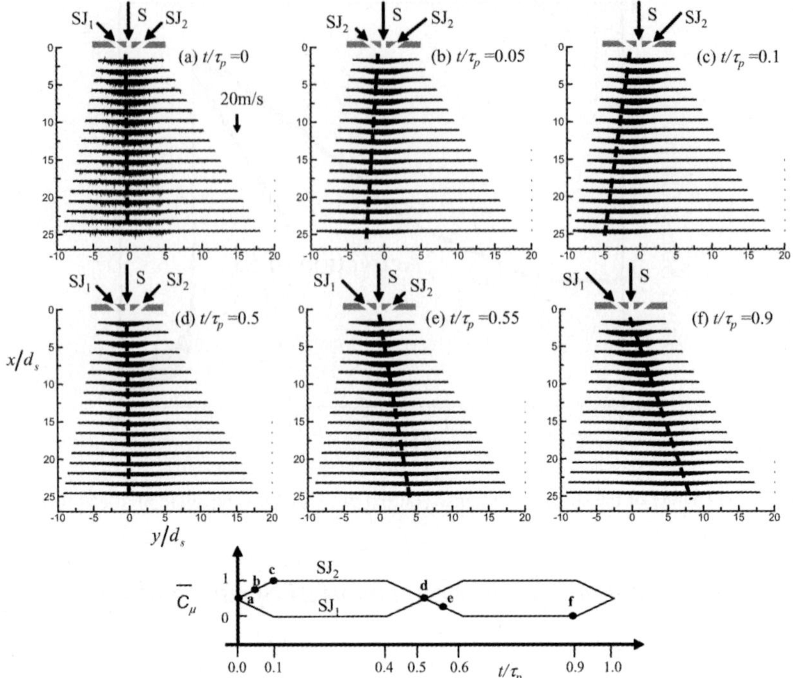

Fig. 8 Velocity vector fields at different times during the flapping motion; $Q_a/Q_w = 166$ and $L_w/d_s = 288$.

angle of $\sim 7°$. Finally, as the left jet is at its maximum strength, and the right jet is off, the spray is further vectored to the right with the angle of $\sim 12°$ (Figure 8f). The slight difference in the vectoring angle from side to side might be due to a slight difference in the jets' strength. Note that another mode of flapping was tested (on-off flapping) where the two opposing jets are driven with a step function, such that when SJ_1 is on, SJ_2 is off, and vice versa. The resulting vector fields are similar to those in Figures 8c and 8f.

Next, the effect of the synthetic jet on the droplet size and distribution at 45 measurement locations (at three x-z planes) were examined. Here, data at one plane (along the spray centerline, $z/d_s = 0$) are presented as a representative case. Figure 9 presents the histograms of the percentage change (with respect to the baseline case) in the number of droplets, at a given droplet diameter, for $Q_a/Q_w = 166$ and $C_\mu = 0.425$, where N_{d_b} and N_{d_f} are the number of droplets (for a given group size) in the interrogation window for the baseline and forced cases, respectively. The synthetic jet was issued from the left side of the spray (i.e., $-y/d_s$). Data are presented at three downstream locations at five cross-stream locations each.

At all three downstream locations, there is a decrease in the number of small droplets on the left side of the spray (i.e., the side closer to the synthetic jet) and an increase of those same small droplets on the right side of the spray, indicating

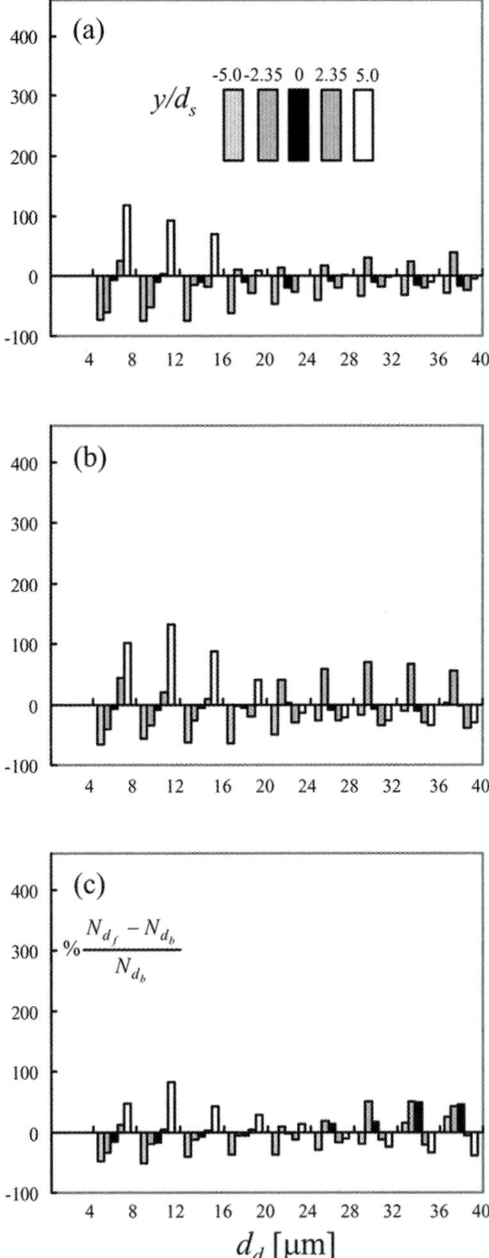

Fig. 9 The change of the droplet size distribution at (a) $x/d_s = 8.8$, (b) $x/d_s = 14.7$ and (c) $x/d_s = 23.5$, at $z/d_s = 0$ for $Q_a/Q_w = 166$ and $C_\mu = 0.425$.

that small droplets ($4\mu m \leq d_d \leq 16$ μm) were pushed away from the synthetic jet due to the impulse of the jet on the droplets. Also, there are more droplets of larger diameters to the left of the spray centerline and fewer to the right side, compared to spray without flow control. This might be due to coalescence of small droplets as they were pushed away by the synthetic jet, creating larger droplets.

Next, the effect of the synthetic jet on the efficiency of the spray to remove heat from a hot surface was investigated, where the normalized distance between the spray and the surface was $H/d_s = 23.5$, 35.3, and 47.0 for a spray with $Q_a/Q_w = 166$ and $L_w/d_s = 288$. Three modes of operation were tested: four synthetic jets activated simultaneously, ramped sweeping, and on-off flapping. All three modes of flow control had an effect on heat transfer from the surface for the three H/d_s tested, as illustrated in Figures 10a–10c. At $H/d_s = 23.5$ (Figure 10a) operating four jets together has the greatest enhancement of heat removal (e.g., 14% lower ΔT for the same heat flux of 30 W/cm^2 out of the top), with the two flapping modes showing smaller enhancement (Figure 10a).

When the distance between the spray and the surface increases to $H/d_s = 35.3$ (Figure 10b) the heat removal for all modes of flow control is enhanced (for example, using on-off flapping yields a reduction in ΔT by up to 36%, at a heat flux of 18 W/cm^2, compared to the baseline). Other forms of flow control also show a cooling improvement, but to a lesser degree. At $H/d_s = 47$ (Figure 10c) flow control yields only a slight cooling enhancement, with all three modes of flow control producing about the same effect. These results illustrate that synthetic-jet-based active flow control of sprays has a noticeable effect on the heat transfer, and augments the performance of spray cooling. This is currently under investigation in the author's research group.

4 Conclusions

The present paper discusses the synthetic jet actuator and shows a couple of applications in fluid/thermal systems. Experiments, using PIV, showed that finite span synthetic jet actuator develops well organized spanwise vortical structures, where the downstream development of the jet is affected by the stroke length, the Strouhal number and the Reynolds number.

The application of synthetic jets actuators for separation control and roll control was tested in wind tunnel experiments on a scaled Cessna 182 model. Using synthetic jets, mounted in the wingtips, yields a delay of separation with a corresponding increase in the maximum lift coefficient. It was shown that synthetic jets can provide similar control authority as conventional ailerons, where proportional roll control was obtained by changing the momentum coefficient. Furthermore, the RMS output of a shear stress sensor was used to detect separation on the wingtips. This detection technique was integrated with a computer and synthetic jet actuators to create a simple closed-loop stall suppression system, which was successfully implemented.

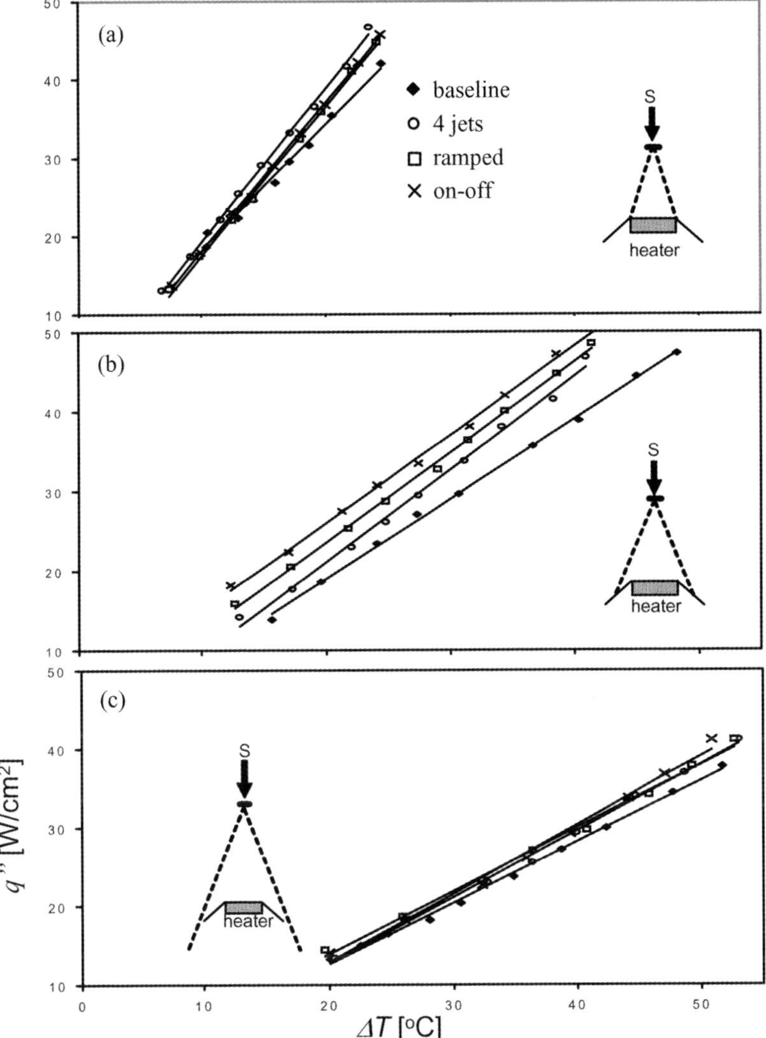

Fig. 10 Changes in heat removal due to flow control at (a) $H/d_s = 23.5$, (b) $H/d_s = 35.3$ and (c) $H/d_s = 47.0$, and $Q_a/Q_w = 166$.

Finally, the effect of synthetic jets on the evolution of a full cone water spray was explored. Flapping motion of the spray was accomplished when two synthetic jets were used directly opposite each other across the spray centerline, in an alternating manner, in both ramped and on-off manner and explored with PIV. Detailed spray parameters of droplet size, concentration, and distribution were measured at 45 locations throughout the spray, with and without flow control, using shadowgraphy, illustrating that microscopic spray characteristics were significantly affected when the synthetic jets were activated. Moreover, the application of the flow control resulted in improved performance of spray cooling, in terms of heat transfer enhancement.

Acknowledgments

The contribution of my graduate students: F. Cannelle, M. Ciuryla, J. Farnsworth, A. Pavlova, and D. Tamburello, is greatly appreciated.

References

1. Amitay, M. and Glezer, A., Controlled transients of flow reattachment over stalled airfoils, *International Journal of Heat and Fluid Flow* **23**(5), 2002, 690–699.
2. Amitay, M., Pitt, D. and Glezer, A., Separation control in duct flows, *Journal of Aircraft* **39**(4), 2002, 616–620.
3. Fung, P. and Amitay, M., Active flow control application on a mini ducted fan UAV, *Journal of Aircraft* **39**(4), 2002, 561–571.
4. Amitay, M., Washburn, A.E., Anders, S.G. and Parekh, D.E., Active flow control on the stingray UAV: Transient behavior, *AIAA Journal* **42**(11), 2004, 2205–2215.
5. Smith, B., and Glezer, A.: The formation and evolution of synthetic jets, *Physics of Fluids* **31**, 1998, 2281–2297.
6. Mallinson, G., Hong, G. and Reizes, J.A., Some characteristics of synthetic jets, *AIAA 30th Fluid Dynamics Conference*, Norfolk, VA, 1999, AIAA Paper 99-3651.
7. Crook, A. and Wood, N.J., Measurements and visualizations of synthetic jets, AIAA Paper 2001-0145, 2001,
8. Rediniotis, O.K., Ko, J., Yue, X. and Kurdila, A.J., Synthetic jets, their reduced order modeling and applications to flow control, 1999, AIAA Paper 99-1000, 1999.
9. Chen, J., Yao, C., Beele, G.B., Bryant, R.G. and Fox, R.L., Development of synthetic jet actuators for active flow control at NASA Langley, 2000, AIAA Paper 2000-2405, 2000,
10. Cater, J.E. and Soria, J., The evolution of round zero-net-mass-flux jets, *Journal of Fluid Mechanics* **472**, 2002, 167–200.
11. Cannelle, F. and Amitay, M., Synthetic jets: Spatial evolution and transitory behavior, *AIAA Fluid Dynamics Conference*, Reno, NV, 2005, AIAA Paper 2005-0102.
12. Amitay, M. and Cannelle, F., Evolution of finite span synthetic jets, *Physics of Fluids* **18** 2006, 054101.
13. Kral, L.D., Donovan, J.F., Cain, A.B. and Cary, A.W., Numerical simulation of synthetic jet actuators, *28th AIAA Fluid Dynamics Conference*, Reno, NV, 1997, AIAA Paper 97-1824.
14. Rizzetta, D.P., Visbal, M.R. and Stanek, M.J., Numerical investigation of synthetic jet flowfields, *AIAA Journal* **37**, 1999, 919.
15. Guo, D. and Kral, L.D., Numerical simulation of the interaction of adjacent synthetic jet actuators, *AIAA Fluids 2000 Meeting*, Denver, CO, 2000, AIAA Paper 2000-2565.
16. Muller, M.O., Bernal, L.P., Miska, P.K., Washabaugh, P.D., Chou, T.K.A., et al., Flow structure and performance of axisymmetric synthetic jets, *AIAA 39th Aerospace Sciences*, 2001, AIAA Paper 2001-1008.
17. Cui, J. and Agarwal, R.K., 3D CFD validation of a synthetic jet in quiescent air (NASA Langley workshop validation: Case 1), AIAA Paper 2004-2222, 2004.
18. Utturkar, Y., Holman, R., Mikttal, R., Carroll, B., Sheplak, M., and Cattafesta, L., A jet formation criterion for synthetic jet actuators, AIAA Paper 2003-0636, 2003.

19. Fugal, S.R., Smith, B.L. and Spall, R.E., Displacement amplitude scaling of a two-dimensional synthetic jet, *Physics of Fluids* **17**, 2005, 045103–045103-10.
20. Glezer, A. and Amitay, M., Synthetic jets, *Annual Review of Fluid Mechanics* **34**, 2002, 503–529.
21. Davis, S.A., The manipulation of large- and small-scale flow structures in single and coaxial jets using synthetic jet actuators, Doctoral Thesis, Georgia Tech., 2000.
22. Smith, B.L. and Glezer, A., Jet vectoring using synthetic jets, *Journal of Fluid Mechanics* **458**, 2002, 1–34.
23. Tamburello, D. and Amitay, M., Three dimensional interaction of a free jet with a perpendicular synthetic jet, *Journal of Turbulence*, 2007, in press.
24. Amitay, M., Smith, D.R., Kibens, V., Parekh, D.E. and Glezer, A., Modification of the aerodynamics characteristics of an unconventional airfoil using synthetic jet actuators, *AIAA Journal* **39**(3), 2001, 361–370.
25. Fung, P. and Amitay, M., Active flow control application on a mini ducted fan UAV, *Journal of Aircraft* **39**(4), 2002, 561–571.
26. Amitay, M., Washburn, A.E., Anders, S.G. and Parekh, D.E., Active flow control on the stingray UAV: Transient behavior, *AIAA Journal* **42**(11), 2004, 2205–2215.
27. Mahalingam, R., Rumigny, N. and Glezer, A., Thermal management using synthetic jet ejectors, *IEEE Transaction on Components and Packaging Technologies* **27**(3), 2004, 439–444.
28. Pavlova, A. and Amitay, M., Electronic cooling using synthetic jet impingement, *Journal of Heat Transfer* **9**, 2006, 897–907.
29. Mumford, J.C., The structures of large eddies in fully developed shear flows. Part 1. The plane jet, *Journal of Fluid Mechanics* **118**, 1982, 241–268.
30. Antonia, R.A., Browne, L.W.A., Rajagopalan, S. and Chambers, A.J., On the organized motion of a turbulent plane jet, *Journal of Fluid Mechanics* **134**, 1983, 49–66.
31. Bernal, L.P. and Roshko, A., Streamwise vortex structure in plane mixing layers, *Journal of Fluid Mechanics* **170**, 1986, 499–525.
32. Nygaard, K.J. and Glezer, A., Evolution of streamwise vortices and generation of small-scale motion in a plane mixing layer, *Journal of Fluid Mechanics* **231**, 1991, 257–301.
33. Pavlova, A., Otani, K. and Amitay, A., Active flow control of sprays with synthetic jets, AIAA Paper 2007-322, 2007.

Is Helmholtz Resonance a Problem for Micro-Jet Actuators?

Duncan A. Lockerby[1], Peter W. Carpenter[1] and Christopher Davies[2]
[1]*School of Engineering, University of Warwick, Coventry CV4 7AL, U.K.;*
E-mail: duncan.lockerby@warwick.ac.uk
[2]*School of Mathematics, Cardiff University, Cardiff CF24 4AG, U.K.*

Abstract. Numerical-simulation studies are undertaken to investigate how Helmholtz resonance affects the interaction of nominally inactive micro-jet actuators with
a laminar boundary layer. Two sets of numerical simulations are carried out. The
first set models the response of an actuator in ambient conditions to a small jump
in its internal pressure. These results verify our theoretical criterion for Helmholtz
resonance. In the second set of simulations, two-dimensional Tollmien–Schlichting
waves, with frequency comparable with, but not particularly close to, the Helmholtz
resonant frequency, are incident on a nominally inactive micro-jet actuator. The simulations show that under these circumstances the actuator acts as a strong source of
3D Tollmien–Schlichting waves. It is surmised that in the real-life aeronautical applications with turbulent boundary layers, broadband disturbances of the pressure
field would cause nominally inactive actuators to act as strong disturbance sources.
Should this be true, it would probably be disastrous for engineering applications of
such massless microjet actuators for flow control.

Key words: Helmholtz resonance, micro actuator, synthetic jet.

1 Introduction

A popular generic model of many micro-jet actuators currently being studied for
flow-control applications [1–4] is depicted in Figure 1. The mode of operation is
straightforward. An electric field is applied to a piezoceramic driver which contracts
to force the diaphragm upwards. This reduces the volume of the cavity, thereby
raising the pressure, which then drives a jet through the exit orifice into the boundary
layer. If an oscillating electric field is applied to the driver, the diaphragm oscillates
up and down, thereby producing a synthetic jet [5]. Alternatively, if a unidirectional
electric field is applied, the diaphragm only deflects upwards to produce a kind of
pulse jet [2–4]; we will follow [3] and call this mode of operation a pressure-jump

J.F. Morrison et al. (eds.), IUTAM Symposium on Flow Control and MEMS, 95–101.
© 2008 *Springer. Printed in the Netherlands.*

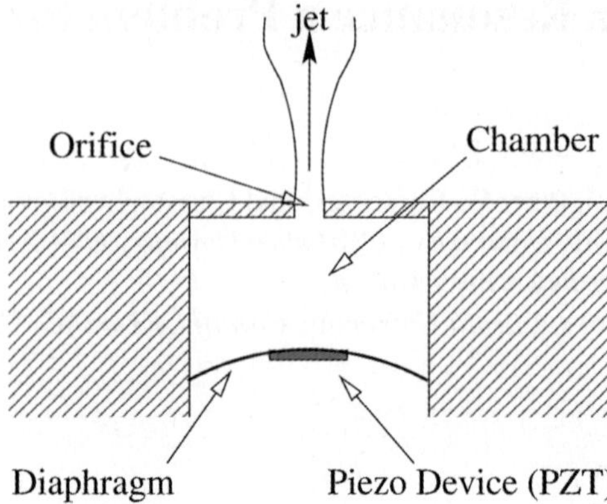

Fig. 1 Schematic of a microjet actuator. Design proposed by Coe et al. [1].

actuator. Of course, when the electric field is removed, the diaphragm will relax to its equilibrium position. Necessarily, this must be accompanied by inflow of air. Thus, both modes of operation produce massless jets in the sense that zero net mass flux of air leaves the actuator over a cycle.

Concerns have been raised that synthetic-jet actuators would be prone to ingest environmental particles. This would make them unattractive for practical flow control. When the pressure-jump mode of operation was first conceived, it was thought that this problem would be overcome. True there was necessarily an inflow phase to replace the air expelled, but this could take place much more gradually than in the case of the synthetic jet. However, our first numerical-simulation study [2] produced a surprise result. There was an inflow into the cavity proceeding the outflow phase, despite the fact that the diaphragm deflected in one direction only (see [2, figure 2]). This puzzling result can be explained by the phenomenon of Helmholtz resonance — a form of intrinsic instability of the cavity [6]. Perversely, it turns out that the very cavity and orifice dimensions that are optimum for actuator performance in terms of maximizing mass flux are close to optimum for Helmholtz resonance [3]. This is true also for synthetic-jet actuators [3, 7].

This paper is structured as follows. In Section 2, we propose a criterion for Helmholtz resonance in terms of fluid and actuator parameters, and verify this using a numerical model developed for devices of the type shown in Figure 1. In Section 3, using the same actuator model coupled to a three-dimensional boundary-layer code, we demonstrate how Helmholtz resonance can be induced in nominally inactive actuators by external pressure fluctuations in a laminar boundary layer.

2 Criterion for Helmholtz resonance

A theory for Helmholtz resonance in massless-jet actuators was given in [3]. However, it was assumed that the orifice flow was fully developed quasi-steady Poiseuille flow (see also [7]). This is a poor assumption because the Helmholtz frequency is well above the limit of validity for quasi-steady Poiseuille theory. Accordingly, we have extended this analysis to include the unsteady pipe-flow theory of Sexl [8]. We also relax the assumption that the flow is fully developed throughout. This results in a criterion for the onset of Helmholtz resonance as follows:

$$\Pi > 12, \tag{1}$$

where the resonant frequency, $\Omega = \sqrt{0.75\Pi}$ and

$$\Pi = \frac{\lambda \pi R_e^6 p_e}{\rho_e \ell V v^2}, \tag{2}$$

where R_e is the radius of the exit orifice, p_e is the mean pressure, ρ_e is the mean density, ℓ is the orifice length, V is the chamber volume, v is the kinematic viscosity and $\lambda < 1$ is a factor that allows for the entry length of the orifice flow.

By comparison, the classic Helmholtz resonant frequency (see, for example, [6]) is $\Omega = \sqrt{\Pi}$, and the condition for Helmholtz resonance obtained using quasi-steady theory, as in [3,7], is $\Pi > 16$.

Here we perform a series of numerical simulations of the micro-jet actuator in Figure 1 in order to verify the criterion stated above. The methods we have adopted to model the microjet actuator are described in some detail in [3,4]. The diaphragm is modelled using thin-plate theory, with the dynamics and stiffness of the attached piezoceramic driver incorporated. For numerical economy, the fluid motion within the plenum chamber is not modelled, and instead the pressure is calculated using the ideal gas law. Within the orifice, where viscous forces are more dominant, the one-dimensional Navier–Stokes equations are solved numerically (this corresponds to $\lambda = 1$).

In these simulations, the actuators (of varying geometry) are given an instantaneous and small internal pressure increase. This induces an initial outflow in all cases, but only when Helmholtz resonance occurs is there a proceeding inflow phase (and in extreme cases enduring jet oscillation). In Figure 2, the inflow velocity (which is zero for sufficiently damped cases) is plotted against Π. There is an inflow, and therefore Helmholtz resonance, at values of Π greater than or equal to 12 – this exactly coincides with the theoretical prediction given by Equation (2).

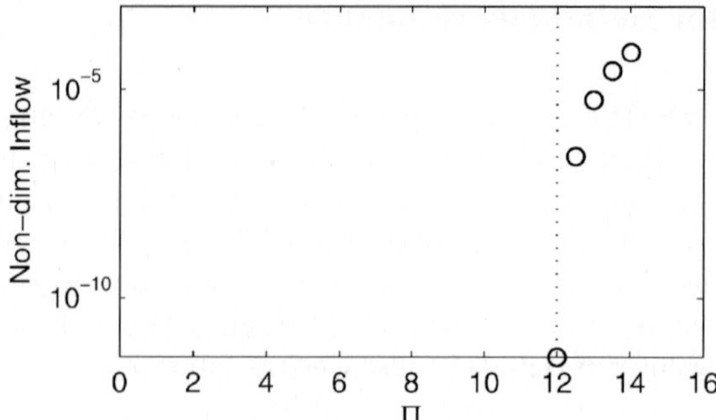

Fig. 2 Non-dimensional inflow velocity against Π. The Helmholtz resonance criterion predicted in Equation (2) is shown by the vertical dashed line.

3 Boundary-layer Induced Helmholtz Resonance

In this section, we investigate the potential for nominally inactive actuators to be resonantly driven by boundary-layer disturbances. Importantly, such a numerical study requires a coupled solution of models for both the actuator and the boundary layer.

For calculating the boundary-layer disturbances generated by the microjet actuator, a novel form of the velocity-vorticity method has been adopted [2,4,9]. This has all the advantages of conventional velocity-vorticity methods (notably, the absence of pressure from the governing equations), but has only three dependent variables rather than the usual six, which reduces the computational expense. This method has been fully validated over a wide range of established results in flow stability and other topics; full details can be found in [9]. In our simulations, we adopt a parallel flow assumption and solve for linear perturbations from a mean Blasius flow profile.

The numerical experiment we have constructed consists of a nominally inactive actuator located within a laminar Blasius boundary layer. Well upstream of the actuator, a Tollmien–Schlichting wave is generated by an inflow-outflow driver located at the wall. As the simulation progresses the Tollmien-Schlichting wave travels downstream and over the actuator's orifice. If the frequency of the wave is fairly close to the Helmholtz frequency, the disturbance pressure fluctuations provoke Helmholtz resonance in the actuator. Figures 3a and 3b give side and plane views of the simulation some time after it has been initiated. These clearly show a strong disturbance has been generated by the 'inactive' actuator that acts as a source of 3D Tollmien–Schlichting waves of much increased amplitude. Probably more important for practical purposes is Figure 3c which shows the variation of centreline (or jet) velocity with time at the orifice exit. This shows that under the influence of Helmholtz resonance there is rapid growth in the amplitude of the jet produced by the actuator. In practice, the nonlinear effects ignored in our simulations would cause the jet amplitude to saturate. In this example the frequency of the incident waves is

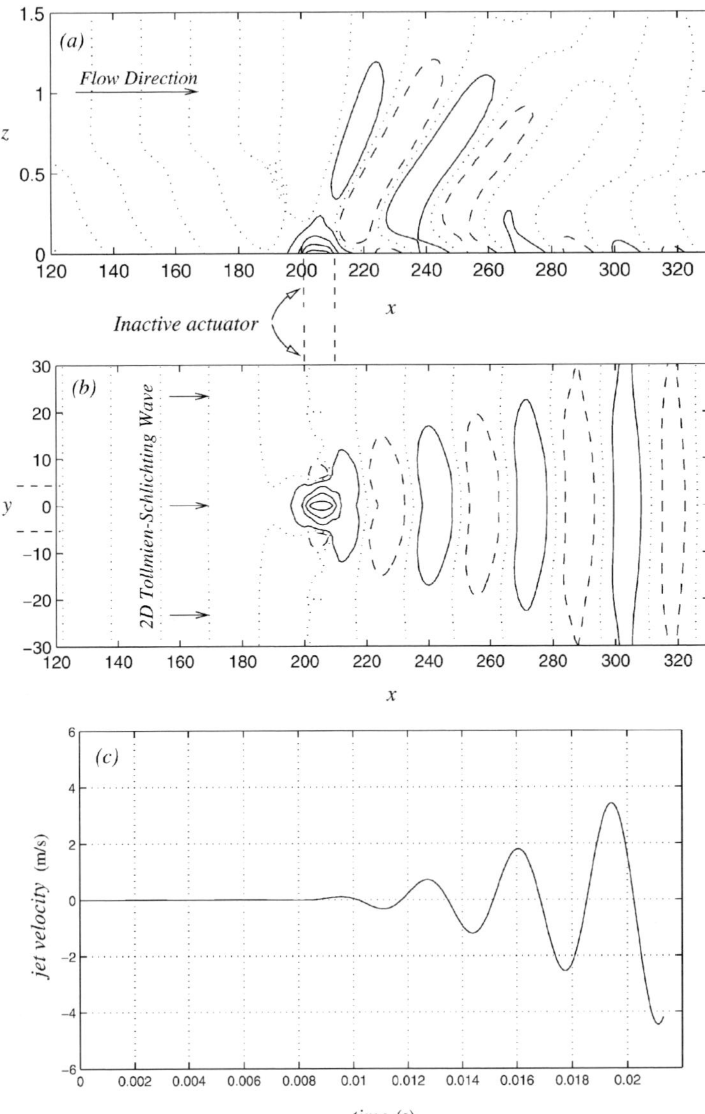

Fig. 3 A nominally inactive macro-scale massless-jet actuator exhibiting Helmholtz resonance excited by incident two-dimensional Tollmien–Schlichting waves. (a) Side view of contours of dimensionless spanwise vorticity perturbation at $y = 0$. (b) Plane view of the same at $z = 0$. Where full and broken lines denote positive and negative values respectively; dotted lines denote zero-level contours. (c) The corresponding variation with time of the centre-line orifice exit (or jet) velocity from the nominally inactive actuator. Boundary-layer parameters: displacement thickness $\delta^* - 1$ mm; free-stream velocity $U_\infty = 30$ m/s; Reynolds number $R - 2000$; ambient pressure $p_e = 1$ bar, ambient density $\rho_e - 1.2$ kg/m^3; and kinematic viscosity $\nu = 15 \times 10^{-6}$ m^2/s. The total dimensionless time of the simulation is 640, and the dimensionless frequency of the incident Tollmien–Schlichting waves is 0.06 (non-dimensionalised with boundary-layer values). Actuator geometry: $\ell = 10$ mm, diaphragm radius is 30 mm, cavity height is 50 mm and orifice radius is 5 mm.

comparable with the Helmholtz frequency, but fine tuning the two frequencies was not necessary to get this resonant effect.

For simplicity we have used Tollmien–Schlichting waves as representative incident waves on 'inactive' actuators. In practice such disturbances could take many forms, e.g. acoustic waves. However, to carry out a numerical simulation of receptivity to acoustic waves would be much more demanding. There is no reason, however, to doubt that a similar resonant response would be seen in these other cases. As remarked above, detuned waves still produce a resonant response. Furthermore, in applications on real aircraft one would expect to find many sources of broadband noise, particularly in turbulent boundary layers.

4 Conclusions

We have proposed and numerically verified a criterion for Helmholtz resonance in terms of fluid and actuator parameters. If these conditions for Helmholtz resonance are satisfied a strong inflow is possible, irrespective of the chosen form of diaphragm forcing. In the simulation of Section 3, two-dimensional Tollmien–Schlichting waves, with frequency comparable with, but not particularly close to, the Helmholtz resonant frequency, are incident on a nominally inactive micro-jet actuator. The results show that under these circumstances the actuator acts as a strong source of 3D Tollmien–Schlichting waves. It is surmised that in the real-life aeronautical applications with turbulent boundary layers, broadband disturbances of pressure field, including acoustic waves, would cause nominally inactive actuators to act as strong disturbance sources. Should this be true, it would probably be disastrous for engineering applications of such microjet actuators for flow control. One would either be faced with the great difficulty of making the actuators inactive or having to design actuators that are not subject to Helmholtz resonance, but as a consequence, far from optimized for performance. These conclusions apply whether the actuator is used in synthetic or pressure-jump mode and may also apply to inactive pulsed-jet actuators.

Acknowledgments

The research presented here was undertaken as part of the AEROMEMS project (an investigation into the viability of MEMS technology for boundary-layer control on aircraft), which was a collaboration between British Aerospace; Dassault Aviation; Centre National de la Recherche Scientifique; and the Universities of Warwick, Manchester, Berlin, Madrid, Athens, Lausanne, and Tel-Aviv. The project was managed by British Aerospace and was partially funded by the CEC under the IMT initiative (Project Ref. BRPR CT97-0573).

References

1. Coe, D.J., Allen, M.G., Trautman, M.A., Glezer, A.: Micromachined jets for manipulation of macro flows. In: *Procedures of Solid-State Sensor and Actuator Workshop* (1994) 243–247.
2. Lockerby, D.A., Carpenter, P.W., Davies, C.: Numerical simulation of the interaction of micro-actuators and boundary layers. *AIAA J.* **40**(1) (2002) 67–73.
3. Lockerby, D.A., Carpenter, P.W.: Modeling and design of microjet actuators. *AIAA J.* **42**(2) (2004) 220–227.
4. Lockerby, D.A., Carpenter, P.W., Davies, C.: Control of sublayer streaks using microjet actuators *AIAA J.* **43**(9) (2005) 1878–1886.
5. Glezer, A., Amitay, M.: Synthetic jets. *Annual Rev. Fluid Mech.* **34** (2002) 503-529.
6. Dowling, A.P., Ffowcs Williams, J.E.,: *Sound and Sources of Sound*, Ellis Horwood Publishers, Chichester, England (1983).
7. Kook, H., Mongeau, L., Franchek, M.A.,: Active control of pressure fluctuations due to flow over Helmholtz resonators. *J. Sound Vibration* **255**(1) (2002) 61–76.
8. Sexl, Th.: Annulareffekt. *Z3. Phys.* **61** (1930) 349–362.
9. Davies, C., Carpenter, P.W.: A novel velocity-vorticity formulation of the Navier–Stokes equations with applications to boundary-layer disturbance evolution. *J. Comput. Phys.* **172** (2001) 119–165.

Passive Scalar Mixing Downstream of a Synthetic Jet in Crossflow

Glen Mitchell[1], Emmanuel Benard[1], Václav Uruba[2] and Richard Cooper[1]

[1]*School of Mechanical and Aerospace Engineering, Queen's University Belfast, Belfast, U.K.; E-mail: glen.mitchell@qub.ac.uk*
[2]*Institute of Thermomechanics, The Academy of Sciences of the Czech Republic, Prague, Czech Republic*

Abstract. An experimental investigation on passive scalar mixing due to the interaction of a synthetic jet with a thermal boundary layer is presented. From velocity measurements, performed by particle image velocimetry, two jet behaviours were identified. For jet to crossflow velocity ratios less than 1.2, the velocity fluctuations due to the jet/crossflow interaction stayed close to the wall. At higher ratios, the fluctuations moved away from the wall. The thermal mixing was examined using laser induced fluorescence. During expulsion, the thickness of the downstream thermal boundary layer increased whilst the thermal boundary layer was annihilated immediately downstream of the jet during entrainment.

Key words: Passive scalar mixing, synthetic jet, crossflow.

1 Introduction

Surface heat transfer is important in many engineering applications, such as energy, aerospace and automotive. For a crossflow, it is possible to increase surface heat transfer through the use of vortex generators upstream of the heat transfer surface [1, 2] and by injecting continuous jets into the flow [3]. In both cases, heat transfer enhancement is obtained by altering the thermal mixing near the surface. It is possible that a synthetic jet could also be used to improve crossflow surface heat transfer.

The synthetic jet actuator consists of a cavity with an oscillating surface acting as one wall and an orifice in another wall. During the forced oscillation of the surface, fluid is alternately entrained into and expelled from the cavity creating a series of vortex pairs that coalesce to form a jet. In a crossflow, the vortex generated travels downstream of the orifice and at low ratios of jet to crossflow velocity, the vortex remains close to the wall. This vortex influences the transport of fluid within the boundary layer and, when the wall is heated, the thermal mixing.

J.F. Morrison et al. (eds.), IUTAM Symposium on Flow Control and MEMS, 103–109.
© 2008 *Springer. Printed in the Netherlands.*

The current study aims at investigating the influence of a 2D synthetic jet on passive scalar mixing in a turbulent boundary layer, with temperature acting as the passive scalar. An initial particle image velocimetry (PIV) study is used to study the hydrodynamics of the jet/crossflow interaction and proper orthogonal decomposition (POD) of the flow field is used to identify the conditions for near wall mixing. A laser induced fluorescence (LIF) study is then conducted in order to examine the thermal mixing due to the jet/crossflow interaction.

2 Water Channel and Synthetic Jet Actuator

The experiments were conducted in a closed circuit horizontal water channel with a test section of length 1000 mm and cross-section 60mm wide by 360 mm high. The channel has a maximum test section velocity of 40 cm/s with a freestream turbulence intensity of less than 1.5%. The flow along the sidewall of the channel was tripped just after the contraction to generate a turbulent boundary layer. The boundary layer can be heated by a 400 mm long copper plate. The plate is located 600 mm downstream from the start of the test section and has two uniform heat flux electrical heaters at the rear. A radiator in the settling chamber of the channel ensures the water temperature fluctuates by less than 0.1°C.

The synthetic jet was generated by a mechanical piston/cylinder arrangement connected to a cavity with a slot type orifice of width 0.75 mm and length 105 mm. The jet orifice is located in the heated copper plate, 130 slot widths downstream from the start of the thermal boundary layer and is orientated so the long axis of the orifice is perpendicular to the flow. An electromagnetic shaker is used to excite the piston with a sinusoidal oscillation. Using this arrangement, jets can be generated with Reynolds number ranging from 20 to 1000 and Strouhal numbers in the range 0.001 to 0.2.

3 PIV/LIF Technique

The PIV/LIF system consists of a 15Hz double pulsed Nd:YAG (532nm wavelength) laser and a CCD camera with a resolution of $2048 \times 2048 \text{pixels}^2$ and a maximum frame rate of 15 frames per second. The laser light sheet was adjusted so it was located along the orifice centreline parallel to the flow. The flow was seeded with hollow glass spheres for PIV and Rhodamine B fluorescent dye for the LIF temperature measurements.

For both the PIV and LIF experiments phase and ensemble averaged results were obtained. The image acquisition was phase-locked to the actuator motion with images acquired at 24 equally spaced intervals during the actuator cycle. For PIV, 150 double frame images were acquired for each phase of the cycle. These images were analysed using a multi pass cross-correlation technique producing vector fields with

Table 1 Parameters for the three jet cases considered.

Case	Re_{U_0}	St_{U_0}	U_∞ [cm/s]	U_0/U_∞
1	278	0.0277	40.0	0.83
2	556	0.0089	40.0	2.22
3	278	0.0089	40.0	1.11

a resolution of 0.22×0.22 mm^2. For LIF, 40 single frame images were taken at each phase. For each image, the fluorescence intensity was averaged over cells of 10×10 pixels^2 to maximise the signal to noise ratio resulting in a spatial resolution of 0.16×0.16 mm^2. The intensity distribution across the light sheet varies for each laser shot. This intensity variation was corrected by a procedure similar to the correction method of Seutiëns et al. [4]. The final temperature field was calculated from the fluorescence intensity using a predetermined calibration relationship.

4 Proper Orthogonal Decomposition

The POD provides an energy efficient decomposition of the fluctuating part of the velocity field. The instantaneous velocity field is decomposed into a time averaged component and a fluctuating component:

$$\mathbf{u}(\mathbf{x},t) = \mathbf{U}(\mathbf{x}) + \sum_{n=1}^{N} a_n(t)\phi_n(\mathbf{x}), \tag{1}$$

where the fluctuating component is described by a linear combination of modes, ϕ. The modes are organised so the first mode contains the most energy and the last mode the least energy. This makes POD a useful tool for identifying dominant coherent flow patterns. In this paper, the POD has been performed on the instantaneous velocity fields by the method of snapshots [5].

5 Hydrodynamic Effects

To investigate the hydrodynamic effects of the interaction of the synthetic jet with a crossflow, eight different cases were studied. Only two representative cases are presented here. Case 1 has a low jet to crossflow velocity ratio and case 2 has a high velocity ratio. The parameters for both jets are given in Table 1. For both cases, the boundary layer had the following parameters at the jet orifice; $Re_\theta = 760$, $\delta = 21$ mm and $H = 1.3$. To confirm that the boundary layer was turbulent, the velocity profile was compared to the logarithmic law of the wall and the turbulence intensity profile was examined.

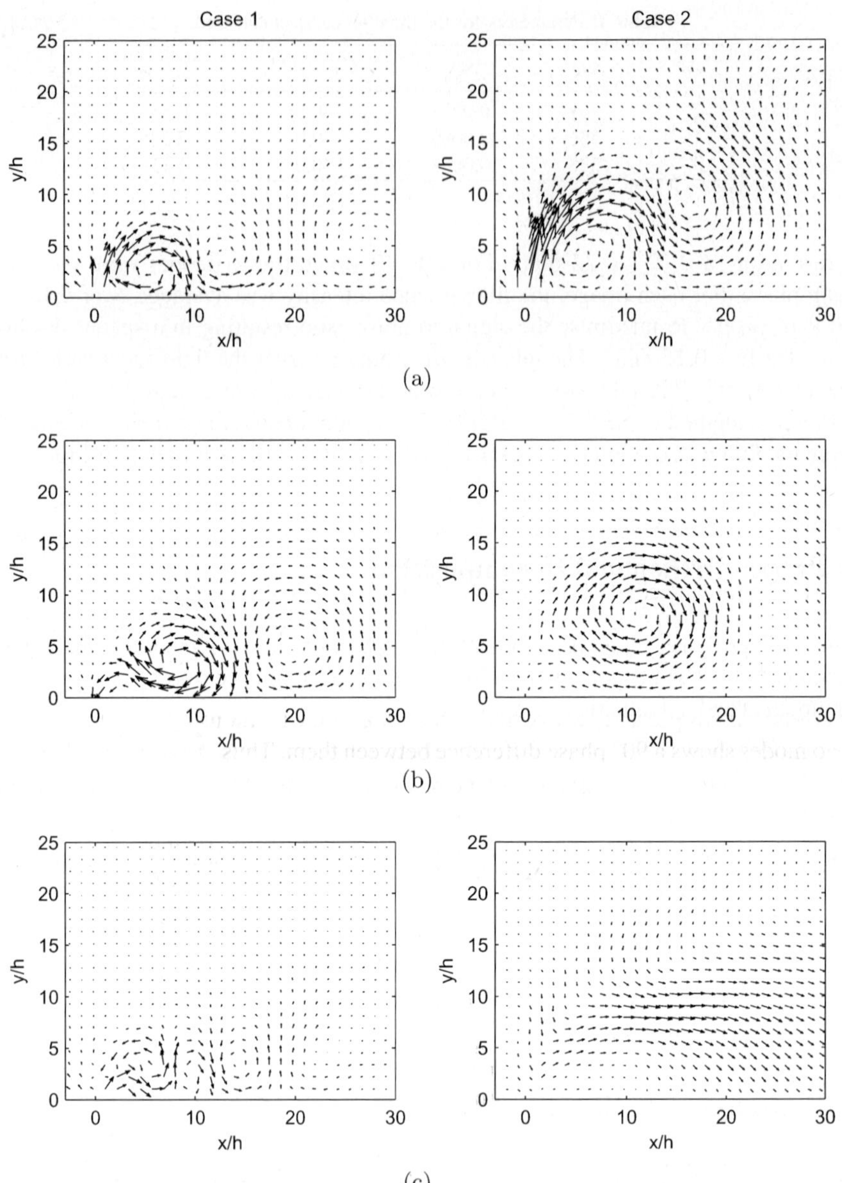

Fig. 1 Visualisation of the first three POD modes for cases 1 and 2: (a) mode 1, (b) mode 2, (c) mode 3.

The first three POD modes for each case are shown in Figure 1. It can be seen that modes 1 and 2 are similar for each case. The first mode has a structure similar to the ejection portion of the jet cycle, with a large velocity magnitude at the orifice exit

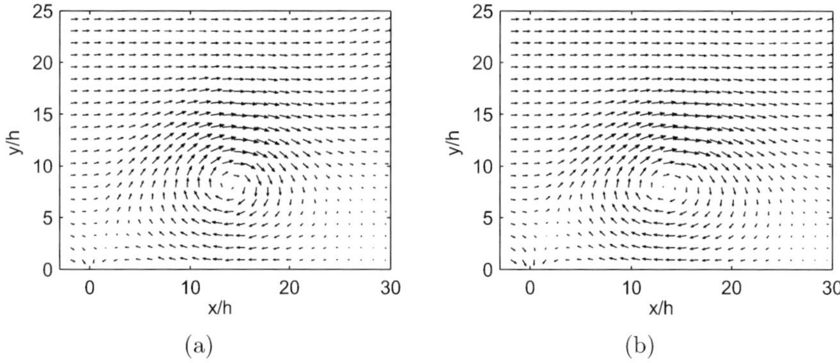

Fig. 2 The synthetic jet flow field during the suction cycle for case 2: (a) phase-averaged, (b) velocity field reconstruction using modes 1 and 2.

and a vortex immediately downstream. The structure of the second mode resembles the suction portion of the cycle, with a smaller velocity around the orifice and the vortex further downstream. Reconstructions of the flow field using these two modes are similar to the phase averaged flow, as demonstrated in Figure 2 for case 2. It should be noted that this is not valid for the start of the expulsion cycle, as modes 1 and 2 do not contain enough detail to reconstruct the complex flow field around the jet orifice at this point in the cycle. A closer look at the relationship between the first two modes shows a 90° phase difference between them. Thus, modes 1 and 2 do not explicitly represent the expulsion and suction cycles respectively, but work together to represent the main structures of the flow field. The first two modes combined contain over 30% of the energy, with subsequent modes having less than 5% each.

After the first two modes, the modes for the two jet cases start to deviate as shown in Figure 1c. For case 1, the third mode consists of a series of vortical structures located close to the wall, whereas case 2 has a horizontal fluctuation located away from the wall. Although not presented here, subsequent modes for case 2 tend to consist of a series of vortical motions moving away from the wall, while for case 1, the fluctuations in higher modes remain close to the wall.

For all eight of the jet cases studied, the POD modes followed the same pattern as cases 1 and 2. The first two modes were similar for all cases. For mode 3 and higher, the two different behaviours described above were observed. The type of behaviour the jet displays was found to be dependant on the jet to crossflow-velocity ratio. For $U_0/U_\infty < 1.2$ the fluctuations remain close to the wall, and they move away from the wall for $U_0/U_\infty > 1.6$.

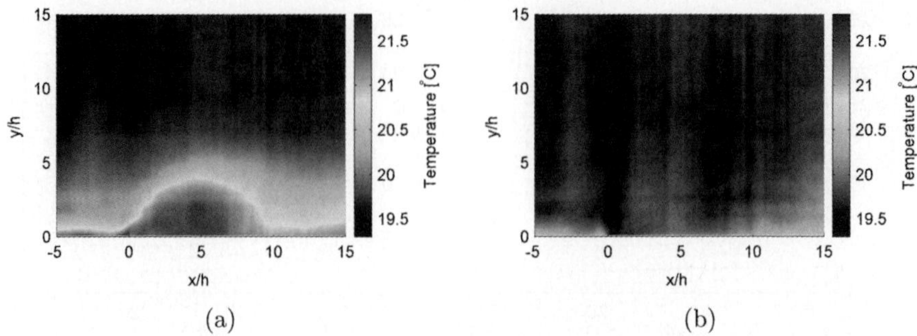

Fig. 3 Thermal mixing between the synthetic jet and a heated crossflow: (a) $t/T = 1/4$, (b) $t/T = 3/4$.

6 Passive Scalar Effects

To investigate the scalar effects, a jet similar to case 1 is considered, so that the fluctuations in the higher modes of the POD remain close to the wall. The jet parameters are given in Table 1 (case 3). For these tests, the copper plate had a mean surface temperature of 25°C. The results of the LIF study are shown in Figure 3 for two different phases of the jet cycle: half way through the ejection cycle and the midpoint of the suction cycle. In this figure, the jet orifice is at the origin.

During the expulsion cycle, the jet pushes warm fluid away from the wall, increasing the thickness of the thermal boundary layer downstream of the jet. Upstream of the orifice, the thermal boundary layer is unchanged from the no-jet condition. During the suction cycle, the thermal boundary layer downstream of the orifice is annihilated as the jet sucks in the warm fluid close to the wall upstream of the orifice as observed in Figure 3b. Comparing the upstream thermal boundary layers in Figures 3a and 3b, it can be seen that the entrainment of the near-wall fluid into the cavity also causes the upstream thermal boundary layer to become thinner.

7 Conclusions

The velocity fields of the interaction of the synthetic jet with a crossflow have been analysed by POD. For all cases considered, the first two POD modes have the same features, but the behaviour of the higher modes is dependant on the jet to cross flow velocity ratio. For $U_0/U_\infty < 1.2$, the velocity fluctuations due to the interaction are located near the wall, whereas the fluctuations are further away from the wall for $U_0/U_\infty > 1.6$. The effect of the jet on thermal mixing varies with the phase of the jet cycle. Thermal penetration is improved during the ejection portion of the cycle. However, during entrainment the thermal boundary layer is terminated at the jet orifice.

References

1. Tiggelbeck, S., Mitra, N., Fiebig, M.: Experimental investigations of heat transfer enhancement and flow losses in a channel with double rows of longitudinal vortex generators. *International Journal of Heat and Mass Transfer* **36** (1993) 2327–2337.
2. Jacobi, A., Shah, R.: Heat transfer surface enhancement through the use of longitudinal vortices: A review of recent progress. *Experimental Thermal and Fluid Science* **11** (1995) 295–309.
3. Zhang, X., Collins, M.: Flow and heat transfer in a turbulent boundary layer through skewed and pitched jets. *AIAA Journal* **31** (1993) 1590–1599.
4. Seutiëns, H., Kieft, R., Rindt, C., van Steenhoven, A.: 2D temperature measurements in the wake of a heated cylinder using LIF. *Experiments in Fluids* **31** (2001) 588–595
5. Sirovich, L.: Turbulence and the dynamics of coherent structures. Part I: Coherent structures. *Quarterly of Applied Mathematics* **45** (1987) 561–571.

Towards a Practical Synthetic Jet Actuator for Industrial Scale Flow Control Applications

Luis Gomes[1] and William Crowther[2]
[1] *Goldstein Research Laboratory,, The University of Manchester, Manchester M30 7RU, U.K.; E-mail: tcs_Luis@yahoo.co.uk*
[2] *School of MACE, George Begg Building, Sackville Street, University of Manchester, M60 1QD, U.K.; E-mail: bill.crowther@manchester.ac.uk*

Abstract. The global evolution of the aerospace market is driving flow control research towards full industrial scale applications. In this approach, technologies need to demonstrate effectiveness, as well as compliance with the aircraft performance constraints. The design of a synthetic-jet-based system for a civil transport aircraft would provide an early understanding on the viability and potential applications of the technology. This study characterises experimentally an optimised piezoelectric-based synthetic jet actuator (SJA). Three full scale systems were developed for an A321: a flap, a slat and a cruise, both for separation and shock control. The systems were designed based both on experimental results and on an extensive hardware research. The laboratory optimisation of SJAs has led to the achievement of peak velocities of 130 m/s and peak conversion efficiencies of around 15%. All systems presented power and weight requirements within the aircraft performance budget, where the flap system resulted in the lowest values. The question that still remains is whether or not the performance benefits will outweigh the system costs.

Key words: Synthetic jet actuator, flow control, aircraft systems, mass, power.

1 Aim of Work

This work intends to describe the laboratory development of optimised synthetic jet actuators and to predict the likely challenges for full scale application to civil transport aircraft.

J.F. Morrison et al. (eds.), IUTAM Symposium on Flow Control and MEMS, 111–118.
© *2008 Springer. Printed in the Netherlands.*

2 Introduction

2.1 Definition and Evolution of Flow Control

Flow control can be defined in numerous ways, however for the purpose of this study it shall be defined as: the achievement of making a flow behave in a way that it would not ordinarily by the application of a minimum amount of energy or effort. As flow control research matures it is becoming more important to consider not just the question of whether it works (technical filter), but rather whether or not it can be made to work successfully in practice (technological filter). Practical success is typically based on obtaining levels of effectiveness and efficiency such that the benefits of application exceed the overall cost of implementation.

2.2 Synthetic Jet Actuators (SJAs)

Synthetic Jet Actuators (SJAs) are electrically powered momentum injection devices. Their fundamental operating principle is that an oscillating diaphragm produces a useful net flow of momentum within an external flow. These devices have been demonstrated to be effective in delaying separation in a variety of applications [1–3]. Previous work [4] has also shown that maximum effectiveness of air jet-like vortex generators occurs around a velocity ratio (ratio of jet to local free stream velocity) of the order of 1. This implies that for peak effectiveness in an industrial application on civil transport aircraft actuators, will typically require high subsonic jet velocities - a significant challenge. Furthermore, efficiency levels should be such that the cost of power provision is acceptable within the overall aircraft performance budget.

3 Laboratory Characterisation and Optimisation of SJAs

Current Airbus-funded work is looking to optimise SJAs for maximum jet velocity and power conversion efficiency. As such, a study was conducted on key geometric variables (i.e. chamber height, H and orifice height, h) for optimisation of the internal physics [5]. The 'standard' configuration SJAs used were based on a 25 mm diameter cylindrical chamber and a piezoceramic driven diaphragm.

The current laboratory state-of-the-art actuator is 5 mm in depth and 30 mm in diameter, weighing $\sim$20 g. It outputs synthetic jets with maximum peak velocities of 130 m/s when driven at an excitation amplitude and frequency of $250V_{pp}$ and 2.5 kHz (see Figure 1). These high-velocity jets exit 1.2 mm diameter orifices, representing a mean fluid power of 0.2 W for a mean electric power consumption of 3 W (i.e. $\simeq$ 7% electric-fluid power conversion efficiency – refer to Figure 2). A

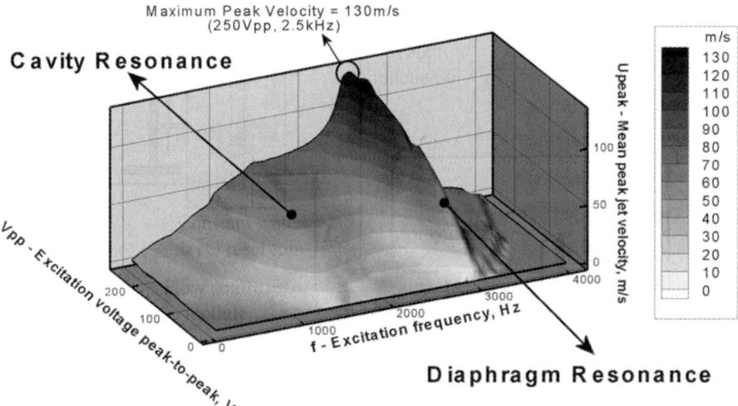

Fig. 1 Peak velocity response map for the geometrically optimised SJ actuator, under the full range of excitation conditions [5].

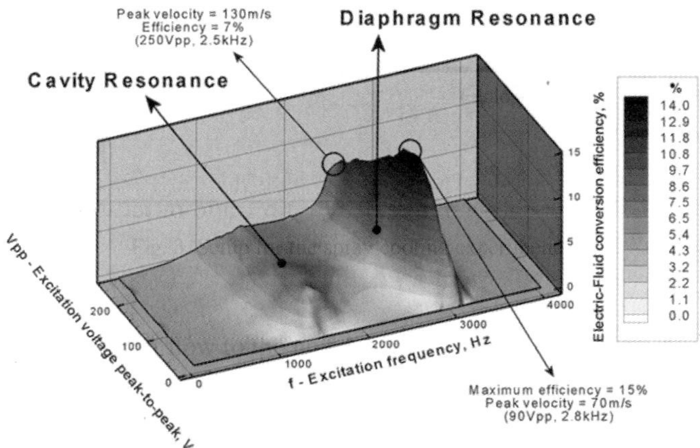

Fig. 2 Electric-fluidic conversion efficiency map for the geometrically optimised SJ actuator, under the full range of excitation conditions.

peak efficiency of $\simeq 15\%$ was reached at a lower excitation amplitude of $90V_{pp}$, resulting in a peak velocity of 70 m/s.

The results from the experimental characterisation of the optimised actuator were used as the building blocks to design the SJA systems, under different operating conditions, for each of the case studies that will be presented in Section 4.

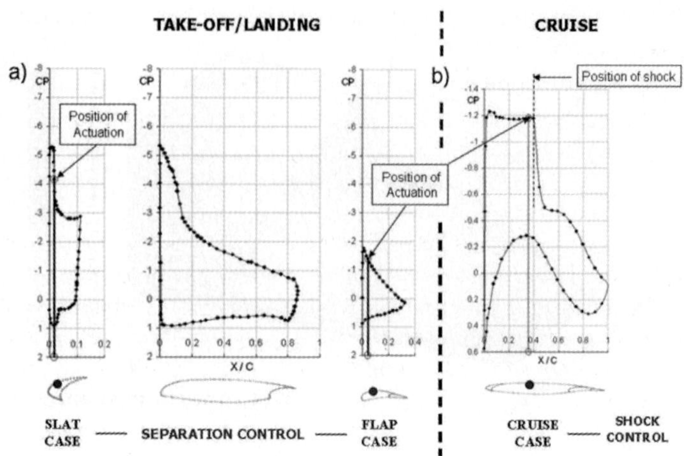

Fig. 3 Experimental pressure distribution around a basic A300 Airbus type aerofoil section in (a) its high lift configuration ($M = 0.2$, $Re = 3.5 \times 10^6$) [6] and (b) its cruise configuration ($M = 0.7$, $Re = 6 \times 10^6$) [7].

4 Industrial Scale Case Study – Application of SJAs to an A321

Three case studies were conducted in order to evaluate the mass and power requirements for a SJA system applied to an A321 aircraft.

Two flight conditions were considered: take off/landing (low speed), and cruise (high speed).

For the low speed condition, the goal is an improvement of maximum lift through the control of separation on the slat and flap. The actuators were positioned at the quarter-chord of the slat/flap, as shown in Figure 3a.

For the high speed condition, the goal is the reduction of wave drag through shock control. The actuators were positioned at $\sim 40\%$ wing chord, i.e. upstream of the shock (see Figure 3b).

Note that the case studies only took into account the cost of implementation. The impact of benefits were not considered.

4.1 Case Study Assumptions Used for Scaling

The A321 specifications can be briefly summarised from [8] as a 2,000 nm range aircraft, with a take-off mass of 90 tonnes, typical flight duration of 2 hours at $M_{\text{Cruise}} = 0.8$, and $M_{\text{TO/Landing}} = 0.2$.

The actuator operating conditions were chosen based on empirical evidence for effectiveness (external physics): jet-to-crossflow velocity ratio of 1; orifice diameter around 1/5th local boundary layer height, and spacing between actuators set to 10 orifice diameters.

4.2 Case Study Results

The SJA system top level break-down structure is composed of 5 main elements: the actuators, the wiring, the power conversion subsystem, the power generation subsystem and the fuel. The fuel is converted into useful electric power through the power generator unit (e.g APU). The raw power is then transformed into a voltage and frequency modulated signal by the power conversion unit. The signal is transported through the wiring to the actuators, and transformed into fluidic power.

All weight and power calculations were based on the data compiled both from laboratory experiments, and from an extensive hardware research (e.g. high power wiring, power amplifiers, etc.) allowing the design of the system to be as realistic as possible.

The case study results presented in Figures 4–6 indicate that as a general rule, all subsystem's weight scale up with fluid power, and hence electric power consumption. The main parameters that drive the fluid power are the jet velocity, the orifice cross sectional area and the number of orifices/actuators in the system, which in turn depend on the local freestream conditions (i.e. crossflow velocity and boundary layer height).

All the subsystems, with the exception of the wiring, present a weight reduction of $\sim 70\%$ when moving from a slat application, i.e. high velocity short duration application (Figure 4), to a flap application, i.e. low velocity short duration application (Figure 5). The system's weight becomes less dependent on the electric power required to drive it, and the element dominating the system's weight changes from the power generator unit to the actuators.

As for the cruise application, i.e. long duration high velocity case (Figure 6), the weight of both the power dependent elements (i.e. power generation and conversion units, actuators and wiring) and the energy storage elements (i.e. fuel) increase when comparing to the short duration type applications. The system's mass penalty increased by $\sim 600\%$ when moving from a flap/slat application to cruise.

Figure 7 presents the overall weight and power penalties of the systems designed from the three case studies. A comparison between them shows that the flap application presents the lowest power and weight penalties, making it the most feasible and attractive solution. An additional advantage comes from the natural sealing of the actuators when the flaps retract, protecting them from the external environmental conditions.

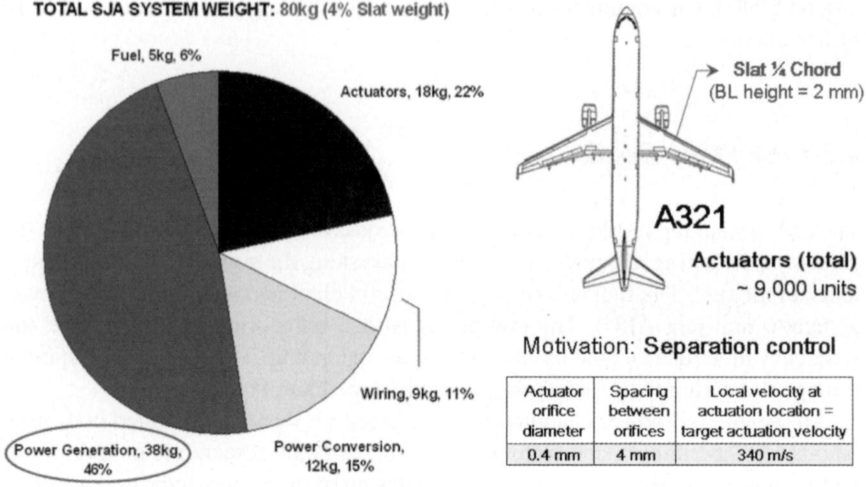

Actuator orifice diameter	Spacing between orifices	Local velocity at actuation location = target actuation velocity
0.4 mm	4 mm	340 m/s

Fig. 4 Slat case – short duration and high velocity application.

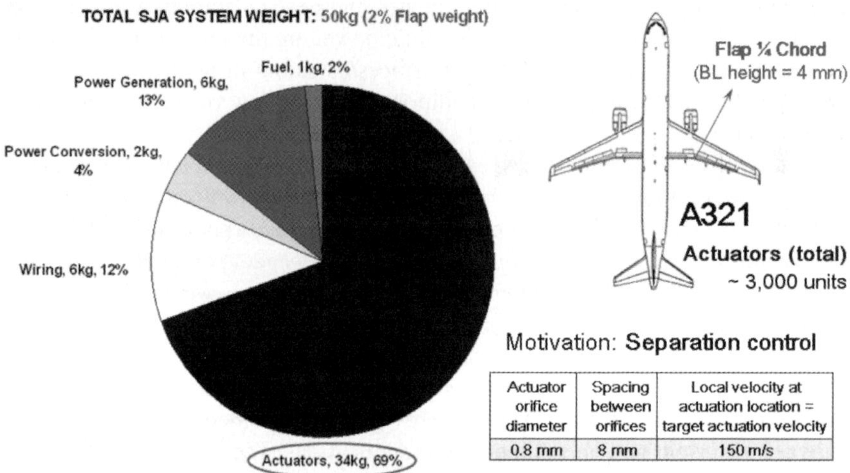

Actuator orifice diameter	Spacing between orifices	Local velocity at actuation location = target actuation velocity
0.8 mm	8 mm	150 m/s

Fig. 5 Flap case – short duration and low velocity application.

5 Conclusions

The laboratory optimisation of SJAs has led to the achievement of a peak velocity of 130 m/s and peak power conversion efficiency of around 15%.

The design of several SJA flow control systems on an A321 (based on empirically developed scaling rules) showed that the power generation subsystem, the actuators and the fuel are the dominant factors determining the system's weight, depending on the application being considered. The weight of the systems considered were still

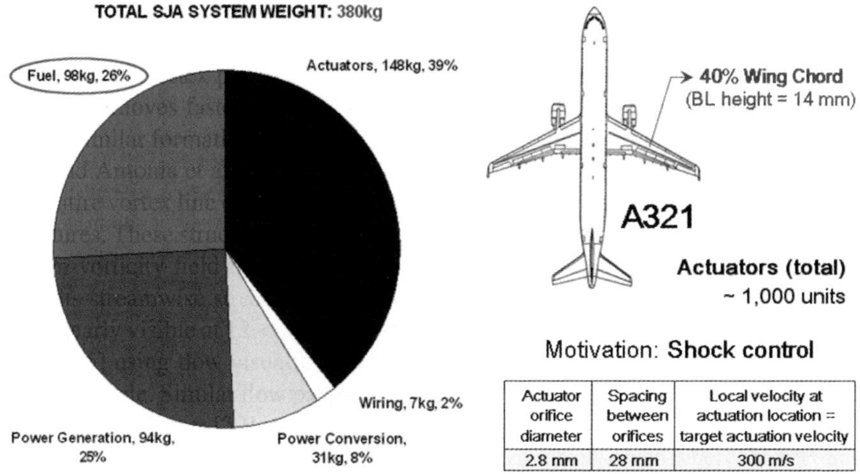

Actuator orifice diameter	Spacing between orifices	Local velocity at actuation location = target actuation velocity
2.8 mm	28 mm	300 m/s

Fig. 6 Cruise case – long duration and high velocity application.

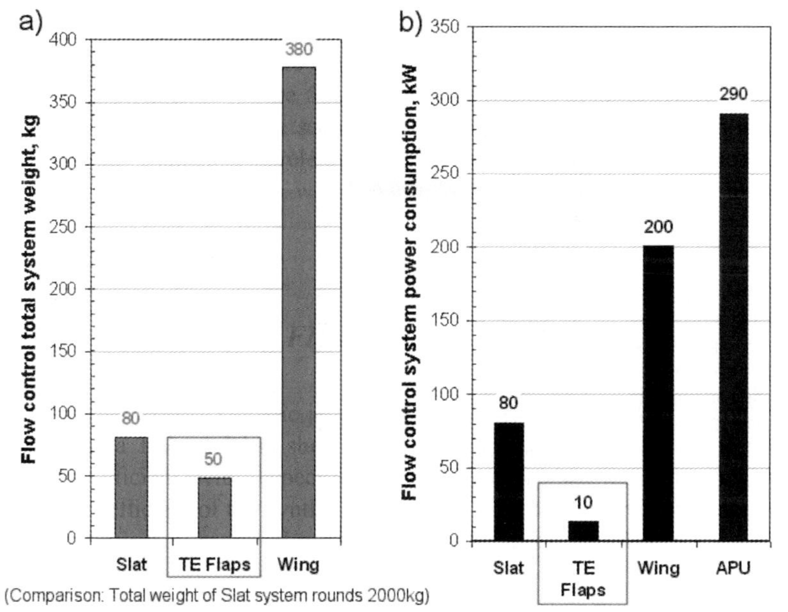

Fig. 7 Comparison between overall SJA system's (a) weight and (b) power penalties.

fractional compared to the weight of the conventional passive flow control system. All of the system's power requirements were within the limit of power rating of an APU typically found in an aircraft of this class. The flap case presented the lowest weight and power penalties, making it the most appealing application. It also presents additional benefits regarding the operation of the actuators.

This investigation showed that it is possible to realistically design a SJ flow control system within the aircraft performance limits, but at a cost of additional power and weight penalties. It becomes therefore apparent that future active flow control systems will need to pay their way. The question that still remains is whether or not the performance benefits will outweigh the system costs.

Acknowledgements

The authors would like to thank Airbus UK and BAE Systems for their financial and technical support of this work, through the Control of Aerodynamic Flows for the Environmentally Driven Aircraft (CAFEDA) programme.

References

1. Amitay, M., Honohan, A., Trautman, M., Glezer, A.: Modification of the aerodynamic characteristics of bluff bodies using fluidic actuators. In *Proceedings 28th AIAA Fluid Dynamics Conference*, Snowmass, CO (1997).
2. Crook, A.: The control of turbulent flows using synthetic jets. Ph.D. Dissertation, The School of MACE (Mechanical, Aerospace and Civil Engineering), The University of Manchester, Manchester (2002).
3. Yehoshua, T., Seifert, A.: Boundary condition effects on oscillatory momentum generators. In *Proceedings 33rd AIAA Fluid Dynamics Conference*, Orlando, FL (2003).
4. Schaeffer, N.: The interaction of a synthetic jet and a turbulent boundary layer. In *Proceedings 41st Aerospace Sciences Meeting and Exhibit*, Reno, NV (2003) AIAA-2003-0643,
5. Gomes, L., Crowther, W., Wood, N.: Towards a practical piezoceramic diaphragm based synthetic jet actuator for high subsonic applications – Effect of chamber and orifice depth on actuator peak velocity. In *Proceedings of 3rd AIAA Flow Control Conference*, San Francisco, CA (2006) AIAA-2006-2859.
6. Wedderspoon, J.: The high lift development of the A320 aircraft, in *Congress of the International Council of the Aeronautical Sciences*, Vol. 1 (1986) pp. 343–351.
7. Greff, E.: The development and design integration of a variable camber wing for long/medium range aircraft, *The Aeronautical Journal* **94** (1990) 301–312.
8. Jackson, P., *Jane's All the World's Aircraft 2004–2005*, Jane's Information Group Ltd (2005) pp. 222–223.

Measurements of Synthetic Jets in a Boundary Layer

Mark Jabbal and Shan Zhong
School of Mechanical, Aerospace and Civil Engineering, The University of Manchester, Manchester M60 1QD, U.K.; E-mail: mark_jabbal@hotmail.com, shan.zhong@manchester.ac.uk

Abstract. PIV measurements along the centerline of a synthetic jet embedded in a flat plate boundary layer were conducted for three types of jet vortex structures identified by the authors in previous flow visualization studies, namely hairpin vortices, stretched vortex rings and tilted vortex rings. The primary purpose of this work was to quantify the near wall effect of these structures in terms of their manipulation of the boundary layer velocity profile. In the near field region, synthetic jets composed of stretched vortex rings, which remain within the boundary layer and tilted vortex rings, which rapidly penetrate the boundary layer produced fuller velocity profiles in comparison to the jet off case. Further downstream, only the velocity profiles manipulated by the hairpin vortices and stretched vortex rings continued to fill out close to the wall, thus suggesting that these embedded structures may offer potential as an optimal configuration for flow separation control.

Key words: Synthetic jets, flow separation control, boundary layer, vortex structures.

1 Introduction

The synthetic jet actuator (SJA) has been demonstrated to have potential for active flow control [1, 2]. A typical SJA [3, 4] consists of a cylindrical cavity bounded by rigid side walls with an orifice plate at one end and an oscillating diaphragm at the other. One of the unique features of the synthetic jet is in its ability to impart additional momentum on a fluid region from which it was originally synthesized without a net mass addition, thereby eliminating the need for air supply and complex piping associated with conventional steady jets.

It is believed that the interaction of the discrete trains of vortices formed out of a SJA with a local boundary layer produces streamwise-aligned vortical structures, which are capable of delaying flow separation by entraining faster moving

J.F. Morrison et al. (eds.), IUTAM Symposium on Flow Control and MEMS, 119–125.
© 2008 *Springer. Printed in the Netherlands.*

fluid from the freestream to the near wall region. The interaction of an embedded SJA array with a turbulent separating flow on a cylinder surface [2] revealed the well-defined footprints of a streamwise vortex pair aft of each orifice, persisting far downstream into the separation line. Subsequent flow visualization experiments of synthetic jets developing over flat plate boundary layers [3, 4] in which the SJA operating conditions (diaphragm oscillation frequency, f and peak-to-peak displacement, Δ) and external conditions (freestream velocity) were varied revealed the formation of three major types of synthetic jet vortex structures – hairpin vortices, stretched vortex rings and tilted vortex rings. The latter of these structures was observed to be coherent at moderate diaphragm displacements [3], becoming fully turbulent at higher displacements [4]. Surface visualization of the near wall effect of these structures [4] showed that for hairpin vortices and stretched vortex rings, both featuring counter-rotating legs that persist close to the wall, the pattern of the footprints resembled those delaying separation on the cylinder model.

In spite of these qualitative assessments, there is a need to quantify the impact of the aforementioned structures on the boundary layer to determine their relative effectiveness for potential flow control. Therefore, this study details the streamwise PIV measurements that were undertaken for a SJA embedded in a flat plate laminar boundary layer towards quantifying the near wall effect on the velocity profiles.

2 Experimental Approach

All tests were conducted in a tilting water flume, which has a length of 4 m and cross-section 0.3 m $\times$ 0.3 m. The SJA embedded in the flat plate has an orifice with a diameter D_o of 5 mm and depth 5 mm and a cylindrical cavity with a diameter of 45 mm and height 25 mm. The SJA orifice was located 140 orifice diameters aft of the leading edge of the plate, thus giving a local boundary layer thickness-to-orifice diameter ratio, δ/D_o of approximately 4, for a fixed working section freestream velocity of 0.05 m/s. The plate was inclined at a small negative incidence of less than 1 degree avoid leading edge separation.

To facilitate PIV measurements, the flow was seeded with hollow glass spheres with a mean diameter of 10 microns. The SJA flow field was illuminated by a 5 W continuous wave Argon ion laser. A lightsheet approximately 1 mm in width was generated, bisecting the orifice along the center plane to produce a two-dimensional slice of the synthetic jet. A Photron Ultima APX CMOS camera with a resolution of 1024×1024 pixels was used to capture images in a field of view of 25 mm $\times$ 25 mm $(5D_o \times 5D_o)$, the vectors of which were resolved using two-frame cross-correlation. A 32×32 pixel interrogation area with an overlap ratio of 50% was chosen giving a spatial separation between adjacent vectors of 0.4 mm. The system was set up such that during experiment, measurements were taken at 60 equally-spaced phases across the diaphragm oscillation cycle in 6-degree increments. For a given phase point, phase-averaged flow fields were obtained by typically averaging

Table 1 Actuator operating conditions and dimensionless flow parameters of the three synthetic jet structures.

Structure Type	f (Hz)	Δ (mm)	VR	L	Re_L	St
Hairpin Vortex	2	0.3	0.32	1.6	131	0.2
Stretched Vortex Ring	1	0.5	0.27	2.7	182	0.1
Tilted Vortex Ring	2	0.5	0.54	2.7	364	0.2

100 instantaneous image pairs. The data was taken at two streamwise locations: in the near field ($-1D_o$ to $4D_o$) and far field ($20D_o$ to $25D_o$) of the orifice.

Conditions for the generation of the three synthetic jet structures are given in Table 1 according to the key dimensionless parameters that define the interaction of a synthetic jet and a boundary layer [4], namely jet-to-freestream velocity ratio VR (which determines jet trajectory), dimensionless stroke length L (which represents the dimensionless length of the fluid column pushed out of the orifice during the blowing stroke), the Reynolds number Re_L, which is based on the jet velocity and stroke length (characterizing vortex strength) and the Strouhal number St (which represents the non-dimensional spacing of successive vortex structures formed out of the orifice).

3 Results and Discussion

Figure 1 shows a centerline phase-averaged vorticity plot for each of the synthetic jet structures obtained from PIV measurements in the near field of the orifice at arbitrarily chosen phase points. Alongside each of these are the corresponding dye visualizations (showing simultaneous side and surface views of the jet) and the thermal footprints generated by the passing of a flow structure using temperature-sensitive liquid crystals. Further details regarding the set up of these techniques can be found in [4]. In all cases, the flow direction is from right to left with the visualization images extending to $25D_o$ downstream of the orifice.

The formation and development of hairpin vortices in the boundary layer is shown in Figure 1a. From the vorticity contour, it can be seen that the emergence of the upstream branch of the vortex is significantly suppressed by the resident vorticity in the boundary layer, whilst the downstream vortex (which has the same sense of rotation as the boundary layer) is strengthened. Due to the low stroke length for this case, the structure does not move sufficiently far from the orifice before the suction stroke begins. Consequently, the weak upstream vortex branch is drawn into the orifice during suction resulting in an asymmetric structure. As the hairpin vortices move downstream, their counter-rotating legs experience significant stretching due to the structures residing within the boundary layer (marked by the dash line). The vorticity plot shows the "head" of one of these hairpins at approximately $1.5D_o$ downstream of the orifice and is seen by the u-velocity component streamlines (from which 50% of the freestream velocity has been subtracted to pick out the vortex).

The thermal footprint reveals the presence of two streamwise streaks of high heat transfer (depicted by the dark regions) created by the motion of the counter-rotating legs that bring in higher momentum fluid from the surroundings to the wall. These streaks are separated by a thin region of low heat transfer (light region), which corresponds to the low momentum fluid accumulating between the legs of the hairpins.

Figure 1b shows the advection of stretched vortex rings in the boundary layer. As in the previous case, the vorticity plot in Figure 1b shows a weakened upstream branch of the vortex forming out of the orifice. However in comparison to the hairpin vortex, the downstream branch appears stronger due to an increase in Re_L and from the contour it is possible to see the vorticity in the trailing legs (note that 70% of the freestream velocity has been subtracted from the u-velocity component). The pattern of the streamlines appear to show that the trajectory of the jet in the near field is confined within $2D_o$ or half the boundary layer height, which is similar to that seen in Figure 1a due to similarities in VR. Although the structures eventually penetrate the boundary layer in the far field, their counter-rotating legs remain close to the wall, thus generating a similar thermal footprint to the hairpin vortices.

Figure 1c shows the advection of tilted vortex rings. These structures have a steep trajectory due to high VR, confirmed by the pattern of the streamlines. Consequently, the structures penetrate the boundary layer within a short distance downstream of the orifice. Due to the short time in which these structures are exposed to the boundary layer shear, there is little evidence of vortex stretching that was seen for the near wall structures. Additionally, with a high Re_L the vortex exhibits a strong roll up on both upstream and downstream branches resulting in a symmetrical structure, as observed in the vorticity magnitude and streamlines (70% of the freestream velocity has been subtracted from the u-velocity). Vortex tilting is prevalent, the mechanism of which is described in [3]. The footprint is distinctly different from Figures 1a and 1b and is likely to be produced by an induced vortex close to the wall.

Figure 2a compares the impact of the synthetic jet structures on the undisturbed centerline zero-pressure gradient laminar boundary layer profile in the near field of the orifice at $x/D_o = 4$. For the passing hairpin vortices, an inflexion in the profile occurs at $y/\delta = 0.15$, with a subsequent velocity deficit extending to $y/\delta = 0.35$. It is believed that this velocity deficit is created by the lift up of low momentum fluid between the legs of the hairpins, as marked by the corresponding thermal footprint (Figure 1a). Conversely, both the stretched and tilted vortex rings generate fuller profiles in the near wall region. The profile manipulated by the tilted vortex ring is seen to have a slightly greater velocity gradient near the wall, which is conducive to higher shear stress and fluid mixing. However, a large velocity deficit is created by these passing structures in the outer part of the boundary layer beyond $y/\delta = 0.4$. Further downstream at $x/D_o = 24$ (Figure 2b), it is seen that the velocity profiles are more fully developed. It is worth noting that compared to their respective profiles in Figure 2a, the velocity gradients near the wall influenced by the stretched vortex rings and tilted vortex rings have increased and decreased respectively. This indicates that the influence of the tilted vortex rings is confined to the near field due to their steep trajectory out of the boundary layer, whereas the stretched vortex rings continue to increase mixing due to their close proximity to the wall. The profile in-

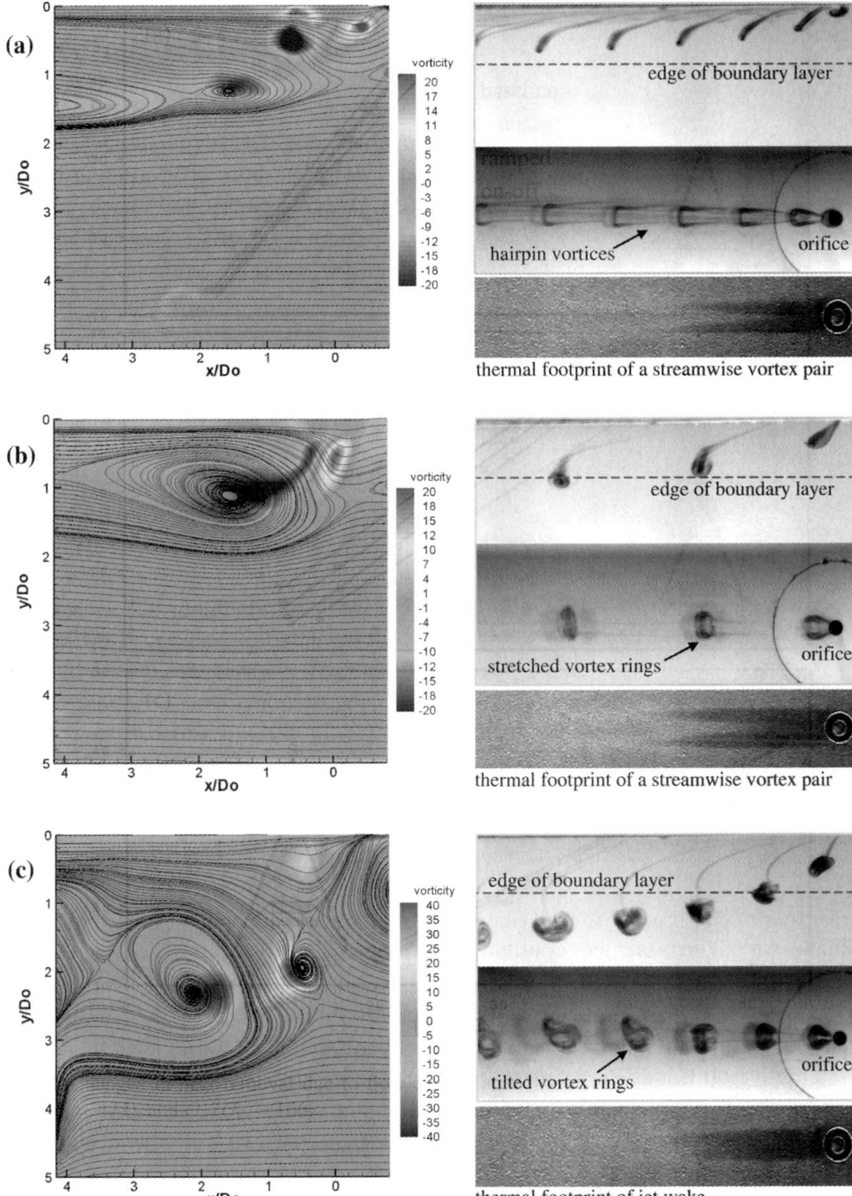

Fig. 1 Corresponding near field centerline phase-averaged vorticity plots, stereoscopic dye visualization and surface thermal footprints for (a) hairpin vortices, (b) stretched vortex rings and (c) tilted vortex rings. The flow visualization images extend to $25D_0$ downstream of the orifice and the dash line represents the edge of the boundary layer.

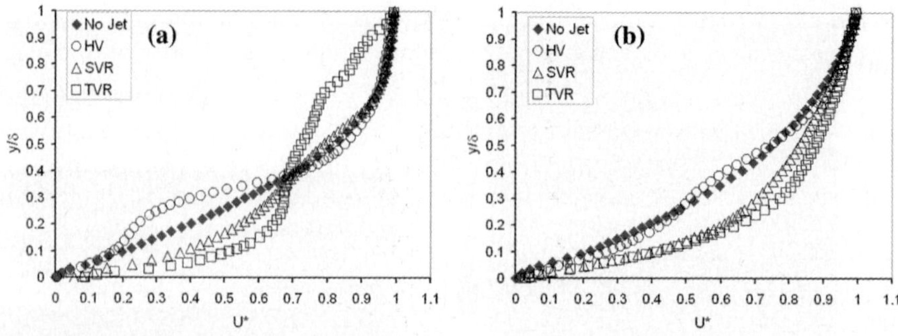

Fig. 2 Centerline boundary layer velocity profiles of the synthetic jet structures at (a) $4D_o$ and (b) $24D_o$ aft of the orifice (HV = hairpin vortex, SVR = stretched vortex ring and TVR = tilted vortex ring).

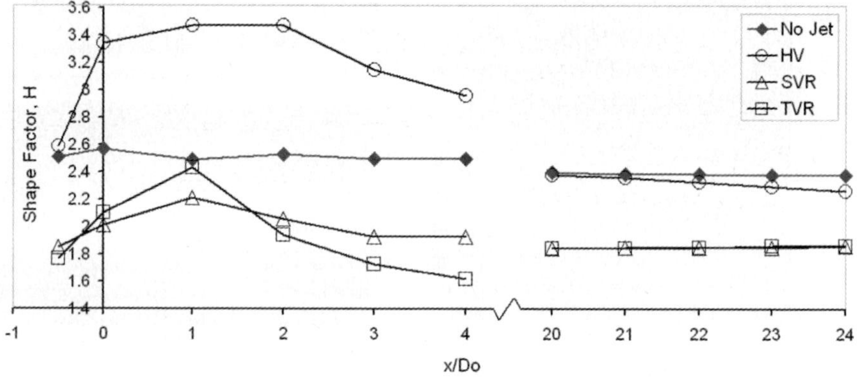

Fig. 3 Streamwise variation of boundary layer shape factor along the centerline.

fluenced by hairpin vortices continues to become fuller, with the inflexion moving to $y/\delta = 0.3$. It is anticipated from this trend that the profile will relax further downstream and exhibit a similar near wall gradient as that generated by the stretched vortex rings.

The overall trends depicted in the boundary layer profiles are mirrored in the streamwise variations of shape factor, H, along the centerline (Figure 3). The shape factor provides an indication of the resistance of a boundary layer to flow separation with a decreasing H being favorable for resisting separation. Between $x/D_o = 4$ and $x/D_o = 20$, the shape factor increases for tilted vortex rings following an initial sharp reduction in the near field region, but decreases for both hairpin vortices and stretched vortex rings. Further downstream between $x/D_o = 20$ and $x/D_o = 24$, whilst the shape factor for the stretched vortex rings appears to have reached a steady value ($H \approx 1.85$), that for the hairpin vortices continues to decrease suggesting that they may exhibit a comparable level of fluid mixing close to the wall in the far field.

4 Conclusions

PIV measurements of different vortex structures emanating from a round synthetic jet in a boundary layer were conducted with a view to quantitatively assessing their impact on the undisturbed centerline boundary layer velocity profile. Hairpin vortices and stretched vortex rings, which remain close to the wall, continue to generate fuller profiles in the far field region and thus may offer potential as an optimal flow control configuration. Further measurements in streamwise planes adjacent to the centerline will be analyzed to assess the lateral influence of these synthetic jet structures in the near wall region.

References

1. Glezer, A. and Amitay, M.: Synthetic Jets. *Annu. Rev. Fluid Mech.* **34**, 503–529 (2002).
2. Crook, A. and Wood, N.J.: Measurements and Visualisations of Synthetic Jets. AIAA Paper 2001-0145 (2001).
3. Zhong, S., Millet, F. and Wood, N.J.: The Behaviour of Circular Synthetic Jets in a Laminar Boundary Layer. *Aeronaut. J.* **110**(1100), 461–470 (2005).
4. Jabbal, M. and Zhong, S.: The Near Wall Effect of Synthetic Jets in a Laminar Boundary Layer. AIAA Paper 2006-3180 (2006).

Large-Eddy Simulations of Synthetic Jets in Stagnant Surroundings and Turbulent Cross-Flow

Don K.L. Wu and Michael A. Leschziner

Department of Aeronautics, Imperial College, Prince Consort Road, London SW7 2BY, U.K.; E-mail: d.wu@imperial.ac.uk

Abstract. Large-Eddy Simulations (LES) are used to investigate the physical processes involved in the injection of a synthetic (zero-net-flux) jet into a zero-pressure-gradient turbulent boundary layer at conditions corresponding to experimental data obtained by others. The boundary layer ahead of the jet is generated by a separate precursor simulation at a momentum-thickness Reynolds number of $Re_\theta = 920$, providing the main simulation with a full and accurate description of the unsteady conditions. Phase-averaged results and time-averaged integral quantities are presented and compared against experimental data. This main study is preceded by a simulation of a synthetic-jet injected into stagnant surroundings, mainly in order to verify the computational framework, but also to gain insight into the behaviour of the vortex rings injected through a square orifice, in accord with corresponding experimental conditions.

Key words: Flow control, synthetic jets, turbulent boundary layer, LES.

1 Introduction

Synthetic jets have a demonstrable ability to delay or prevent the onset of separation of an adverse-pressure-gradient boundary layer into which the jets are injected. While the flow field of a synthetic jet has been studied extensively by experimental means [1, 6], only relatively few numerical studies have been conducted, with most of these being of the URANS variety, as is demonstrated by the compilation of Rumsey et al. [4]. However, it is arguable that only scale-resolving simulations are capable of capturing the complex dynamics associated with the unsteady injection, especially the interplay between the vortex rings and the turbulent structure in the boundary layer approaching the injection region. URANS simulations are seen to return unsatisfactory predictions of the flow, even at a 2D representation of a spanwise homogeneous slot jet [5]. Also, only numerical computations are able to provide de-

J.F. Morrison et al. (eds.), IUTAM Symposium on Flow Control and MEMS, 127–134.
© 2008 *Springer. Printed in the Netherlands.*

tailed information on the properties of fluid ejected through the jet orifice. This is particularly important when the jet is operating in a highly turbulent environment and the interaction of the near wall turbulence and the cavity flow is complex.

In the present study, the flow of a single synthetic jet, both in stagnant surroundings and in a turbulent cross-flow, has been investigated by means of Large-Eddy Simulation (LES). The simulation domains include the cavity and the neck of the orifice connecting this with the external domain. The jet orifice is treated as being square. This simplification not only offers numerical advantages (e.g. accuracy) in applying the structured multi-block procedure used in this study, but is also compatible with the orifice geometry adopted by Garcillan et al. [8] for the case of injection into stagnant surroundings that is used for comparison purposes.

2 Numerical Framework

The simulations were performed with the in-house finite-volume code STREAMLES. The algorithm is second-order accurate in space and time. The fractional-step method is employed. The first step is fully explicit, with the convection and diffusion terms being approximated by means of the Adam–Bashforth scheme. In the second step, a multi-grid Poisson-solver is employed to solve for the pressure field in an implicit fashion. The velocities are then subsequently updated to satisfy global conservation, prior to the advancement of the solution in time by a second-order backward scheme.

All computations presented in this document utilise structured hexagonal meshes. Parallelisation is achieved by splitting the grid into sub-domains with structured block-interfaces and assigning each block to individual processors. Sub-grid scale modelling in all simulations is based on the standard Smagorinsky SGS model in conjunction with the van Driest wall-damping.

3 Synthetic Jet in Stagnant Surroundings

A synthetic jet operating in a quiescent environment has but very few practical applications (e.g. as a cooling jet). Yet, it is important to analyse this flow in order to understand the underlying physics of the jet. In the present study, this test case serves as a validation of the computational framework and as a means towards studying the macroscopic structures of the jet.

A LES of a synthetic jet with a square orifice in stagnant surroundings has been performed by reference to experimental set-up from [8]. The external computational domain spans $20D_0 \times 10D_0 \times 10D_0$ in the streamwise and the two lateral directions respectively, with D_0 denoting the orifice-edge length. The complete numerical grid, including the external domain, the orifice and the cavity, consists of approximately 1.8 million cells. The simulation has been performed at $Re = 750$, based on D_0 and

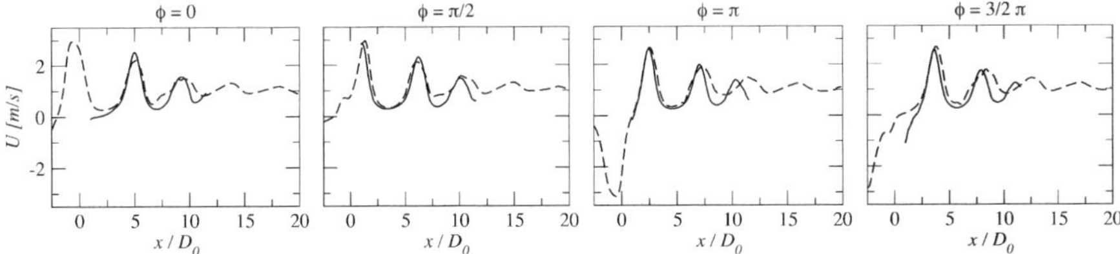

Fig. 1 Phase-averaged centreline velocity of an isolated synthetic jet at various phase of the actuation cycle; —: PIV data from [8]; − −: LES phase-averaged data.

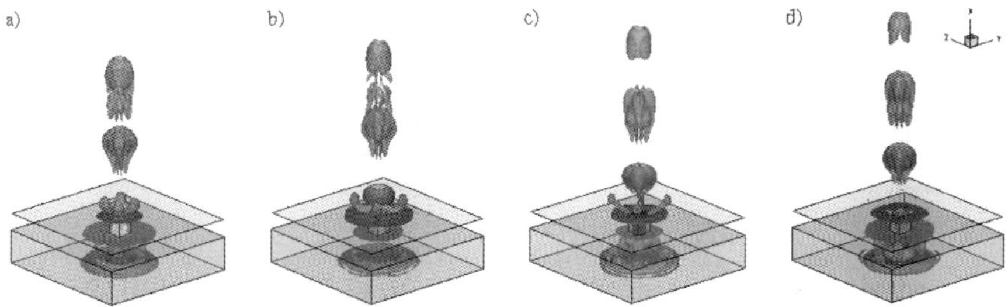

Fig. 2 Iso-surfaces of phase-averaged vorticity magnitude at various phases: (a) $\phi = 0$; (b) $\phi = \pi/2$; (c) $\phi = \pi$; (d) $\phi = 3/2\pi$.

the spatial-averaged peak orifice velocity $\hat{U}_0$. The actuation at the lower end of the cavity is imposed as a spatially uniform, sinusoidal velocity-transposition condition at $St = 0.089$ (based on the same scales). Phase-averaged results over 10 actuation cycles are compared against PIV data from [8].

Figure 1 compares the centre-line velocities as returned by LES and PIV measurements at four phases of the actuation. A series of peaks is observed, each indicating the position of a vortex ring. Both the position and the magnitude of the two initial peaks at $0 \leq x/D_0 \leq 7.5$ predicted by LES agree well with the experimental data throughout the actuation cycle. The discrepancy between the predicted third peak at $x/D_0 \approx 10$ and the measurements can be attributed to the comparably short duration of phase-averaging (10 cycles) associated with the computational results. Moreover, the non-smooth shape of the curve indicates that the vortex ring is less stable and undergoes transitional breaks-up.

This is confirmed by Figure 2, which displays iso-surfaces of the phase-averaged vorticity magnitude at four different phases of the flow as returned by LES. It also illustrates the evolution of the transitional vortex ring. Upon formation, the vortex ring is subsequently advected away from the orifice. It eventually merges with vorticity patches that are produced by the break-up of the ring from the preceding cycle. This process of merging and mixing presumably triggers the break-up of the current vortex ring in turn, prior to its dissipation in the far field.

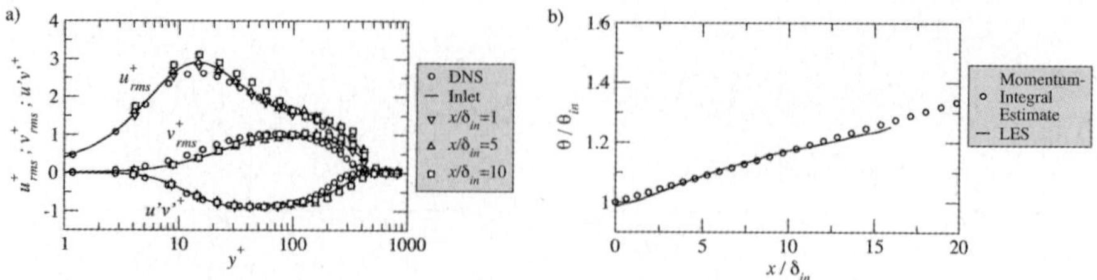

Fig. 3 Various properties of the baseline flow of a turbulent boundary layer; (a) profiles of time-averaged streamwise-turbulence intensity u^+_{rms}, wall-normal-turbulence intensity v^+_{rms} and shear stress $u'v'^+$ at various streamwise locations; (b) streamwise evolution of momentum thickness $\theta/\theta_{\text{in}}$ (normalised by its inlet value).

4 Synthetic Jet in Turbulent Boundary Layer

4.1 Baseline Boundary Layer

For the study of the turbulence physics of a jet-in-cross-flow interaction, it is important that the baseline flow of a turbulent boundary layer is physically realistic in terms of its temporal and spatial scales. In a LES environment, this means that a time-dependent inlet condition is required to impose realistic fluctuations that can be sustained and advected into the downstream domain. In the present study, this is achieved by means of a precursor LES using a pseudo-periodic recycling scheme for a flat-plate turbulent boundary layer at zero pressure gradient [3]. The baseline turbulent cross-flow has been computed at $Re = 2240$ (based on free-stream velocity U_∞ and orifice-edge length D_0) and inlet momentum-thickness Reynolds number $Re_{\theta,\text{in}} = 920$. The utilised numerical grid consists of ca. 1.18 million cells and resolves the near-wall regime down to $y^+ = 0.5$. The profiles of the Reynolds stresses at various streamwise locations are displayed in Figure 3a and compared against DNS data from [7] at a slighty lower inlet momentum-thickness of $Re_{\theta,\text{in}} = 670$. The overall agreement with the DNS data is fair and self-similarity along the streamwise direction is established for all stresses. Also, the streamwise evolution of the momentum thickness (normalised by its inlet value) $\theta/\theta_{\text{in}}$, as shown in Figure 3b, agrees well with the estimation returned by the momentum-integral analysis as proposed by Lund et al. [3].

4.2 Injection into Boundary Layer

With the successful simulation of a physical and realistic baseline, uncontrolled boundary layer, attention is focused next on the simulation of a synthetic jet emerging from a square into the turbulent cross-flow. The centre point of the orifice is located at $10D_0$ downstream of the domain's inlet. The inlet boundary-layer thick-

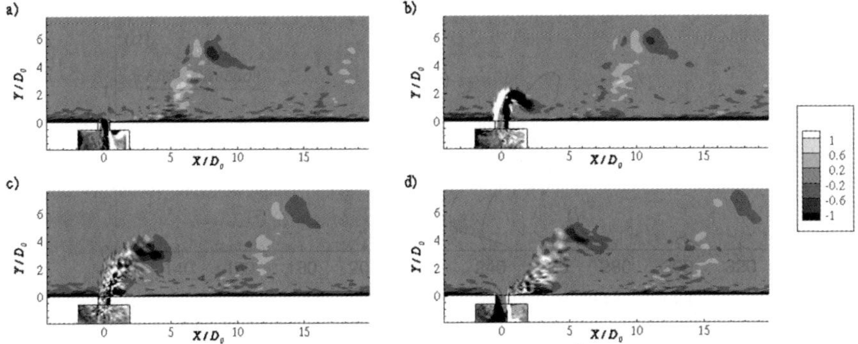

Fig. 4 Contours of phase-averaged out-of-plane vorticity ω_z at various phases of the actuation; (a) $\phi = 0$; (b) $\phi = \pi/2$; (c) $\phi = \pi$; (d) $\phi = 3/2\pi$.

ness is approximately $\delta \approx 4D_0$, and the jet-to-freestream/velocity ratio U_0/U_∞ is 2. The Strouhal number of the actuation is $St = 0.07$, based on U_∞ and D_0. These operational parameters are in accord with PIV and hot-wire measurements of a circular jet in a turbulent cross-flow by Garcillan et al. [2]. The computational domain includes the orifice and the cavity and is resolved by 1.96 million cells. Phase-averaged results are obtained over 10 actuation cycles after the initial transients have been removed.

Figure 4 displays the contours of phase-averaged out-of-plane vorticity ω_z at various phases of the actuation as returned by LES. At $\phi = \pi/2$, when the outward stroke is strongest, two distinct patches of spanwise vorticity with opposing rotational directions are visible just above the orifice, representing the 3D vortex ring that has been formed. As the vortex ring is penetrating further into the outer region of the boundary layer with higher cross-flow velocity, it is substantially deflected towards the downstream direction. At $\phi = \pi$, when the outward stroke has come to an end and the suction half-cycle is about to begin, the previously distinct vortex ring has already fragmented into small-scale vorticity patches, indicating that the laminar roll-up process has evolved into transitional break-down under the influence of the turbulent cross-flow. The injected vorticity is subsequently advected downstream by the cross-flow and gradually diffuses in the far field. Also, at $\phi = 3/2\pi$, when the suction is strongest, a significant amount of vorticity is being entrained into the cavity. With a rapid break-down into small-scale vorticity patches during the outward stroke, the reintroduction of these structures into the external domain might also contribute towards triggering the transitional break-down of the vortex ring between $\phi = \pi/2$ and $\phi = \pi$.

In order to identify the most energetic structures in the presented flow, Proper Orthogonal Decomposition (POD) has been employed to filter the velocity field by reference to its energy spectrum. Figure 5 visualises the flow in the near field of the orifice at $\phi = \pi$, as returned by the POD analysis, in which the mean flow is superposed with the four most energetic modes of the flow. The vorticity lines are overlaid with contours of the local vorticity magnitude ω, where light and dark

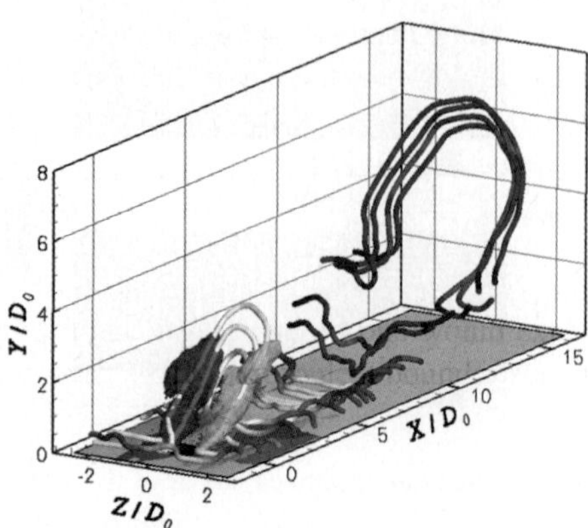

Fig. 5 Visualisation of the flow field at $\phi = \pi$ as returned by the POD analysis; stream-traces: vorticity lines overlaid with contours of vorticity magnitude ω; iso-surfaces: iso-surfaces of streamwise vorticity normalised by D_0/U_∞ for $\omega_x = -1$ and $\omega_x = 1$.

Fig. 6 Time-averaged streamwise evolution of the momentum thickness θ/θ_{in} and the shape factor H at the centre-plane; o: hot-wire data [2]; —: baseline; – –: jet-in-cross flow.

shades indicate high and low values of ω, respectively. Additionally, iso-surfaces of the streamwise vorticity ω_x for the values $\omega_x = -1$ and $\omega_x = 1$ have been included.

Two Ω-shaped vortices of two consecutive injections puffs are visible in Figure 5, one that is being formed at the orifice and the other in the far field. During the injection, the expelled fluid acts in a manner similar to an obstacle in a crossflow. As a result, the boundary layer flows around the puff and rolls up to form an Ω-vortex with a high vorticity magnitude, as indicated by the light shade of the vorticity lines in Figure 5. Further downstream, the vorticity lines are associated with the Ω-vortex of the preceding puff and feature a darker shade, indicative for a significantly lower vorticity magnitude. Vorticity lines located between the two Ω-vortices are deflected towards the wall as a result of the downward movement of the fluid, which is possibly related to the suction stroke and the presence of the horseshoe vortex.

A quantitative view on the effect of the actuation is provided by Figure 6, which compares the time-averaged streamwise evolutions of the momentum thickness θ/θ_{in} and the shape factor H in the centre-plane $z/D_0 = 0$, as returned by the baseline flow and the jet-in-cross-flow configuration. The momentum thickness is significantly increased as a result of the actuation, indicating the increased mixing induced by the injection. The shape factor only features a local peak close to the orifice before dropping back to the standard value indicative of a fully turbulent boundary layer. The large differences between the computed and measured values of shape factor and momentum thickness in the vicinity of the orifice may be due to limitations of the hot-wire measurements in this region. Specifically, the hot-wire probe is reported to spatially average the flow effectively over a distance of approximately $1D_0$. In any event, at the location of injection, the state of the flow is far from

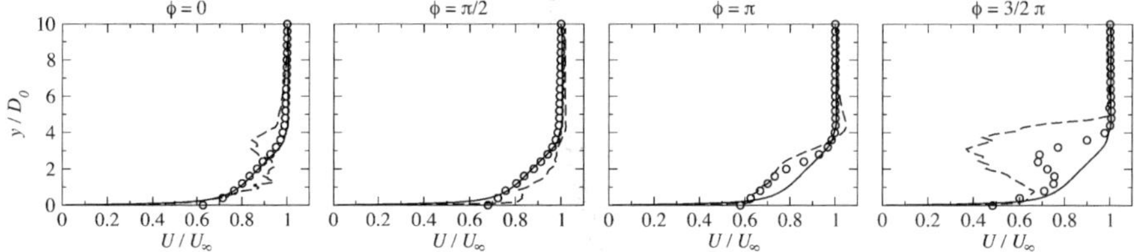

Fig. 7 Phase-averaged streamwise-velocity profiles in the centre plane at $x/D_0 = 5$ at various phases; ○: hot-wire data [2]; —: baseline; −−: jet-in-cross flow.

that of a standard boundary layer, so that large excursions in integral quantities from the standard values are not unexpected.

A more detailed comparison of the experimental and numerical results is offered in Figure 7 by reference to phase-averaged velocity profiles. The figure compares computed and experimental profiles in the centre-plane for $x/D_0 = 5$ at four different phases of the cycle against each other and against the velocity of the uncontrolled boundary layer. Qualitatively, agreement is good, indicating that the LES has captured the main control effect. The rather 'noisy' profiles returned by the simulation reflect the relatively small number of cycles over which integration has been performed. At the beginning of the cycle, the puff related to the previous cycle is located at $x/D_0 \approx 5$, as suggested by Figure 4a. The U-profile returned by the LES indicates substantial variations throughout the boundary layer as a result of the locally present turbulent structures. In contrast, the experimental data show virtually no differences relative to the baseline flow, suggesting a discrepancy in the location of the puff in question, a difference possibly aggravated by insufficient phase-averaging time. A quarter cycle later, the velocity profile is similar to the uncontrolled baseline counterpart, although the numerical results show the stronger impact of the suction stroke. At $\phi = \pi$, the effects of the upstream injection begins to show, as the streamwise velocity is reduced throughout a large portion of the boundary layer in the experimental and numerical results. The impact of the reverse flow that is formed in the wake of the injection column on the velocity profile is clearly visible at $\phi = 3/2\pi$, as the streamwise velocity is significantly reduced throughout the entire boundary layer. This trend is also confirmed by the hot-wire measurements, although the magnitude of streamwise-velocity loss is less prominent.

References

1. Cater, J.E., Soria, J.: The evolution of round zero-net-mass-flux jets. *J. Fluid Mech.* **472** (2002) 167–200.
2. Garcillan, L., Little, S., Zhong, S., Wood N.J.: Time evolution of the interaction of synthetic jets with a turbulent boundary layer. CEAS/KATnet Conference on Key Aerodynamic Technologies, Bremen, Germany (2005).

3. Lund, T.S., Wu, X., Squires, K.D.: Generation of turbulent inflow data for spatially-developing boundary layer simulations. *J. Comput. Phys.* **140** (1998) 233–258.
4. Rumsey, C.L., Gatski, T.B., Sellers, W.L., Vatsa, V.N., Viken, S.A.: Summary of the 2004 CFD Validation Workshop on Synthetic Jets. *AIAA J.*, Vol. 44(**2**) (2004) 194–207.
5. Rumsey, C.L., Manceau, R., Gatski, T.B.: Flow over 2-D wall mounted hump using URANS. In *Proceedings of 12th ERCOFTAC/IAHR Workshop on Refined Turbulence modelling* (2006).
6. Smith, B.L., Glezer, A.: The formation and evolution of synthetic jets. *Phys. Fluids* 10(**9**) (1998) 2281–2297.
7. Spalart, P.R.: Direct simulation of a turbulent boundary layer up to $Re_\theta = 1410$. *J. Fluid Mech.* **187** (1988) 61–98.
8. Garcillan, L., Zhong, S., Pokusevski, Z., Wood, N.J.: A PIV study of synthetic jets with different orifice shape and orientation. AIAA Paper 2004-2213 (2004).

Characteristics of Small-Scale Synthetic Jets – Numerical Investigation

Hui Tang and Shan Zhong
School of Mechanical, Aerospace and Civil Engineering, The University of Manchester, Manchester, U.K.; E-mail: hui.tang@postgrad.manchester.ac.uk, shan.zhong@manchester.ac.uk

Abstract. In synthetic jet actuators with an orifice diameter typically less than 1 mm (termed as small-scale synthetic jet actuators), the compressibility and viscosity effects often become significant causing them behave differently from large-scale synthetic jets. In this paper, a numerical study of the synthetic jet from an orifice diameter of 0.5 mm is undertaken and the results are compared with the synthetic jet from an orifice diameter of 5 mm. The results reveal that, given the same dimensionless parameters L and Re_L, the appearance and circulation of vortex rings produced from synthetic jets of different scales are identical in the near field. It is also found that although the linear relationships between (L and Re_L) and actuator operating conditions observed for large-scale synthetic jets are no longer valid for small-scale synthetic jets, the linear relationships between the dimensionless jet performance parameters and (L and Re_L) still exist. This finding is very useful to support the development of low-dimensional predictions models.

Key words: Synthetic jet, CFD simulation.

1 Introduction

Synthetic jets (SJ), also known as zero-net-mass-flux jets, have shown to have potential in active flow control [1]. As illustrated in Figure 1, a typical synthetic jet actuator (SJA) consists of a cavity bounded by rigid side walls, with an orifice plate at one end and an oscillating diaphragm at the other. To implement the selection of SJAs for successful flow control applications, it is essential to understand the fluid mechanics of SJs.

Based on incompressible flow assumption, the behaviour of SJs can be predicted from the actuator parameters [2]. However, in practical applications, the incompressible flow assumption usually breaks down when the diaphragm oscillates near the Helmholtz resonant frequency, or when the potential core in the exit jet dimin-

J.F. Morrison et al. (eds.), IUTAM Symposium on Flow Control and MEMS, 135–140.
© 2008 Springer. Printed in the Netherlands.

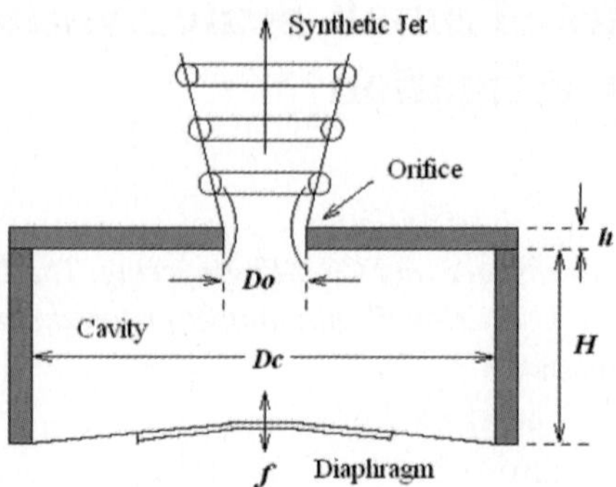

Fig. 1 Schematic of SJA.

ishes such that a further increase in the input power to the actuator will not produce a proportional gain in jet velocity due to the significant viscous effect inside the orifice duct. The latter is more likely to occur in actuators with a very small orifice diameter. For this case, the jet behaviour has not been fully understood and is not easy to predict. In this paper, therefore, the behaviour of small-scale SJs will be investigated. The objectives of the current study are: (1) to provide a suitable CFD model for SJAs with orifice diameter of small scales, where the compressibility effect in the cavity is significant; (2) to provide information for supporting the development of low-dimensional prediction models, by examining the relationship between the dimensionless jet flow parameters and the actuator operating parameters, i.e. the diaphragm oscillating displacement and frequency, as well as the relationship between the jet performance parameters and the dimensionless jet flow parameters. Here the dimensionless jet flow parameters are the dimensionless stroke length L, which represents the dimensionless length of the fluid column pushed out of the orifice during the blowing stroke, and the Reynolds number Re_L, which is based on the jet velocity and the stroke length. The jet performance parameters include the jet velocity U_o, mass flux rate Q_o, momentum flux rate M_o, and vortex circulation Γ.

2 Method and Validation of Simulations

The SJA studied in this paper has an orifice of diameter $D_o = 0.5$ mm, depth $h = 0.5$ mm, and a cavity of diameter $D_c = 20$ mm, height $H = 10$ mm. By carefully selecting the diaphragm oscillating displacement Δ and frequency f, a series of operating cases were simulated. Unsteady axisymmetric compressible numerical simulations were performed using a commercial CFD solver, FLUENT. For simplification, a velocity-inlet boundary condition instead of a moving boundary condition was applied at the neutral position of the diaphragm, which produces a negligible difference in the jet performance. A sensitivity study has also been conducted and

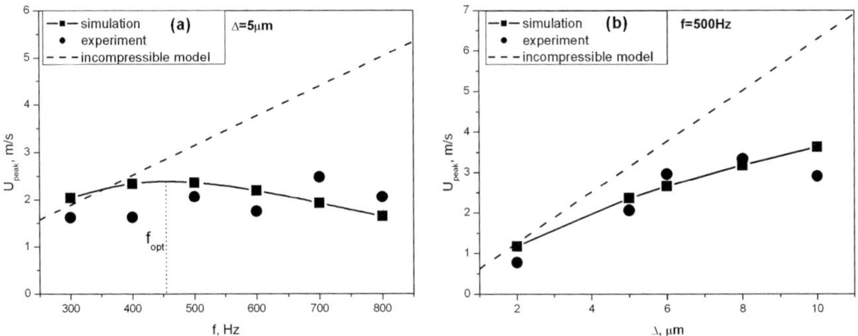

Fig. 2 Peak jet velocity U_{peak} against (a) frequency f for cases with $\Delta = 5$ μm, and (b) displacement Δ for cases with $f = 500$ Hz.

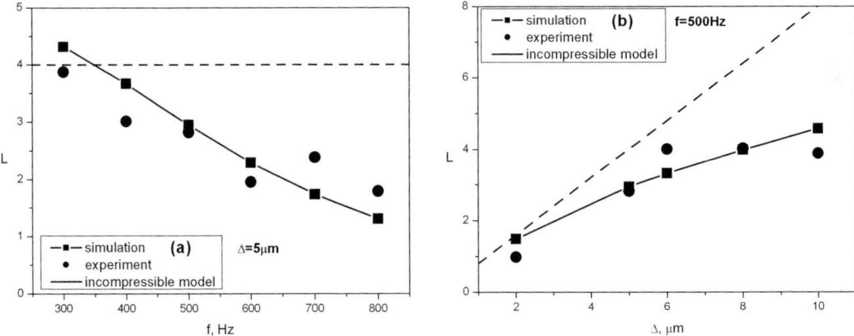

Fig. 3 Dimensionless stroke length L against (a) frequency f for cases $\Delta = 5$ μm, and (b) displacement Δ for cases with $f = 500$ Hz.

showed that the choices of mesh size and time-step used in the present study are adequate [3].

The simulation results were validated with existing PIV data. The simulated peak jet velocities (defined as the maximum space-averaged velocity in a duty cycle), the dimensionless stroke lengths L and the Reynolds numbers Re_L are reasonably close to the experimental data as shown in Figures 2 and 3. Although not shown here, the simulated vortex ring has similar shape and size as that obtained in the experiment. Moreover, the distributions of the vorticity levels are quite similar.

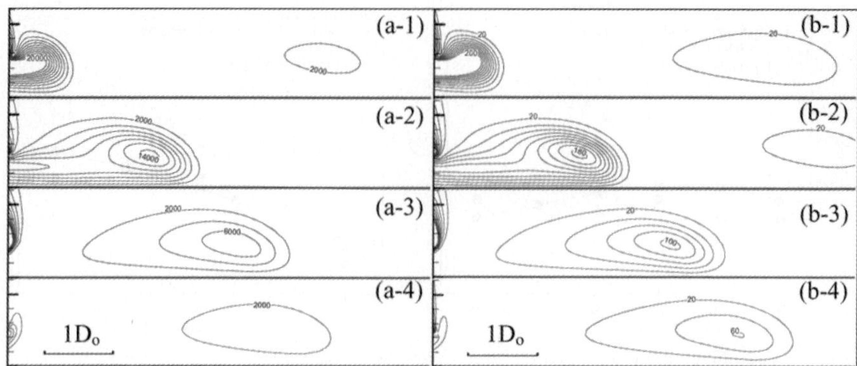

Fig. 4 Time sequences of vorticity contour for SJs with (a) $D_o = 0.5$ mm and (b) $D_o = 5$ mm. Ten levels of vorticity are presented in the simulation. The left side of each figure is the orifice exit and the right side is $6D_o$ distance from the orifice.

3 Results and Discussion

3.1 Comparison of Vortices Produced by SJAs of Different Scales

Figure 4 shows the vortex rings produced from SJAs with $D_o = 5$ mm and $D_o = 0.5$ mm. It is seen that the vortex structures of the two SJs have identical shape and convective velocity. Nevertheless, the value of the highest vorticity level in the vortex centre for the $D_o = 0.5$ mm case (Figure 4a) decreases more rapidly than that of the $D_o = 5$ mm case (Figure 4b), indicating that, even though no turbulence model is included in the simulation, the jet expelled from the orifice with $D_o = 0.5$ mm has a higher dissipation rate.

3.2 Performance of SJAs with $D_o = 0.5$ mm

Two groups of cases were carefully selected in this study. It is shown that, with the diaphragm displacement held at $\Delta = 5$ μm, the peak jet velocity firstly increases and then decreases with the frequency f, revealing a maximum value at about $f = 450$ Hz (see Figure 2a). This is inconsistent with the prediction of the previous incompressible model, in which the peak velocity is proportional to both the oscillating frequency f and the amplitude Δ [2]. On the other hand, with the frequency held at $f = 500$ Hz, the peak velocity increases with Δ monotonically (see Figure 2b). However, unlike the prediction of the incompressible model, the increase is not linear.

The incompressible model predicts that the dimensionless stroke length L is proportional to Δ but independent of f [2]. However, for the SJA with $D_o = 0.5$ mm, L

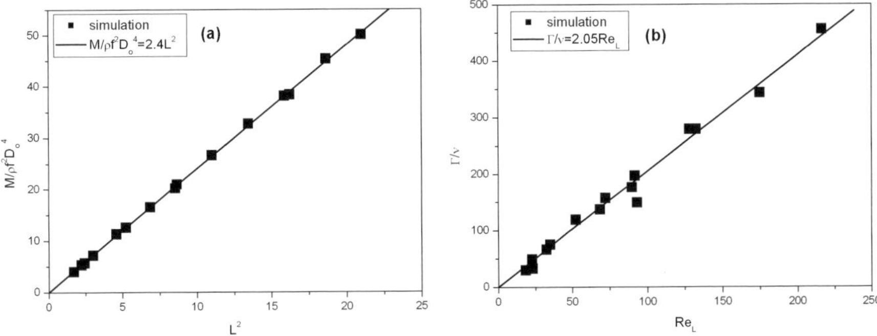

Fig. 5 Correlations between (a) dimensionless momentum flux rate and L^2, and (b) dimensionless vortex circulation and the Reynolds number Re_L.

is dependent on both Δ and f. As shown in Figure 3a, with the diaphragm displacement held at $\Delta = 5$ μm, L decreases monotonically with f. On the other hand, L increases with Δ at a constant oscillating frequency (see Figure 3b). The Reynolds number ReL has the similar variation trends to L. Therefore, the relations, predicted by the incompressible model, between the operating parameters and the dimensionless operating parameters are no longer valid when compressibility effects in the actuator are significant, which usually happens for SJAs with small orifice diameters.

The relations between the dimensionless operating parameters and jet performance parameters were also investigated. Since L is calculated by integrating the exit velocity with respect to the blowing time, it is a necessary result that L is linearly proportional to the exit velocity U_o, or equivalently, the mass flow rate Q_o. Interestingly, the linear relationship between the dimensionless momentum flux rate at the orifice exit and L^2, as predicted in the incompressible model [2], still holds for SJs of $D_o = 0.5$ mm (see Figure 5a), except that the slope 2.4 is greater than 1.94 for the incompressible model. Also, the dimensionless vortex circulation shows a good linear relation with Re_L, with slope 2.05 much greater than 1.23 of the incompressible model (Figure 5b). The higher slopes for SJs of $D_o = 0.5$ mm are caused by viscosity effects in the small orifice duct.

4 Conclusion

Numerical results in the present study reveal that, given the same L and Re_L, the appearance and level of circulation of vortex rings produced from SJAs of different scales are identical. But the vortex rings of smaller scales dissipate more quickly after they come out of the orifice exit. It is also found that, for small-scale SJs, while the linear relations predicted by the incompressible model [2] between the dimensionless jet flow parameters and the actuator operating parameters are no longer valid, the linear relations between the dimensionless jet performance parameters and the dimensionless jet flow parameters still hold.

References

1. Glezer, A. and Amitay, M.: Synthetic jets. *Annu. Rev. Fluid Mech.* **34** (2002) 503–529.
2. Tang, H. and Zhong, S.: Incompressible flow model of synthetic jet actuators. *AIAA J.* **44**(4) (2006) 908–912.
3. Tang, H. and Zhong, S.: 2D numerical study of circular synthetic jets in quiescent conditions. *Aeronautical J.* **109** (2005) 89–97.

Large Eddy Simulations of Transitional and Turbulent Flows in Synthetic Jet Actuators

Sanjay Patel and Dimitris Drikakis
Fluid Mechanics and Computational Science Group, Department of Aerospace Sciences, Cranfield University, Cranfield, Bedfordshire MK43 0AL, U.K.; E-mail: {s.patel, d.drikakis}@cranfield.ac.uk

Abstract. We examine the evolution and transition-to-turbulence of synthetic jets into quiescent air by performing three-dimensional numerical investigations using a high-resolution scheme for solving the compressible Navier–Stokes equations in the context of Implicit Large Eddy Simulation (ILES). The computational results show a good comparison to experimental data obtained from the NASA Langley workshop on CFD Validation of Synthetic Jets and Turbulent Separation Control.

Key words: Synthetic jets, large eddy simulation, high-resolution methods.

1 Introduction

Synthetic jets sometimes referred to as zero-net-mass-flux jets, have been investigated both computationally and experimentally in the context of flow control for a number of years. Synthetic jets differentiate themselves from continuous or pulsed jets by being synthesized from the working fluid of the flow system in which they are positioned. This enables synthetic jets to transfer linear momentum to the flow without net mass injection across the flow boundary. Synthetic jets are able to provide momentum flux, alter pressure distribution and to introduce arbitrary scales to another flow.

This investigation concerns the computational simulation of a synthetic jet issuing into quiescent air using high-resolution numerical methods. High resolution methods in the context of LES have characteristics that tend to mimic the effects of finite viscosity and appear to achieve many of the sub-grid scale (SGS) properties [1–5]. The idea of using high-resolution methods to implicitly model complex transitional and turbulent flows is referred to as Implict Large Eddy Simulation (ILES). The computational model is based on experiments carried out at NASA Langley [6], which were specifically intended to be used for CFD validation. The experimental data was collected using a variety of methods including Particle Image

J.F. Morrison et al. (eds.), IUTAM Symposium on Flow Control and MEMS, 141–144.
© 2008 Springer. Printed in the Netherlands.

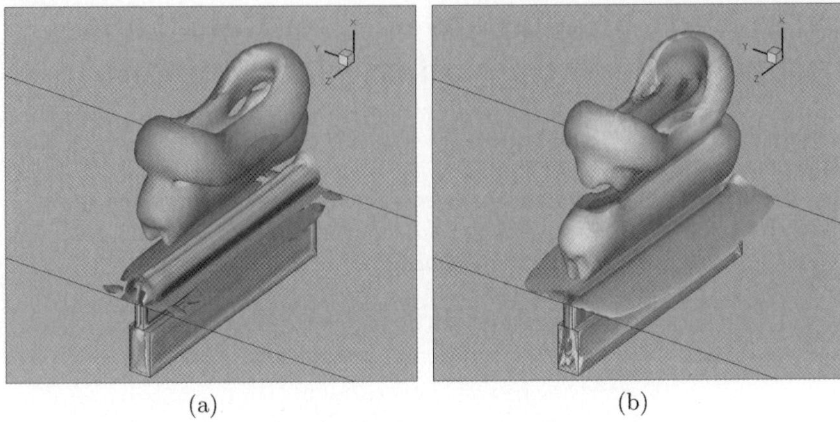

Fig. 1 Isosurfaces of vorticity at (a) 90° and (b) 270°.

Velocimetry (PIV), Laser Doppler Velocimetry (LDV) and Hot Wire anemometry (HW), thus providing a range of data available to use in the validation.

2 Numerical Framework

A three-dimensional parallel compressible flow solver has been used for the simulations. High-resolution methods are non-linear, non-oscillatory methods. which use the local solution to select a technique for approximating the solution. This suggests that high-resolution methods tend to adapt themselves to the particular circumstances so that the solution obtained is accurate and has some physical meaning. It has been shown by Drikakis [1] and Margolin and Rider [3] that high-resolution methods can be used in (under-resolved with respect to grid resolution) turbulent flow computations without the need to resort to a turbulence model.

The high-resolution scheme employed here is the Godunov-type, characteristics-based (CB) scheme by Eberle [7]. It is a Godunov-type method that defines the conservative variables along the characteristics as function of their characteristic values. We used a third-order interpolation scheme [8] to compute the characteristic values from the left or right states depending on the sign of the characteristic speed (eigenvalues). The scheme is not presented here in detail and the reader is referred to the paper by Eberle [7] for further details.

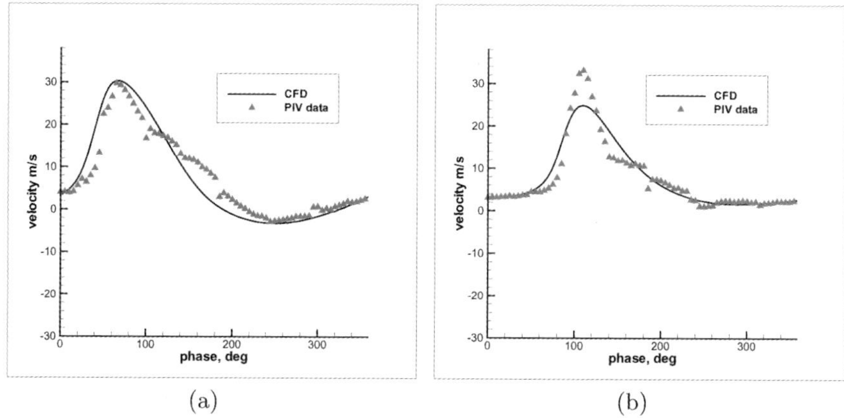

Fig. 2 Phase-averaged u-velocity at (a) 2 mm and (b) 5 mm above the center of the orifice exit.

3 Results and Discussion

We have performed three-dimensional computations of a synthetic jet exiting into quiescent air. The results have been compared to experiments undertaken at NASA Langley [6]. The LDV and HW data contain a number inconsistencies, hence the PIV data has been chosen to be used for the validation purpose.

A sinusoidal blowing/suction boundary condition has been used to model the unsteady behaviour of the the synthetic jet diaphragm. The Reynolds number based on the orifice width and average velocity over the discharge phase of a cycle was 1150. The forcing frequency that the diaphragm oscillates at was 444.7 Hz, taken from the experimental data. The flow was allowed to develop over several cycles and phase averaged data was calculated once the flow was regarded to be fully developed. It was found that the synthetic jet flow close to the orifice exit was dominated by time-periodic formation, streamwise advection and the interaction of counter-rotating vortex pairs, which breakdown over the course of the actuator cycle. The reversal of the flow on the suction phase is because the time-averaged static pressure at the slot exit was lower than the ambient pressure, resulting in a reversal of both the streamwise and cross-stream velocity components.

Figure 1 shows isosurfaces of vorticity at maximum expulsion (90°) and maximum suction (270°). It can be clearly seen that, as the flow separates at the edges of the orifice, a ring type vortex is formed. This vortex travels in the streamwise direction over the course of the actuator cycle and starts to breakdown with increasing streamwise distance from the orifice exit.

Phase averaged streamwise velocity profiles were obtained at various streamwise positions above the center of the orifice. Figure 2 shows velocity profiles at two positions (2 mm and 5 mm) above the orifice exit. The CFD results agree very well with the PIV data with regards to both phase angle and velocity profile. The magnitude of velocity predicted by CFD close to the orifice exit (2 mm) matches

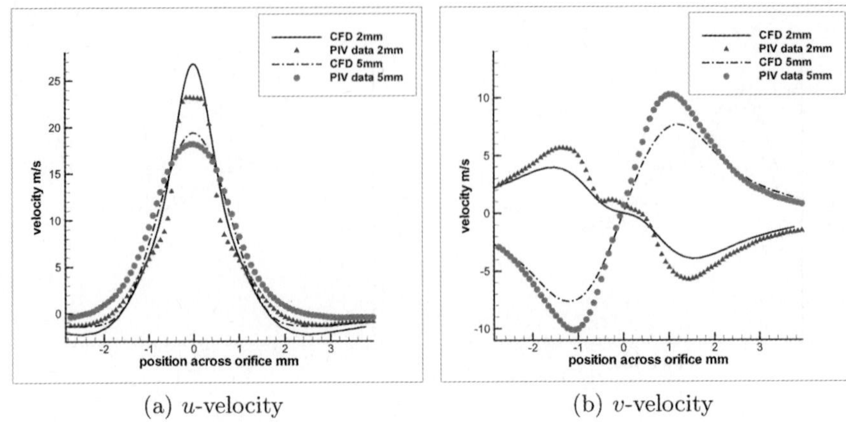

(a) u-velocity   (b) v-velocity

Fig. 3 Phase-averaged velocity profiles across the orifice exit at 2 mm and 5 mm above the orifice for 90° phase angle.

well with the PIV data, but further away (5 mm), the maximum velocity is slightly under predicted.

Phase-averaged velocity profiles across the orifice exit for horizontal positions equal to 2 mm and 5 mm above the orifice at a phase angle of 90° are shown in Figure 3. The cross-stream maximum and minimum velocities are slightly under-predicted by the CFD but the general trend is very good at both distances away from the orifice exit. Overall, the results obtained compare well, both qualitatively and quantitatively, with the experimental PIV data obtained from NASA Langley.

References

1. Drikakis, D.: Advances in turbulent flow computations using high-resolution methods. *Progress in Aerospace Sciences* **39** (2003) 405–424.
2. Fureby, C., Grinstein, F.F.: Large eddy simulation of high Reynolds number free and wall bounded flows. *Journal of Computational Physics* **181** (2002) 68–97.
3. Margolin, L.G., Rider, W.J.: A rationale for implicit turbulence modeling. *International Journal for Numerical Methods in Fluids* **39** (2002) 821–841.
4. Margolin, L.G., Smolarkiewicz, P.K., Wyszogrodzki, A.A.: Implicit turbulence modeling for high reynolds number flows. *Journal of Fluids Engineering* **124** (2002) 862–867.
5. Youngs, D.: Application of miles to Rayleigh–Taylor and Richtmyer–Meshkov mixing. AIAA Paper 2003-4102 (2003).
6. Yao, C., Chen, F., Neuhart, D., Harris, J.: Synthetic jet flow field database for cfd validation. In: *2nd AIAA Flow Control Conference* (2004).
7. Eberle, A.: 3-D Euler calculation using characteristic flux extrapolation. AIAA Paper 85-0119 (1985).
8. Zóltak, J., Drikakis, D.: Hybrid upwind methods for the simulation of unsteady shock-wave diffraction over a cylinder. *Computer Methods in Applied Mechanics and Engineering* **162** (1998) 165–185.

SEPARATION CONTROL

Model Reduction and Control of a Cavity-Driven Separated Boundary Layer

Espen Åkervik[1], Jérôme Hœpffner[2], Uwe Ehrenstein[2] and Dan S. Henningson[1]
[1]*Linné Flow Centre, KTH Mechanics, SE-100 44 Stockholm, Sweden;*
E-mail: henning@mech.kth.se
[2]*IRPHÉ, Université de Provence, F-13384 Marseille Cedex 13, France*

Abstract. The control of a globally unstable boundary-layer flow along a two-dimensional cavity is considered. When perturbed by the worst-case initial condition, the flow exhibits a large transient growth associated with the development of a wave packet along the cavity shear layer followed by a global cycle related to the least stable global eigenmodes. The flow simulation procedure is coupled to a measurement feedback controller, which senses the wall shear stress at the downstream lip of the cavity and actuates at the upstream lip. A reduced model for the control optimization is obtained by a projection on the least stable global eigenmodes. The LQG controller is run in parallel to the Navier–Stokes time integration. It is shown that the controller is able to damp out the global oscillations.

Key words: Control, model reduction, global eigenmodes, separated flows.

1 Introduction

During the last decade modern control theory has increasingly been applied to fluid flow problems, given the available computer capacities and sensor/actuator developments. Linear optimal control theory has been introduced to flow systems governed by linear instability mechanisms [1], for instance, spatially developing boundary-layers [2]. It may also be relevant for nonlinear flows, such as turbulent boundary-layers [3]. Optimal control of fluid flow based on full state-space representation of the flow field necessitates manipulation of very high-dimensional dynamical systems. In weakly non-parallel flow configurations, the problem may become tractable by determining control and estimation kernels for individual wavenumbers in the homogeneous space directions [2]. In most practical situations the flow state must be estimated based on a limited amount of measurements. We use the linear quadratic Gaussian (LQG) synthesis, where external disturbances and measurement noise are modeled as random processes. This gives rise to a measurement feedback

J.F. Morrison et al. (eds.), IUTAM Symposium on Flow Control and MEMS, 147–155.
© 2008 *Springer. Printed in the Netherlands.*

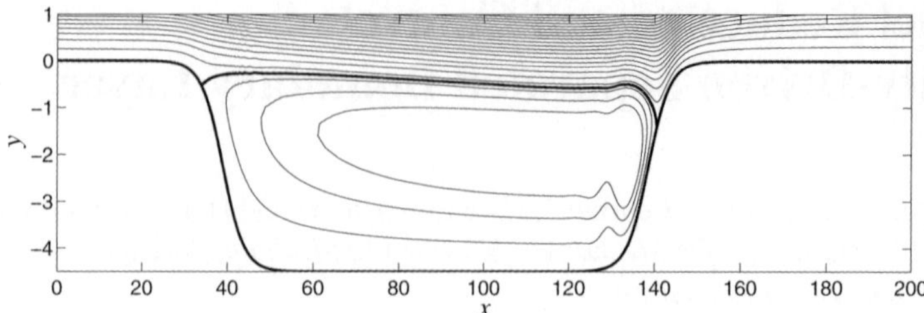

Fig. 1 Streamlines of the steady state base flow used for stability analysis at Re $= 350$. The thick line represents the dividing streamline. Note that the length to depth ratio of the cavity is ≈ 20.

controller which aims at minimizing the flow kinetic energy. The design of the controller is intimately related to model reduction and the usual procedure is that of projecting the equations onto a subspace spanned by a set of vectors, followed by truncation. One common approach is to use the Proper Orthogonalized Decomposition modes of the excited flow, thereby capturing the high-energy content of the flow. Balanced truncation provides a more attractive basis by selecting vectors that are equally controllable and observable (see e.g. [4]).

2 Direct Numerical Simulation

The approach taken here is to use the global eigenmodes of the linearized Navier–Stokes system as the projection basis, the flow considered being a non-parallel cavity-driven incompressible boundary-layer flow. See Figure 1 for a sketch of the geometry. The Navier–Stokes equations are solved in the domain $0 \leq x \leq 400$, $\eta(x) \leq y < y_{\max}$, for $y_{\max} = 80$ large enough to recover freestream uniform flow, with $\eta(x)$ describing the wall. The Reynolds number is based on the displacement thickness at inflow $x = 0$ where a Blasius profile is prescribed. The streamlines in a subset of the computational domain for the steady state at Re $= 350$ are depicted in Figure 1. Note that the main effect of the smooth large aspect-ratio cavity is the generation of a recirculation zone and a shear layer. The numerical solution procedure has previously been used for instability investigations of boundary-layer separation in [5] and it will be referred to as the DNS. To account for wall curvature, a mapping transforms the physical coordinates into the computational ones $(\bar{x}, \bar{y})$, which are discretized using fourth-order finite differences in $\bar{x}$ (with 2048 grid points) and Chebyshev-collocation in $\bar{y}$ (with 97 collocation points). Note that by transforming the partial differential operators in the Navier–Stokes equations, extra-terms arise which depend on the first and second derivatives of η so that sharp corner cavities are not allowed.

Cavity-flow is known to be subject to self-sustained oscillations (cf. the review in [6]). Therefore the technique proposed in [7] is used to recover steady states for Reynolds numbers above criticality. The Navier–Stokes equations are forced by adding a term proportional to the difference between the flow state and a filtered solution. If $\dot{q} = \mathrm{NS}(q)$ represents the nonlinear Navier–Stokes system, the modified system reads

$$\dot{q} = \mathrm{NS}(q) - \chi(q - \bar{q}), \quad \dot{\bar{q}} = (q - \bar{q})/\Delta, \tag{1}$$

where the rightmost equation represents the differential form of a causal low-pass temporal filter with an exponential kernel. The steady state of Equation (1) is also a steady state of the Navier–Stokes system. A filter width of $\Delta = 15$ has been chosen such that the frequencies of the instability are targeted and a damping coefficient $\chi = 0.02$ was found to be appropriate [7].

3 Eigenmodes and Optimal Growth

The global instability modes are computed by linearizing the Navier–Stokes system about the steady state $\mathbf{U}(x,y) = (U(x,y), V(x,y))$. The disturbance flow field $\mathbf{u}(x,y,t) = \hat{\mathbf{u}}(x,y)\, e^{-i\omega t}$ and pressure $p(x,y,t) = \hat{p}(x,y)\, e^{-i\omega t}$ satisfy the partial differential system

$$-i\omega\hat{\mathbf{u}} = -(\mathbf{U}\cdot\nabla)\hat{\mathbf{u}} - (\hat{\mathbf{u}}\cdot\nabla)\mathbf{U} - \nabla\hat{p} + \frac{1}{\mathrm{Re}}\nabla^2\hat{\mathbf{u}}, \tag{2}$$

$$0 = \nabla\cdot\hat{\mathbf{u}}. \tag{3}$$

After discretization this is written as

$$-i\omega_l\mathbf{B}\mathbf{q}_l = \mathbf{A}\mathbf{q}_l \quad \text{with adjoint} \quad i\omega_l^*\mathbf{B}\mathbf{q}_l^+ = \mathbf{A}^+\mathbf{q}_l^+ \tag{4}$$

for the modes $\mathbf{q}_l$, $\mathbf{B}$ is the projection of the total disturbance field on the velocity components; $\mathbf{A}^+$ is the adjoint operator and the bi-orthogonality condition $\langle \mathbf{q}_k, \mathbf{B}\mathbf{q}_l^+ \rangle = \delta_{kl}$ applies. The discrete adjoint has been considered, for the associated finite-dimensional inner product on a discretized domain of extent $0 \le x \le 300$, $\eta(x) \le y \le H$, used for the stability computations. The height $H = 75$ has been chosen for the perturbation to vanish and the cavity eigenmodes are not affected by the Neumann boundary condition at outflow $x = 300$ (at inflow homogeneous Dirichlet conditions are used). The domain is mapped into $[-1,1] \times [-1,1]$ and a Chebyshev–Chebyshev collocation discretization is used. The basic steady flow is then interpolated on the new grid. A similar procedure has been used in [8] for the computation of global modes in the flat plate boundary layer. A collocation grid with up to 350×65 points was used to obtain converged stability results. The resulting system is far too large to be solved by standard QZ algorithms. However Krylov subspace projections with dimension $m = 800$ together with the Arnoldi algorithm (see [9]) proved suitable to recover the part of the spectrum relevant for our analysis. For the steady state shown in Figure 1, the spectrum is depicted in Figure 2. For the present parameters, there are two unstable eigenvalues (only half of the spectrum with $\omega_r > 0$ is shown). The eigenfunction corresponding to the least stable eigen-

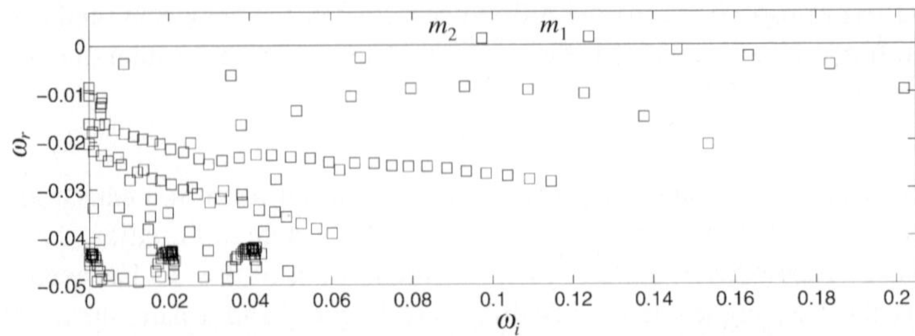

Fig. 2 Eigenvalues of the direct problem (4). There are two unstable modes labeled m_1 and m_2. The mode m_1 is depicted in Figure 3.

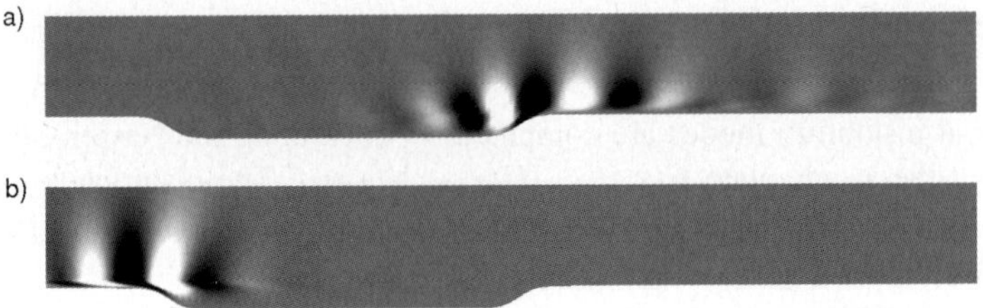

Fig. 3 Unstable eigenmode, corresponding to the eigenvalue labeled m_1 in Figure 2. (a) Vertical velocity, (b) adjoint vertical velocity.

value is depicted in Figure 3. We observe a clear separation in space between the direct and adjoint eigenmode, indicating a strong streamwise non-normality (see [13]).

For sufficiently small flow perturbations $\mathbf{q}(t)$, an eigenmode expansion $\mathbf{q}(t) = \sum_{j=1}^{N} k_l(t)\mathbf{q}_l$, can be used to describe the flow dynamics. The Navier–Stokes equations are initiated by superimposing the optimal initial condition $\mathbf{q}_0$ onto the steady state, leading to the maximum energy growth $\|\mathbf{q}(t)\|$ at a given time t. The procedure to compute the optimal initial condition is outlined in [10] and the subsequent energy evolution for the present flow case is depicted in Figure 4. Using one mode, we observe the exponential growth of the most unstable mode. Transient energy growth, due to non-normality of the eigenmodes, results in a much faster growth up to $t = 200$, followed by a global cycle of approximately 300 time units. This cycle is associated with the least stable eigenvalues in Figure 2. Since the real parts of these modes are a distance of $\delta \approx 0.02$ apart, and the corresponding eigenfunctions have a very similar structure, they have the ability to cancel each other, giving rise to a cycle with a period of $2\pi/\delta$ (see [11]).

Contours of the vertical velocity component of the optimal initial condition are shown in Figure 6. It corresponds to time $t = 1000$ of the envelope in Figure 4 and has been superimposed on the steady state in the DNS as the initial condition. The spatio-temporal diagram of the resulting dynamics is depicted in Figure 5, where one sees the convection and growth of the wave packet along the shear layer, and regeneration at the upstream cavity lip. In addition, a global pressure change, visible in the form of vertical rays, occurs when the wave packet reaches the downstream

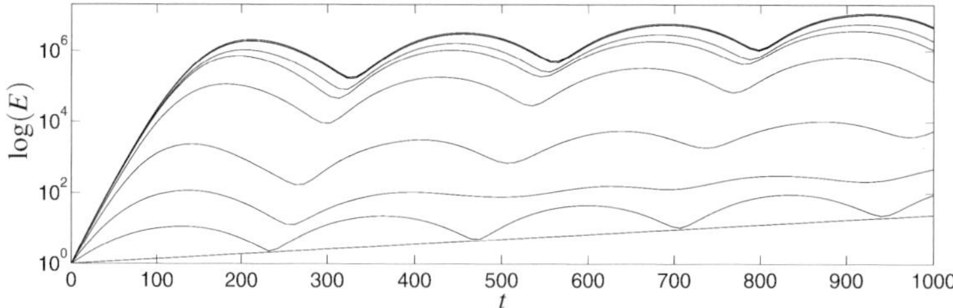

Fig. 4 Envelope of maximum energy growth from initial conditions. The different lines correspond to increasing number of eigenmodes included in the optimization, 1 to 184 from bottom to
top.

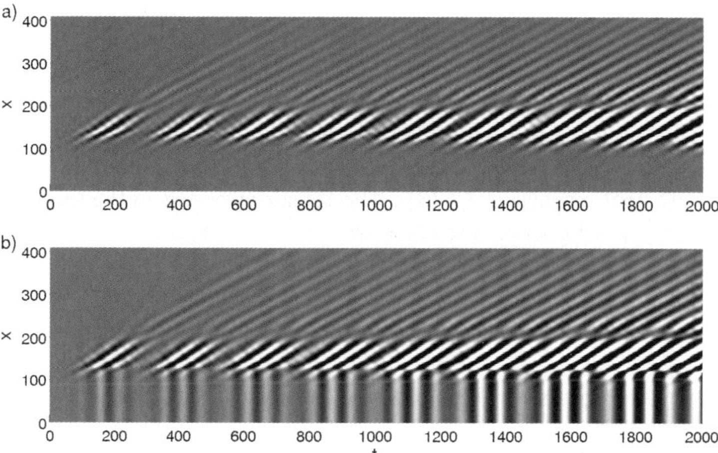

Fig. 5 x/t diagram for (a) the vertical flow velocity and (b) the pressure. The flow initial condition
is the optimal initial condition.

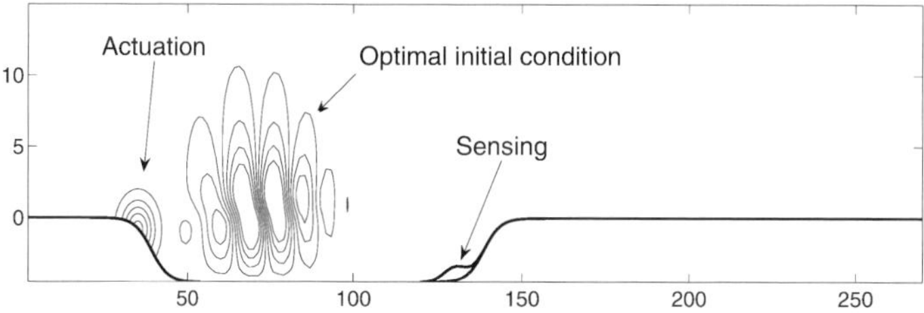

Fig. 6 Sketch of the control setting, with a volume forcing actuator, and a wall skin friction sensor.

cavity lip. The energy evolution for the flow is shown in Figure 7, exhibiting a
cycle with the same period as observed in Figure 4 for the system of eigenmodes.
A similar instability of a compressible shear layer above a rectangular cavity and its
control has recently been studied in [12].

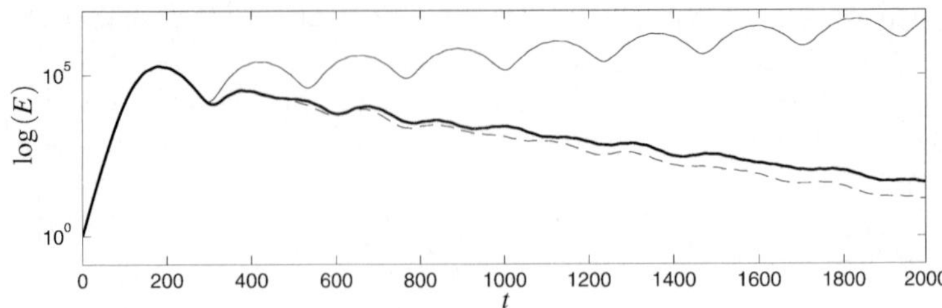

Fig. 7 Energy of the uncontrolled flow (*thin solid line*), controlled flow using model with 4 modes (*thick solid*) and 25 modes (*dashed*).

4 Control

To control the cavity flow, we introduce one sensor and one actuator as sketched in Figure 6. The actuator is located at the upstream limit of the cavity, where the unstable adjoint eigensolution, shown in Figure 3, has its maximum, so as to trigger the most efficient response (cf. [13]). The sensor is placed in the vicinity of the downstream cavity lip where the eigenfunctions have large amplitude. The sensor $R(x)$ is a Gaussian-shaped function at the wall with a width of ≈ 20 and measures the wall shear stress $\int R(x)(\partial u/\partial y)dx$. This operation may formally be written as $r = Cq$ for the flow state q. The actuator is a volume forcing of Gaussian type on the vertical velocity component located close to the wall about the upstream cavity lip, with a width of ≈ 20 and a height of ≈ 2.

A dynamic model for the cavity flow is built by using the eigenmode expansion. Based on this model, a LQG control procedure gives rise to the system

$$\begin{cases} \dot{k} = A^M k + B_1^M w + B_2^M \phi, & r = C^M k + g, \\ \dot{\hat{k}} = A^M \hat{k} + B_2^M \phi - L(r - \hat{r}), & \hat{r} = C^M \hat{k}, \quad \phi = K\hat{k}. \end{cases} \tag{5}$$

The vector $k(t)$ of the expansion coefficients of the flow obeys the model dynamics, where A^M is the diagonal matrix of the eigenvalues. The external disturbances are modeled as white noise stochastic input $w(t)$ with variance W, and B_1^M is the projection on the eigenmodes of the Gaussian-shaped spatial forcing function centered at $x = 50$. The projection of the actuator is represented by B_2^M, and $\phi(t)$ is the actuation signal. These projections are achieved by performing the inner product with the adjoint modes. The measurement is denoted r, and C^M is the measurement operator. The measurement is corrupted by a stochastic sensor noise $g(t)$ with variance G. An estimator is built, with estimated state $\hat{k}$, obeying the model dynamics, with an estimation feedback forcing $L(r - \hat{r})$. The estimation gain L is designed such that the estimated state $\hat{k}$ converges to the flow state k, i.e. minimizes the mean kinetic energy of the estimation error $k - \hat{k}$. The control actuation ϕ is a feedback of the estimated flow state, with control feedback gain K, designed such as to minimize a weighted sum of the flow mean kinetic energy and the actuation effort. The kinetic energy norm is represented in the basis of the eigenmodes by the matrix Q^M, and the weighting on control effort is denoted ℓ^2.

The optimal feedback gains K and L that minimize the flow and estimation error mean kinetic energy are found by the solution of two algebraic Riccati equations (see [14, 15])

$$0 = (A^M)^H X_c + X_c A^M - X_c B_2^M \ell^{-2} (B_2^M)^H X_c + Q^M,$$

$$0 = A^M X_e + X_e (A^M)^H - X_e (C^M)^H G^{-1} C^M X_e + B_1^M W (B_1^M)^H,$$

for the matrix unknowns X_c and X_e, and the feedback gains can be obtained as $K = -l^{-2} B_1^M X_c$ and $L = -X_e (C^M)^H G^{-1}$. In our computations, we have assumed an external disturbance w with unit variance ($W = 1$). The control penalization and sensor noise variance were chosen $\ell = 5 \cdot 10^5$, $G = 5 \cdot 10^{11}$ in order to enforce low amplitude feedback gains.

Once the two Riccati equations are solved and the feedback gains are obtained, we can couple the flow and the controller by extracting the skin-friction measurement from the DNS, using it as input to the second line of Equation (5), and injecting the subsequent actuation signal ϕ back in the DNS

$$\dot{q} = \mathrm{NS}(q) + B_2^{\mathrm{NS}} \phi, \quad r = C^{\mathrm{NS}} q, \tag{6}$$

where B^{NS} and C^{NS} represent the actuator and the sensor in the Navier–Stokes system. In order to assess the performance of the computed control and estimation gains the controller is applied to the same configuration that lead to the development shown in Figure 5. Reduced models consisting of the 25 or the 4 least stable eigenmodes are considered. Figure 7 shows that when control is applied, the exponential energy growth is turned into exponential decay after the first peak. There is an almost equivalent performance for both controller dimensions. The actuator signal shown in Figure 8a) quickly decays in time after the first reflections of the wave packet at $t \approx 300$. The wall measurement with and without control depicted in Figure 8b) illustrates the excellent control performance. It is not possible to control the initial energy growth, before the wave packet has reached the sensor (at $t \approx 100$) located at the downstream cavity lip. The x/t diagram for the controlled flow in Figure 9 is to be compared with Figure 5. When the control is applied, one still observes the vertical rays of the global pressure changes but the wave packet regeneration is prevented.

5 Conclusions

Concluding, the cavity-like flow considered here may be seen as a prototype of non-parallel flow with self-sustained global instability behavior. The direct numerical simulation results provide the *real* flow for the feedback-coupling with the controller constructed using global eigenmodes. The model with only few degrees of

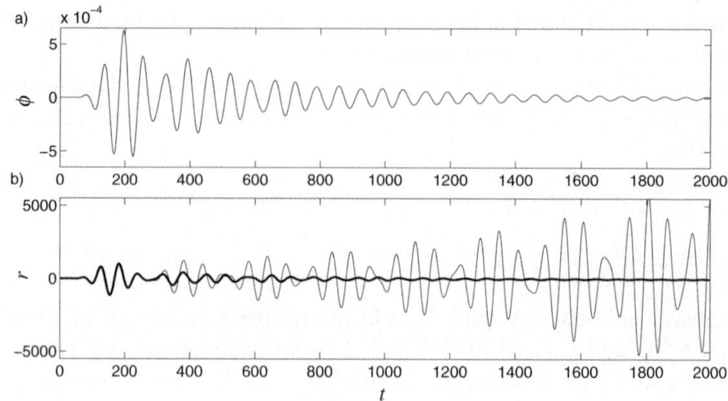

Fig. 8 (a) Actuation signal in time. (b) Wall shear stress measurements at $x = 130$ for controlled flow with four modes (thick line) and uncontrolled flow (thin line).

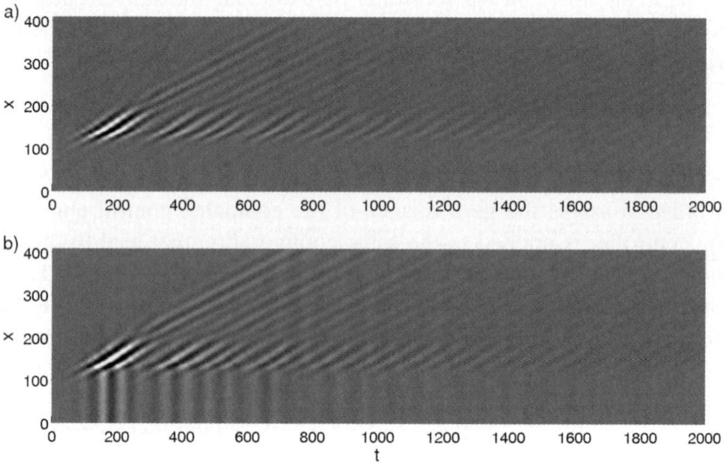

Fig. 9 x/t diagram for (a) the vertical flow velocity and (b) the pressure starting when the control is applied.

freedom satisfactorily controls the flow dynamics and it may easily be run in parallel to the Navier–Stokes time integration. This provides promising perspectives for the possibility of applying closed-loop control to complex flow systems based on global modes.

References

1. T.R. Bewley and S. Liu, Optimal and robust control and estimation of linear paths to transition, *J. Fluid Mech.* **365**, 1998, 305.
2. M. Högberg and D.S. Henningson, Linear optimal control applied to instabilities in spatially developing boundary layers, *J. Fluid Mech.* **470**, 2002, 151.
3. J. Kim, Control of turbulent boundary layers, *Phys. Fluids* **15**, 2003, 1093.
4. C.R. Rowley, Model reduction for fluids, using balanced proper orthogonal decomposition, *Int. J. Bifurcation and Chaos* **15**, 2005, 997–1013.
5. M. Marquillie and U. Ehrenstein, On the onset of nonlinear oscillations in a separating boundary-layer flow, *J. Fluid Mech.* **490**, 2003, 169.
6. D. Rockwell and E. Naudasher, Review-self sustained oscillations of flow past cavities, *J. Fluids Engng.* **100**, 1978, 152.
7. E. Åkervik, L. Brandt, D.S. Henningson, J. Hœpffner, O. Marxen and P. Schlatter, Steady solutions of the Navier–Stokes equations by selective frequency damping, *Phys. Fluids* **18**, 2006, 068102.
8. U. Ehrenstein and F. Gallaire, On two-dimensional temporal modes in spatially evolving open flows: The flat-plate boundary layer, *J. Fluid Mech.* **536**, 2005, 209.
9. M. Nayar and U. Ortega, Computation of selected eigenvalues of generalized eigenvalue problems, *J. Comput. Phys.* **108**, 1993, 8.
10. P.J. Schmid and D.S. Henningson, *Stability and Transition in Shear Flows*, Springer, New York, 2001.
11. P.J. Schmid and D.S. Henningson. On the stability of a falling liquid curtain, *J. Fluid Mech.* **463**, 2002, 163–171.
12. C.W. Rowley, D.R. Williams, T. Colonius, R.M. Murray and D.G. Macmynowski, Linear models for control of cavity flow oscillations, *J. Fluid Mech.* **547**, 2006, 317.
13. J.-M. Chomaz, Global instabilities in spatially developing flows: Non-normality and non-linearity, *Annu. Rev. Fluid Mech.* **37**, 2005, 357.
14. R.E. Skelton. *Dynamics System Control*, John Wiley and Sons, New York, 1988.
15. J. Hœpffner, M. Chevalier, T.R. Bewley and D.S. Henningson. State estimation in wall-bounded flow systems. Part I: Laminar flows, *J. Fluid Mech.* **534**, 2005, 263.

Collaborative Studies on Flow Separation Control

Wei Long Siauw[1], Jean Paul Bonnet[1], Jean Tensi[1], Avi Seifert[2], Oxana Stalnov[2], Vikas Kumar[3], Farrukh Alvi[3], Callum Hugh Atkinson[4], Stephen Trevor[4] and Luis Daniel Gomes[5]

[1]*Laboratoire d'Etudes Aérodynamiques (LEA), Université de Poitiers, ENSMA Téléport 2, 1 Avenue Clement Ader, BP 40109, 86961 Futuroscope, Chasseneuil Cedex, France*
[2]*Department of Fluid Mechanics and Heat Transfer, School of Mechanical Engineering, Tel Aviv University (TAU), Ramat Aviv, Israel*
[3]*Fluid Mechanics Research Laboratory (FMRL), Florida State University, Tallahassee, FL 32310, U.S.A.*
[4]*Laboratory for Turbulence in Aerospace and Combustion (LTRAC), Department of Mechanical Engineering, Monash University, Melbourne, VIC 3800, Australia*
[5]*The University of Manchester, Manchester, M60 1QD, U.K.*

Abstract This paper presents the wind tunnel test results concerning the effects of deploying steady and synthetic jets on a NACA0015 airfoil and describes the design of a multi-orifice-single-chamber synthetic jet actuator. Three steady jets with different configurations were tested. The orifice diameter, orientation and spacing were the varying parameters. Synthetic jets were deployed through a single row of orifices that were orientated normal to the airfoil surface. A single row of each type was positioned at 30% of chord from the leading edge. These jets exhibited varying degree of control authority over the lift and drag coefficients. The timescales of attachment and separation were estimated for the test cases of angled steady and synthetic jets. In view of controlling the flow separation in a dynamic manner, a multi-orifice-single-chamber actuator with a typical response time smaller than that of the afore-mentioned time scales was designed, fabricated and tested in quiescent condition.

Key words: Airfoil, turbulent separation control, characteristic time, normal jet, angled jet, synthetic jet, multi-orifice, piezo-device, pneumatic device.

1 Introduction

Several authors [1–6] have performed experiments related to the control of flow separation over a generic airfoil with the NACA0015 being the most widely utilized test model. Almost all the experiments focused mainly on a particular technique of flow control. Comparison was difficult as the differences in test parameters span across different Reynolds numbers, model aspect ratio and blockage ratio.

J.F. Morrison et al. (eds.), IUTAM Symposium on Flow Control and MEMS, 157–166.
© 2008 *Springer. Printed in the Netherlands.*

The use of splitter plates and application of turbulator further rendered difficulties in comparison. In the current test program, a common airfoil was used; this eliminated the ambiguity of the above-mentioned parameters. The objectives were to: (1) access the general aerodynamics, in terms of lift and drag coefficients, of the airfoil in response to the deployment of jet actuators; (2) estimate the times scales of full/partial flow attachment and separation; (3) design and test a multi-orifice-single-chamber synthetic jet actuator utilizing piezo-electric material as the driver. In the first objective, the baseline aerodynamic behavior between $0°$ to $16°$ incidence was determined from an external force balance; the incidence where transition from two to three-dimensional separation was identified from surface flow visualization using oil mixture and tuffs. In the second objective, incidences where quasi two-dimensional separation occurred were used to estimate the time scale of flow attachment and separation. These estimates were used as criteria to design an actuator that could respond to the flow faster than it could change between attached and separated state. Such information would be required when implementing a closed-loop flow control system. This collaborative study, "Second European Forum on Flow Control (EFFC2)", was conducted within the framework of the Airbus program "Control of Aerodynamic Flows for Environmentally Designed Aircraft (CAFEDA)". During a period of three months, researchers from five different countries were hosted by LEA of the University of Poitiers to achieve the afore-mentioned objectives.

2 Model, Actuator and Test Conditions

The NACA0015 airfoil was measured 2.4 m (span) by 0.35 m (chord), see Figure 1. 88 μm carborandum was placed near to the leading edge. All steady jet configurations were tested at a Reynolds number of 0.96×10^6. The incidences tested ranged from $0°$ to $16°$. Actuator configurations were tabulated in Table 1. In the case of a steady jet, three configurations were tested; two of which were orientated normal to the surface with diameter and spacing as the varying parameters. An angled steady jet, one of the three configurations, was deployed at a pitch and yaw angles of $30°$ and $60°$ (relative to a plane that is tangential at $x/c = 30\%$) respectively as shown in Figure 2a. All jet configurations were implemented on a removeable cover similar to that as shown in Figure 2b. Depending on the pressure loss in the cavity and piping system, a nominal peak velocity of 70 m/s to 100 m/s could be achieved at the outlet of the 1 mm diameter orifice. In the case of the 0.5 mm diameter, it was designed to operate when choked which gave sonic output velocity at the orifice outlet. The synthetic jet, optimized for an output peak velocity of 30 m/s, augmented by an array of piezo-electric actuator was tested at a Reynolds number of 0.25×10^6.

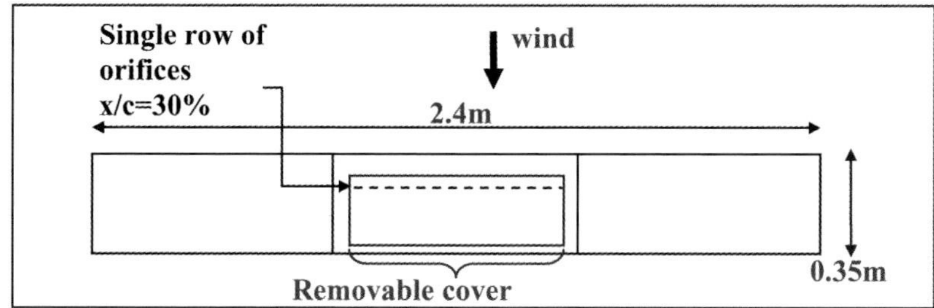

Fig. 1 Plan view of NACA0015.

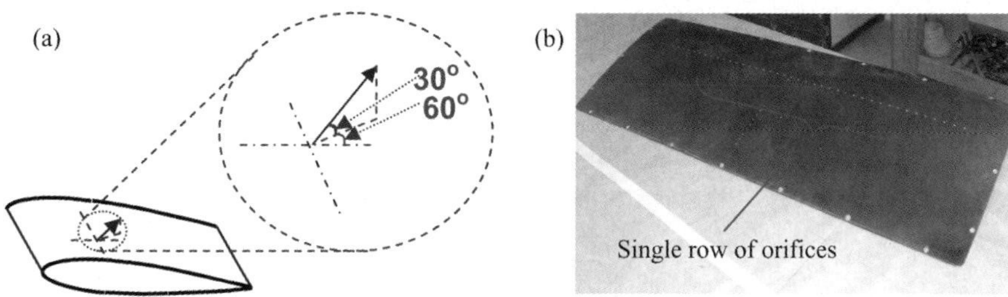

Fig. 2 (a) Schematic of angled steady jet deployed at $x/c = 30\%$. (b) Overview of cover with a single row of orifices machined at $x/c = 30\%$.

Table 1 Configurations of actuator implemented in the current test program.

	Mode of Deployment	Means of Deployment	Jet Orientation	Position (x/c)	Number of Orifices
Steady "Angled" Jets (1mm diameter)	Continuous		30deg (pitch), 60deg (yaw)	0.3	44
Steady "Normal" Jets (1mm diameter)	Continuous	pressurized cavity		0.3	51
Steady "Normal" Jets (0.5mm diameter)	Continuous		normal to surface	0.3	64
"Normal" ZNMF Jet (1mm diameter)	Amplitude, plused modulated	piezo-electric		0.3	56

3 Baseline Characteristics of the NACA0015

With reference to Figure 3, the Cl versus alpha curve indicated that the airfoil started to stall at an incidence of 8° and reached a maximum at 12°. Comparison of the baseline flow with and without orifices exposed revealed that the Cl and Cd were nominally larger. This suggested that the orifices were acting like a mechanical vortex generator and not due to transition to turbulence since the boundary layer was tripped further upstream. Flow separation between incidences of 8° and 11.5° was quasi two-dimensional, and transition to three-dimensional separation starts with the formation of three stall cells at 12°. The physics of this transition had been described in the literature [7]. The observation of the stall cells persists until an incidence of 16°, see Figures 4a and 4b. Since baseline aerodynamic characteristics were differ-

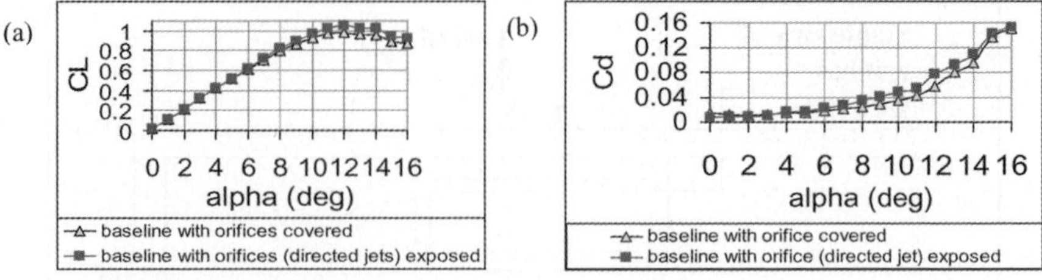

Fig. 3 (a) Lift coefficient (Cl) versus alpha. (b) Drag coefficient (Cd) versus alpha.

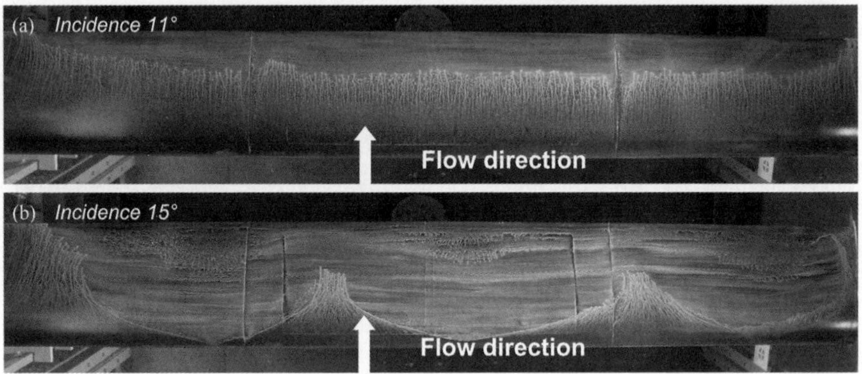

Fig. 4 Surface flow visualization at incidence of (a) 11° that depicts a quasi two-dimensional separation; (b) 15° that depicts a three-dimensional separation with three stall cells.

ent for cases with and without covering the orifices, all comparisons with actuator deployed were made with respect to the baseline of NACA0015 with its orifices exposed.

4 Effects of Steady Jets

The jet characteristics were the momentum coefficient and velocity ratio, defined by

$$C_\mu = (\rho_j U_j^2 A)/(0.5\rho_\infty U_\infty S), \tag{1}$$

$$VR = U_j/U_\infty. \tag{2}$$

Subscripts j and ∞ denote, respectively, parameters associated with the jet and free stream; U, ρ, A and S represent, respectively, the velocity, density, total surface area of jet orifices and planform area of the wing spanned by the row of orifices. Table 2 provides a summary of the effects of steady jets, on Cl and Cd, that were investigated.

Table 2 Summary of the effects of various steady jets on Cl and Cd.

	Steady jet parameters	C_μ	VR	Cl	Cd
1	Angled, 1 mm diameter	0.27% to 0.36%	2.5 to 3.5	5% to 16% (improvement)	30% to 50% (reduction)
2	Normal, 1 mm diameter	0.03% to 0.14%	0.9 to 1.8	3% to 5% (improvement)	Negligible change
3	Normal, 0.5 mm diameter	0.4%	8	3% to 8% (improvement)	15% to 22% (reduction)

From Table 2, it can be deduced that, at maximum $C_\mu \sim 0.4\%$, normal jet deployment through orifices with diameter of 0.5 mm diameter was less efficient as compared to angled jets. At incidences greater than $16°$, there was no improvement in Cl and possibly, a significant increase of C_μ might be necessary to bring about further improvement. Due to the limited range of C_μ, comparison was not possible for the case of a normal jet with a diameter of 1 mm. Its trend suggested that the Cl could be increased by increasing the C_μ. Trends of its Cd versus alpha showed little change. The Cl and Cd characteristics as shown in Figures 5, 6 and 7 would give a detail perspective of the results in Table 2. Note that the percentage improvement for Cl would be three times as much since actuator was deployed over a region of slightly less than 30% of the airfoil span.

5 Effects of Synthetic Jet Actuator

A sinusoidal signal (1.95 kHz) with amplitude modulation (AM, 41 Hz sine wave) was fed into an array of 14 piezo-electric actuators, these signals correspondeing to a non-dimensional frequency (F^+) of 1 and 49 respectively. F^+ was defined as follows:

$$F^+ = fL/U_\infty, \tag{3}$$

where f denotes the frequency of the signal for a pure sine wave and the low frequency component of the AM signal and L denotes the length between actuator and trailing edge of airfoil. Four orifices, orientated normal to the surface, were associated with each piezo-electric actuator. In AM, the high frequency component was set the resonance frequency of the system. This was required to provide maximum output velocity ($\sim$30 m/s) at the orifice; the low frequency was introduced for the purpose of perturbing instabilities that might exist in the flow. The resulting C_μ and VR were 0.32% and 3 respectively. With reference to Figure 8a, there was an improvement in Cl of up to 4.1%. AM was able to provide drag reduction at low incidences, see Figure 8b.

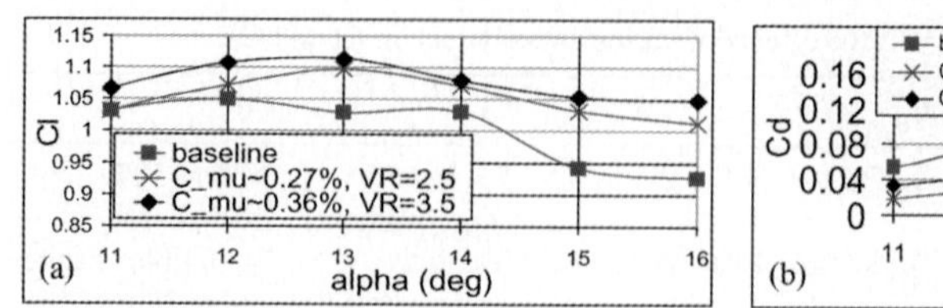
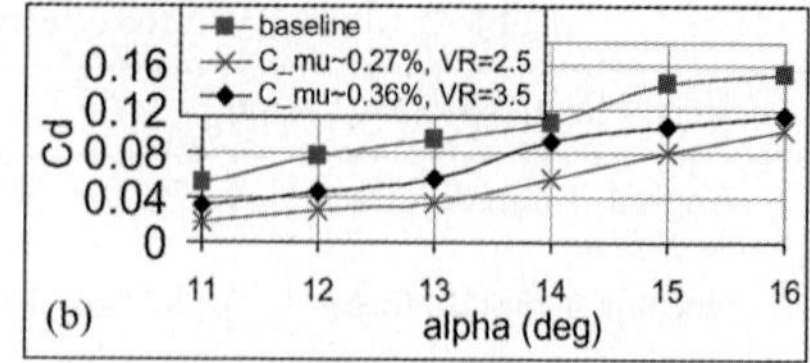

Fig. 5 Effects of steady angled jet with a yaw and pitch angle of 60° and 30°. (a) Cl versus alpha. (b) Cd versus alpha.

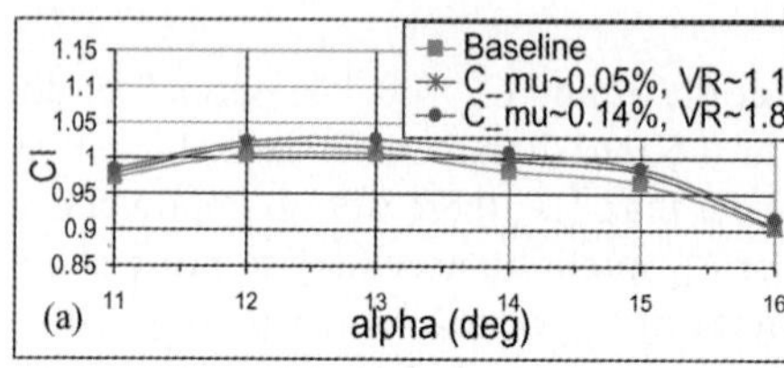
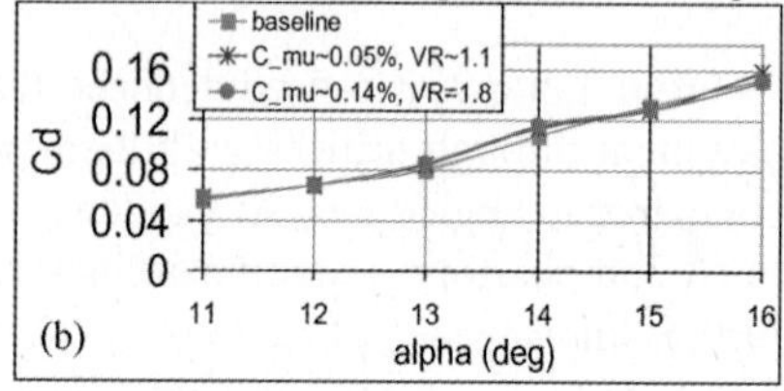

Fig. 6 Effects of normal jet deployed through orifices with diameter of 1 mm. (a) Cl versus alpha. (b) Cd versus.

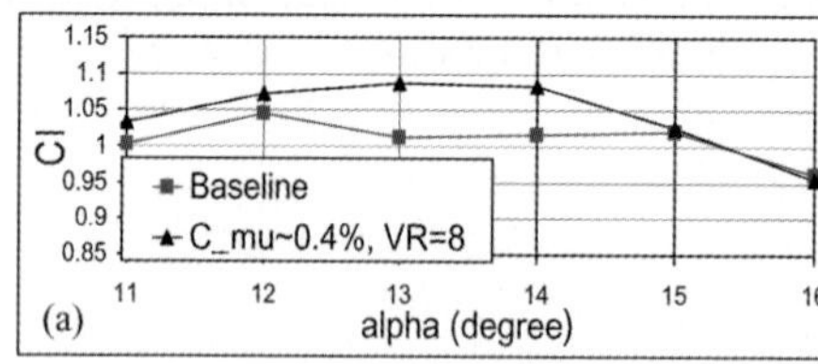
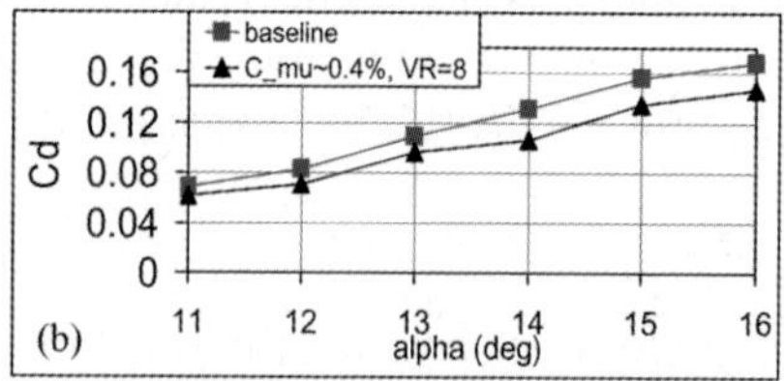

Fig. 7 Effects of normal jet deployed through orifices with diameter of 0.5 mm. (a) Cl versus alpha; (b) Cd versus alpha.

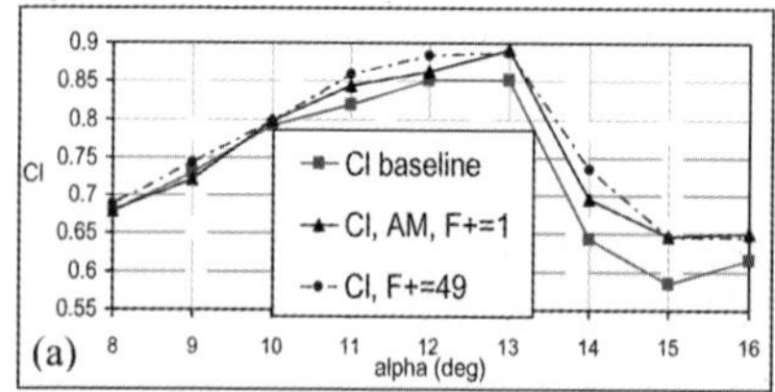
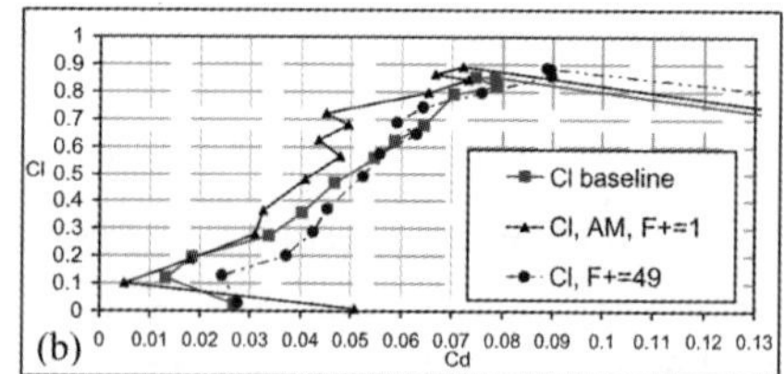

Fig. 8 Comparison of AM and sinusoidal signal with respect to baseline; (a) Cl versus alpha, (b) drag polar.

6 Time Scales of Flow Attachment and Separation

For the case described in Section 5 (synthetic jet study), time response due to a change in state, during actuator "ON" and "OFF", was performed at an incidence of 13°. A Kulite pressure transducer was installed at $x/c = 85\%$ to realize the study.

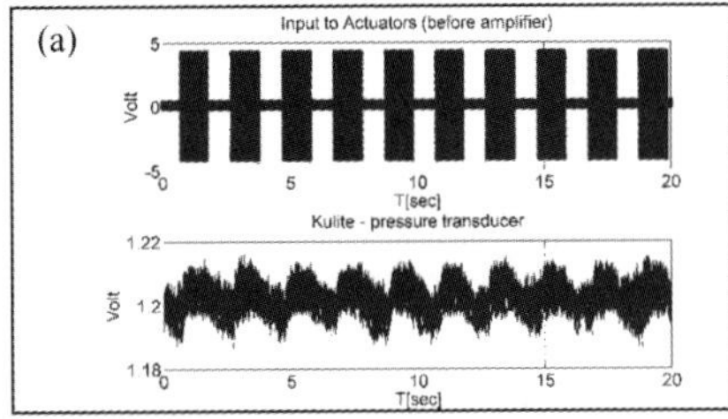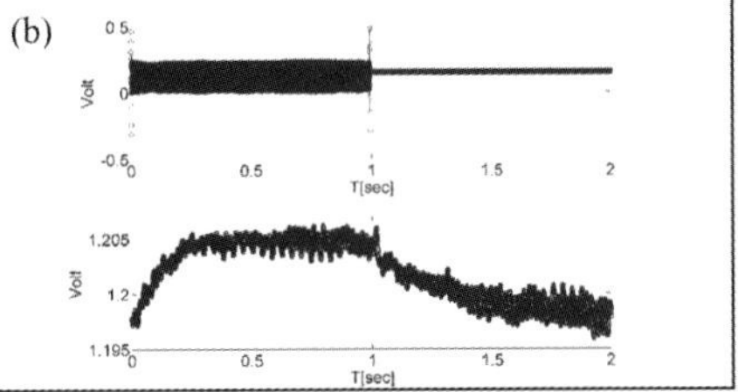

Fig. 9 (a) Response of actuator and Kulite pressure transducer. (b) Ensemble average of ten cycles of the time response in (a).

Ten cycles of ON/OFF state at 0.5 Hz, were acquired. An ensemble average was performed to estimate the time response.

Values of 0.2 sec ($T^+ = 8.1$) and 0.6 sec ($T^+ = 24.3$), corresponding to the actuator ON and OFF state were estimated, see Figure 9. The non-dimensional time T^+ was defined as

$$T^+ = tU_\infty/L,\qquad(4)$$

where t denotes the time taken for a change in state and L distance from actuator to trailing edge of the airfoil. Surface tuft visualization indicated a separation extent of 50% (actuator OFF state) and 20% (actuator ON state) of the airfoil chord. Darabi and Wygnanski [8, 9] determined values of 16 (attachment) and 20 (separation) in the case of a generic flap.

For a steady angled jet as described in Section 4, a hot wire was positioned at $x/c = 97\%$ and 10 mm above the airfoil surface to acquire the time histories during the ON/OFF state as shown in Figure 10. The airfoil was set at an incidence of 11°, at which a quasi two-dimensional separation (15% of chord) occurred. In this case, the ON state corresponded to full flow attachment. As the process of activating/deactivating the actuator was manual and thus without a reference, a cross correlation between ten data sets was performed for the purpose of data alignment before an ensemble average was computed. A curve fit was then performed on the data utilizing the following function:

$$V(t) = b + d\left(\mathrm{erf}\left(\frac{\sqrt{\pi}\,(t+c))}{a}\right)\right),\quad \mathrm{erf}(T) = \frac{2}{\sqrt{\pi}}\int_0^T e^{-t^2}\,dt,\qquad(5)$$

where V and t denote velocity and time respectively; a, b, c and d denote the constants that were determined during curve fitting. The constant a was estimated to be the typical time. Typical time of attachment and separation was both estimated to be 0.11 sec ($T^+ = 16.3$) and 0.21 sec ($T^+ = 33.8$) respectively.

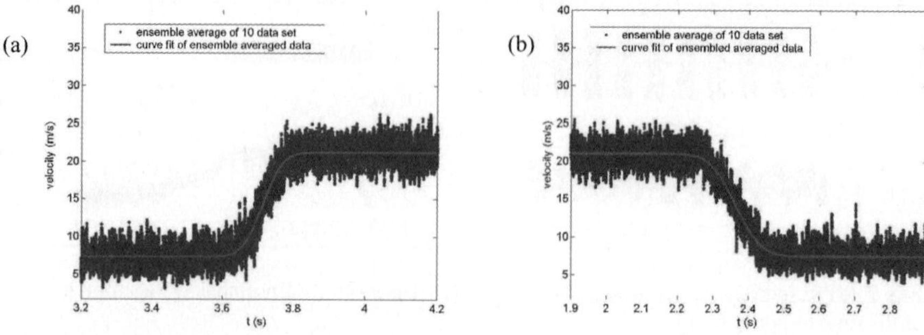

Fig. 10 Time response during a time interval of 1 sec. (a) Attachment process, (b) separation process.

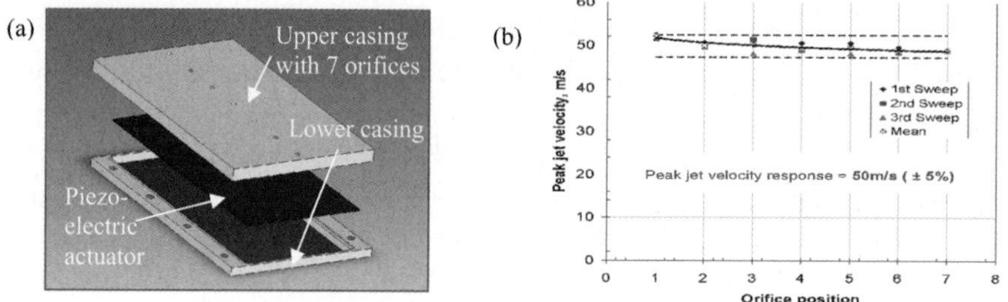

Fig. 11 (a) Overview of the multi-orifice-single-chamber actuator. (b) Plot of peak velocity versus orifice position (sweep indicates the number of experiment performed).

7 Development of Multi-Orifices-Single-Chamber Synthetic Jet Actuator

Two main requirements of an actuator were (1) faster response than flow attachment and separation over the NACA0015 and (2) ease of installing several orifices ($\sim$45) over the span of an airfoil. Thus, a compact multi-orifice-single-chamber actuator was designed and tested in quiescent conditions. The design was based on the extrapolation of the parameters derived from a single-orifice-single-chamber [10]. Seven orifices were associated to a rectangular chamber as in Figure 11a. It was determined that the response time was a few milliseconds during the ON state and approximately four times during the OFF state. This would satisfy the requirement for the case of dynamic flow separation control on a NACA0015 by two orders of magnitude. The maximum velocity achieved for each orifice was approximately 50 m/s, see Figure 11b.

8 Conclusion

At a Reynolds number of 0.96×10^6, the collaborative test program indicated that steady angled jets could provide better efficiency compared to steady normal jets deployed through orifices with diameter of 0.5 mm. In the case of synthetic jet, tested at a Reynolds number of 0.25×10^6, Cl improvement of up to 4.1% was possible and it could provide drag reduction at low incidences. Drawing inference from the results of angled steady jet, further enhancement to the efficiency of synthetic jet would require its orifice to be angled and its outlet velocity be increased. The non-dimensional response time for synthetic jet, T^+, was calculated to be 8.1 (partial attachment) and 24.3 (separation). In the case of steady angled jet, T^+ was estimated to be 16.1 and 33.8 for full attachment and separation respectively. It was noted that different actuators, models, locations, sensing technique and positioning of sensors might give rise to different values. A better estimate could be obtained, in the first step, by acquiring field data via PIV and surface pressure data. Bi-orthogonal-decomposition on the data could then be applied to extract the temporal evolution. Nevertheless, current experience suggested that a response time of one order less than 0.1 sec would be required of an actuator system to control the separation in a dynamic manner. To meet this requirement, a compact multi-orifice-single-chamber actuator was designed, fabricated and tested in the absence of cross flow. Its response time was found to be well below the time scales of the attachment and separation, and was capable of generating an average peak velocity of 50 m/s.

Acknowledgements

The authors acknowledged the valuable fundings from the Airbus "CAFEDA" program and CNRS GDR "Contrôle des Décollements". The assistance rendered by the team of technical personnel at LEA-ENSMA and LEA-CEAT was well appreciated for making the collaborative test program a unique and successful experience. Notably, Jean Marc Breux and Khoo Wee Hon for technical assistance throughout the wind tunnel testing; Jean Pierre Bal for the fabrication of the model covers to accommodate different fluidic actuator.

References

1. Tuck A. and Soria J. (2004), Active flow control over a NACA 0015 airfoil using a ZNMF jet, in *15th Australasian Fluid Mechanics Conference*, 13–17 December 2004.
2. Gilarranz J.L., Traub L.W. and Rediniotis O.K. (2002), Characterization of a compact, high-power synthetic jet actuator for flow separation control, AIAA-2002-0127.
3. Greenblatt D. and Wygnanski I. (2000), The control of flow separation by periodic excitation, *Progress in Aerospace Sciences* **36**, 487–545.

4. Chen F.J. and Beeler G.B. (2002), Virtual shaping of a two-dimensional NACA 0015 airfoil using synthetic jet actuator, AIAA-2002-3273.
5. Bales K., Khoo P. and Jefferies R. (2003), Flow control of a NACA 0015 airfoil using a chordwise array of synthetic jets, in *41st Aerospace Sciences Meeting and Exhibit*, Reno, NV, 6–9 January 2003, AIAA 2003-0061.
6. Tian Y., Cattafesta L.N. and Mittal R. (2006), Adaptive control of separated flow, in *41st Aerospace Sciences Meeting and Exhibit*, Reno, NV, 9–12 January 2006, AIAA 2006-1401.
7. Weihs D. and Katz J. (1983), Celluar patterns in poststall flow over unswept wings, *AIAA Journal* **21**(12), 1757–1759.
8. Darabi A. and Wygnanski I. (2004), Active management of naturally separated flow over a solid surface. Part 1. The forced reattachment process, *Journal of Fluid Mechanics* **510**, 105–129.
9. Darabi A. and Wygnanski I. (2004), Active management of naturally separated flow over a solid surface. Part 2. The separation process, *Journal of Fluid Mechanics* **510**, 131–144.
10. Gomes L.D., Crowther W.J. and Wood N.J. (2006), Towards a practical piezoceramic diaphragm based synthetic jet actuator for high subsonic applications – Effect of chamber and orifice depth on actuator peak velocity, in *3rd AIAA Flow Control Conference*, AIAA-2006-2859.

High Resolution PIV Study of Zero-Net-Mass-Flow Lift Enhancement of NACA 0015 Airfoil at High Angles of Attack

Trevor Stephens and Julio Soria

Laboratory for Turbulence Research in Aerospace and Combustion (LTRAC), Department of Mechanical Engineering, Monash University, Melbourne, Australia 3800; E-mail: julio.soria@eng.monash.edu.au

Abstract. The effect of changing the pitch to diameter ratio (P/D) of a row of round, wall-normal, zero-net-mass-flux (ZNMF) jets located at the leading edge of a NACA 0015 airfoil was investigated. A parametric study and particle image velocimetry (PIV) measurements were conducted on a two-dimensional airfoil in a water tunnel at a Reynolds number of 6.56×10^4. Different optimal forcing frequencies and percentage lift increases between the two P/D cases were observed. It is possible that differences in jet interaction mechanisms may have caused the differences in control effectiveness between P/D cases. Time-averaged streamlines indicate a reduction in the size of a recirculation region over the upper surface of the airfoil may be causing the improved lift.

Key words: Flow control, lift enhancement, separation control, zero-net-mass flux jet, jets, synthetic jet, PIV.

Nomenclature

A_j	ZNMF jet orifice area		$\bar{h}$	Average ZNMF jet orifice height
b_n	Span of ZNMF jet orifices		J	Characteristic jet momentum,
c	Airfoil chord			$J \equiv \rho U_{j\mathrm{rms}}^2 \bar{h}$
C_L	Airfoil lift coefficient		n	Number of ZNMF jet orifices
c_μ	Oscillatory momentum blowing		P	ZNMF jet orifice pitch
	coefficient		q	Dynamic pressure, $q \equiv \frac{1}{2}\rho U_\infty^2$
D	ZNMF jet orifice diameter		$U_{j\mathrm{rms}}$	Root mean square of jet velocity
f	Excitation frequency		U_∞	Freestream velocity
F^+	Non-dimensional frequency		ΔC_{L20}	Percentage increase in C_L at $20°$

J.F. Morrison et al. (eds.), IUTAM Symposium on Flow Control and MEMS, 167–173.

© 2008 *Springer. Printed in the Netherlands.*

1 Introduction

The use of passive flow control devices such as vortex generators and strakes has been accepted by the commercial aviation industry. These fixed objects have the disadvantage of creating unwanted parasitic drag when they are not required.

Active flow control refers to techniques where energy is expended to modify the flow [2] and often involves replacing the fixed surfaces used in passive flow control with a series of jets intended to manipulate the fluid within the boundary layer. Its main advantage is that it can be "switched on" when required. One promising implementation involves using ZNMF jets [6].

ZNMF jets are created from the working fluid of the flow system in which they are deployed, and thus transfer linear momentum without net mass injection into the system [1, 4]. In the case of round jets, as used in this study, they are created from the periodic formation of vortex rings [3].

An airfoil with ZNMF jet active flow control can be characterised by two non-dimensional parameters: The non-dimensional excitation frequency

$$F^+ \equiv \frac{fc}{U_\infty} \tag{1}$$

and the oscillatory momentum blowing coefficient

$$c_\mu \equiv \frac{J}{qc} = 2\frac{\overline{h}}{c}\left(\frac{U_{j\mathrm{rms}}}{U_\infty}\right)^2. \tag{2}$$

The geometry of the row of circular jet orifices used in this study was expressed as the pitch to diameter ratio, P/D, defined in Figure 1b.

The effect of the P/D of dual circular ZNMF jets on their propagation in still air has been previously investigated [7]. The ZNMF jets in that study had $P/D = 5.71$ and 1.71, and different modes of propagation were identified for each case. For the wider spacing, the jets were unaffected by one another but when the spacing was reduced, the jets almost immediately combined into a single, larger jet, indicating that jet interaction relies heavily upon P/D.

Previous studies of ZNMF jets in flow control focused on two-dimensional jets. However, a continuous slot along the leading edge of an aircraft has serious implications to the structure of the wing. The motivation for this study was to investigate the relationship between ZNMF jet geometry and the effectiveness of active flow control and is described in more detail in [5].

2 Experimental Apparatus and Procedure

Two NACA 0015 airfoils with 150 mm chord and 510 mm span were manufactured for this study. Experiments were carried out in a closed-circuit horizontal water tun-

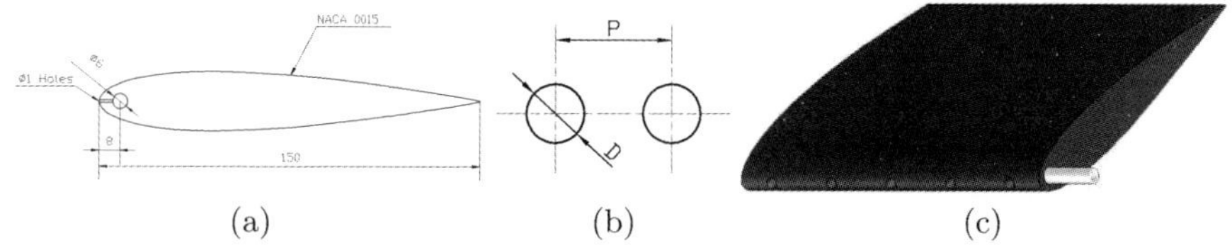

Fig. 1 Drawings of the NACA 0015 airfoils (a, c) and ZNMF jet geometry (b).

Table 1 Specifications of the NACA 0015 airfoils used in this investigation.

P/D	n	A_j (mm^2)	b_n (mm)	$\bar{h}$ (mm)
6.0	76	59.69	456.0	0.1309
1.7	265	208.13	450.5	0.4620

nel, having a 5 m long working section with a cross section of 0.5×0.5 m. The airfoils were mounted vertically, through a rotation stage, to a six-axis force transducer which measured lift forces at 21.33 Hz with a resolution of 0.116 N or $\Delta C_L \approx 0.025$. Force data was acquired for 4 minutes or 5120 samples for each test condition. Analysis of force transducer data found that the error from the transducer's operation and the aerodynamic force fluctuations was $\Delta C_{L20} \approx 2\%$.

Rows of 1 mm diameter, surface normal holes were machined along the leading edges with $P/D = 6.0$ and 1.7; equivalent to the values used by Watson [7]. The holes were fed by a 6 mm diameter cavity inside the leading edge of the airfoil, as shown in Figure 1a, through which flow oscillations were supplied to generate a row of ZNMF jets. The flow oscillations were supplied by a 20 mm diameter piston driven by a computer controlled stepper motor via a Scotch–Yoke mechanism. Table 1 lists the specifications of the two airfoils.

For PIV measurements, the water tunnel was seeded with Potters Sphericell hollow glass particles which have a mean diameter of 11 µm and a density of 1100 kg/m^3. Based on Stokes law, the particles have a relaxation time of 9 µs, indicating the seed particles will follow the fluid motion with high fidelity.

A dual cavity Nd: YAG laser was used to generate coplanar laser sheets, approximately 1 mm in thickness, incident near the mid-plane of the airfoil and aligned with a jet orifice. The pulsed laser sheets were generated at 2 Hz and synchronised with a digital camera used to capture the two single-exposed PIV images. The camera had a resolution of 4008×2672 pixels and using the 105 mm focal length lens, gave a spatial resolution of 56.25 µm/pixel.

An interrogation window size of 32×32 pixels was used, yielding a 125×83 vector field. During the control experiments, phase locking was avoided by offsetting the optimum frequencies, multiples of 0.5 Hz, by 0.04 Hz. The 2 Hz acquisition frequency also ensured the statistical independence of each PIV image pair. Two hundred instantaneous vector fields were averaged to yield time-averaged data.

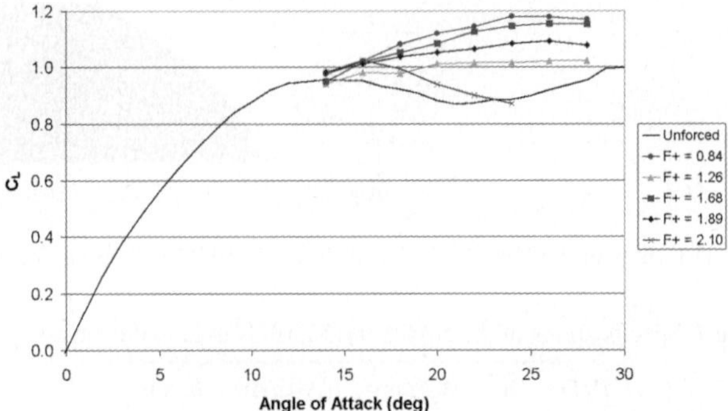

Fig. 2 Sample of results for $c_\mu = 0.053\%$ at $Re = 6.56 \times 10^4$.

3 Parametric Study

The effect of changing the geometry of ZNMF jet orifices was investigated through a parametric study of each airfoil at the same conditions and comparing their behaviour at a Reynolds number of 6.56×10^4. The uncontrolled $C_{L\,\max}$ and C_{L20} were measured for both airfoils. $C_{L\,\max}$ for both airfoils was found to occur at $16°$. For $P/D = 6.0$, $C_{L\max} = 0.987$ and $C_{L20} = 0.911$ and for $P/D = 1.7$, $C_{L\max} = 0.981$ and $C_{L20} = 0.906$. These values are well within the experimental error of each other.

The dependence of the effectiveness of the active flow control on F^+ and c_μ was investigated. This was evaluated by holding c_μ constant and varying F^+ from 0.42 to 2.11. The first stages of this investigation aimed to find the angle of attack, if one existed, at which a controlled airfoil would reach its maximum lift coefficient. Figure 2 shows a sample result for this investigation.

A broad range of F^+ and c_μ were investigated in this manner with $2°$ steps in angle of attack. Forty-four different combinations of the ZNMF jet orifice P/D and the amplitude and frequency of oscillation were tested. For more than 90% of the test conditions examined, the controlled maximum lift angle was in the range of $20° \pm 2°$. Thus, a $20°$ angle of attack was used for all further work.

The effectiveness of the flow control was quantified as the percentage increase in lift at $20°$ and given the notation ΔC_{L20}. Both airfoils were investigated in this way for the same range of F^+ previously examined at $c_\mu = 0.027\%$ and 0.049%. The dependence of c_μ on the experimental set up meant not all forcing frequencies could be investigated due to the piston displacement limits.

Figure 3a shows the parametric study's results for both airfoils. The general shape of this percentage lift increase curve is similar for both airfoils with two peaks of high active flow control effectiveness at two frequencies, F^+_{low} and F^+_{high}. Between and on either side of these peaks, the effectiveness reduces. This is summarised in the schematic in Figure 3b.

It is apparent that over some ranges of F^+, no lift increase due to increased c_μ is measured for the $P/D = 1.7$ case. The $P/D = 6.0$ case however shows a greater

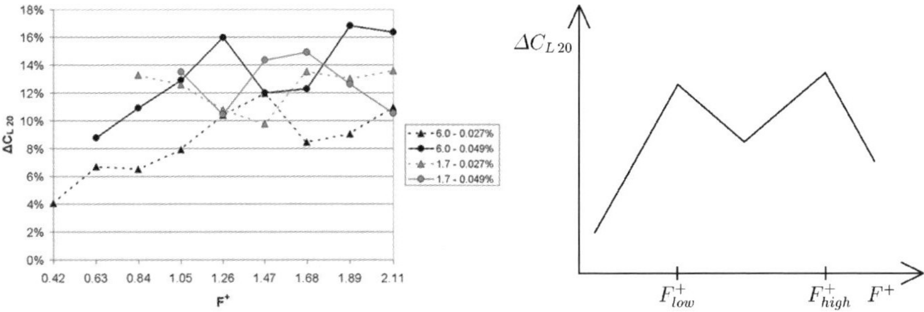

Fig. 3 ΔC_{L20} as a function of F^+ at Re $= 6.56 \times 10^4$ (a) and schematic of ΔC_{L20} curve shape (b).

Table 2 The effect of P/D on ΔC_{L20} at $c_\mu = 0.049\%$ and Re $= 6.56 \times 10^4$.

P/D	F^+_{low}	ΔC_{L20}	F^+_{high}	ΔC_{L20}
6.0	1.26	16.0%	1.89	16.9%
1.7	1.05	13.5%	1.68	14.9%

increase in control effectiveness over the entire spectrum of F^+ investigated. It was also observed that increasing c_μ appeared to shift the peak frequencies to slightly lower values.

For $c_\mu = 0.049\%$, the peak frequencies and their active flow control effectiveness are clearly different between the two different P/D cases. Table 2 summarises the important results at $c_\mu = 0.049\%$. This data indicates that the control effectiveness at F^+_{low} and F^+_{high} is similar, being within the experimental error. The worst performing P/D case was 1.7. It is possible that jet interaction differences, similar to those seen in [7], may have caused the reduced control effectiveness.

4 Particle Image Velocimetry

The conditions found to be optimal for control in the parametric study were investigated further with PIV at the same Reynolds number of 6.56×10^4. The F^+_{high} control frequencies were selected since the peaks were better resolved. In order to obtain the maximum difference between the uncontrolled and controlled cases the angle of attack was reduced to $16°$ with $c_\mu = 0.049\%$.

Figure 4 shows time-averaged streamlines for the uncontrolled case and both controlled cases. The uncontrolled airfoil presents a massive recirculating region with its centre lying at about 80% chord. With control activated, the $P/D = 6.0$ case shows a drastically reduced recirculation region with smooth, attached flow seen for almost the entire chord. A very small recirculation region is seen right at the trailing edge. $P/D = 1.7$ also shows significant improvement. Time-averaged Reynolds stress contours for these cases are reported in [5].

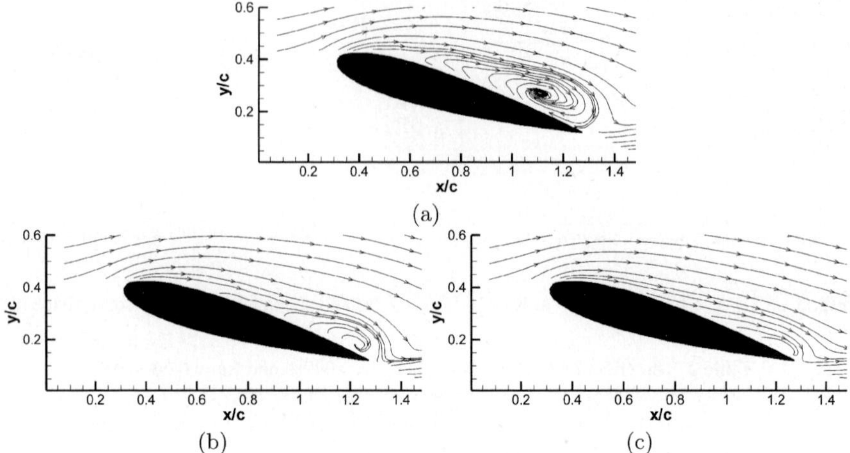

Fig. 4 Time-averaged streamlines for the uncontrolled airfoil (a), the controlled airfoil with $P/D = 1.7$ (b) and $P/D = 6.0$ (c).

5 Conclusions

The geometry of a row of round, wall-normal, ZNMF jets located at the leading edge of a NACA 0015 airfoil has been seen to effect the optimum frequencies of active flow control. These optimum frequencies were identified as $F^+ = 1.89$ for a row of jets with $P/D = 6.0$ and $F^+ = 1.68$ for $P/D = 1.7$ but were seen to rely on the c_μ at which the jets were operated. A second optimum frequency of similar effectiveness was seen at a lower frequency for each P/D case. The effectiveness of each was found to increase with increasing c_μ. The most effective P/D of the two was found to be 6.0. It is possible that differences in jet interaction mechanisms due to their geometry may have caused the different control effectiveness for each P/D case. Time-averaged streamlines indicate a reduction in the size of a recirculation region over the upper surface of the airfoil, which was likely to have caused the improvement in the observed lift.

Acknowledgements

The authors would like to acknowledge the tremendous efforts of Dr. Kamal Parker and Ashley Tuck for setting up much of the apparatus adapted to this experiment, Eric Wirth and Ivor Little for manufacturing the apparatus used, Professor Bijan Shirinzadeh for the generous loan of the force transducer and the financial support of the ARC and AFOSR/AOARD.

References

1. Cater, J.E., Soria, J.: The Evolution of Round Zero-Net-Mass-Flux Jets. *Journal of Fluid Mechanics* **472** (2002) 167–200.
2. Donovan, J.F., Krai, L.D., Gary, A.W.: Active Flow Control Applied to an Airfoil. AIAA Paper 0210 (1998).
3. Gordon, M., Soria, J.: PIV Measurements of a Zero-Net-Mass-Flux Jet in Cross Flow. *Experiments in Fluids* **33**(6) (2002) 863–872.
4. Smith, B.L., Glazer, A.: The Formation and Evolution of Synthetic Jets. *Physics of Fluids* **9** (1998) 2281–2297.
5. Stephens, T.: The Effect of Zero-Net-Mass-Flux Jet Geometry on Active Flow Control of a NACA 0015 Airfoil. Honours Thesis, Laboratory for Turbulence Research in Aerospace and Combustion (LTRAC), Department of Mechanical Engineering, Monash University, Melbourne, Australia (2006).
6. Tuck, A.: Active Flow Control of a NACA 0015 Airfoil Using a ZNMF Jet. Honours Thesis, Laboratory for Turbulence Research in Aerospace and Combustion (LTRAC), Department of Mechanical Engineering, Monash University, Melbourne, Australia (2004).
7. Watson, M., Jaworski, A.J., Wood, N.J.: Contribution to the Understanding of Flow Interactions between Multiple Synthetic Jets. *AIAA Journal* **41**(4) (2003) 747–749.

Separation Control along a NACA 0015 Airfoil Using a Dielectric Barrier Discharge Actuator

Jérôme Jolibois, Maxime Forte and Eric Moreau
Laboratoire d'Etudes Aérodynamiques (CNRS), University of Poitiers, Téléport 2, Bd. Marie & Pierre Curie, BP 30179, 86962 Futuroscope Cedex, France; E-mail: jerome.jolibois@lea.univ-poitiers.fr

Abstract. This paper deals with the control of airflow separation above a NACA 0015 airfoil using a surface plasma actuator. A dieletric barrier discharge plasma is used to bring velocity in the boundary layer, tangentially to the wall. The goal of the actuation is to displace (upstream or downstream) the separation point, in either reattaching a naturally detached airflow or in detaching a naturally attached airflow. The ultimate goal of these experiments is to better understand where one has to act along the profile chord (as a function of the angle of attack) to be the most efficient. These experiments show that the plasma actuator is more efficient when it acts close to the separation point, and that the power consumption can be highly reduced by using a non-stationary actuation.

Key words: Electrofluidodynamic (EFD), AC dielectric barrier discharge, flow separation control, ionic wind, NACA 0015 airfoil.

1 Introduction

Several studies have demonstrated the ability of plasma actuators to control airflow along airfoils, resulting in stall delaying and lift increase, for velocity up to 30 m/s ([1–5] for instance). More, recent works have shown that a 110 m/s detached airflow could be fully reattached [6]. A review has been published recently [7].

However, in all these studies, we cannot be sure that the plasma effect is due to the momentum added by the discharge-induced electric wind, or to the laminar-toturbulent transition induced by the disturbances generated by the plasma actuator. Indeed, the plasma actuator is usually located at the leading edge, and the airfoil chord is low (typically between 9 and 20 cm).

The present study does not have these three drawbacks. First, the airfoil chord is equal to 1 meter. Second, the boundary layer is tripped with a turbulator located at the leading edge to uncouple the effects of the ionic wind and the laminar-to-

J.F. Morrison et al. (eds.), IUTAM Symposium on Flow Control and MEMS, 175–182.
© 2008 *Springer. Printed in the Netherlands.*

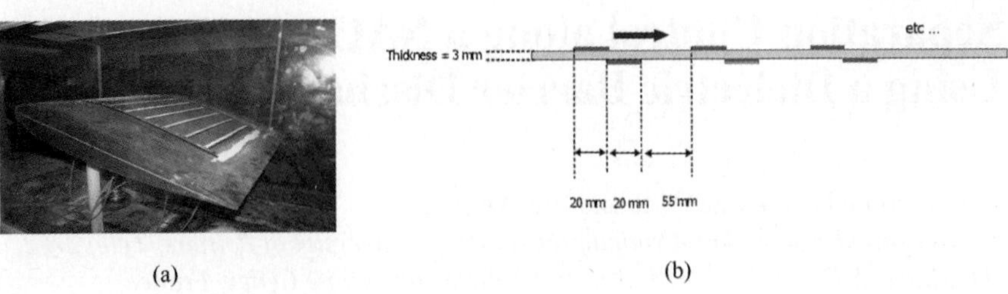

(a)   (b)

Fig. 1 Photo of the NACA 0015 airfoil in the open-wind tunnel (a) and schematic side view of the plasma actuators on the PVC plate (b).

turbulent transition. Then, seven plasma actuators are placed along the airfoil suction side, from 30 to 80% of the chord, in order to investigate accurately where we must act to have the most efficient effect on the airflow. The free stream velocity is always equal to 6 m/s, corresponding to $Re = 0.4 \times 10^6$.

The plasma actuator used here is a dielectric barrier discharge actuator, which is made of two metallic electrodes separated by an insulator. The application of an AC high voltage between both electrodes generates a weakly ionized plasma and the discharge induces an electric wind parallel to the wall. In order to increase the ionic wind velocity, several parametric studies on the dielectric barrier discharge (DBD) were previously realized [8, 9]. In the present paper, the electric wind and the free air stream are in the same direction.

2 Experimental Setup

The measurements have been done in an open-wind tunnel which has a rectangular cross-section ($2.5 \times 1.2 \text{ m}^2$). It can operate at velocities up to 8 m/s. The airfoil model was a NACA 0015. The chord was 1 m and the spanwise was 1.2 m. In order to tripp the boundary layer, a turbulator was placed near the leading edge (at about 1% of the chord). The turbulator was made of silicon carbide (carborundum). Its thickness was equal to 0.3 mm. The airfoil was a hollow body in wood, enabling to fix a 3 mm thick polyvinyl chloride (PVC) plate on its suction side (Figure 1a), where the control device was placed (Figure 1b). It was composed by seven DBD actuators. Each DBD actuator was constituted by two metallic electrodes asymmetrically mounted on each side of the dielectric PVC plate. Each electrode consisted of a 10 µm thick aluminium tape strip. The strips were 20 mm wide and 600 mm long (spanwise). The upper electrode was connected to a power amplifier (Trek Model 30/30C, ±30 kV, ±40 mA). The high voltage had a sine waveform. The seven DBD actuators are placed on the PVC plate, from 30 to 80% of chord, every 8%. The distance between two successive actuator was 55 mm, and then the actuators did not interact between them. They are numbered from actuator #1 (upstream) to actuator #7 (downstream).

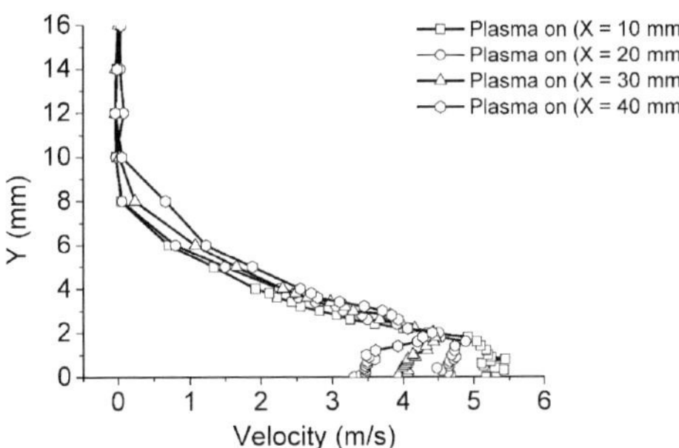

Fig. 2 Velocity profiles of the discharge-induced electric wind, for 23 kV and 1000 Hz, at different X positions ($X = 0$ corresponds to the right side of the HV electrode when $X = 20$ mm corresponds to the right side of the grounded electrode).

The experiments consisted of visualizations and Particle Image Velocimetry (PIV) measurements. They are undertaken at 6 m/s, corresponding to a Reynolds number of 0.4×10^6. Visualizations were carried out with a camera (3072×2048 pixels). The suction side of the airfoil was lighted by a SPECTRA PHYSICS multimode argon Laser (488–512 nm, 5 W) and the main flow was seeded by smoke particles. For PIV measurements, the images were recorded with a CCD camera (LAVISION®, 1280×1024 pixels, 12 bits) using a Yag-Nd double impulsion laser (QUANTEL®, 30 mJ, 532 nm). The images were processed with the cross-correlation algorithm of LAVISION® and the final window size was 32×32 pixels with an overlap of 75%.

3 Flow Control by Barrier Discharge Actuator

3.1 Optimization of the Position of the Actuator

In this section, an AC high voltage of 23 kV ($f = 1000$Hz) was applied at the air-exposed electrode when the lower electrode was grounded. In such conditions, a maximum electric wind velocity of 5.5 m/s was obtained at about 0.5 mm from the wall, for $X = 10$ mm (10 mm downstream the right side of the high voltage electrode, see Figure 2).

The incidence of the airfoil (called α) was modified from 12 to 17°. At $\alpha = 12°$, the airflow begins to separate at the trailing edge of the airfoil. The separation point moved upstream when α increased. At each angle, the effect of each actuator was investigated, one after one, from actuator #1 to #7.

Figure 3 shows the velocity vector fields and the streamlines for an angle of attack equal to 15°, when the actuation is off and on. In the absence of actuation (a), the airflow is naturally detached above the airfoil. Here, the separation point is located at the middle of the airfoil (45% of chord), close to actuator #3. When actuator #6 is switch on (b), the airflow is partially reattached. The separation point is displaced downstream, at 55% of the chord (this has been accurately estimated with the help of a zoomed PIV view at the wall). If actuator #3 is switched on, the airflow is then fully reattached above the airfoil (Figure 3c).

Figure 4a presents velocity profiles, deduced from PIV measurements, at the airfoil trailing edge. On one hand, it clearly shows that actuator #6, placed near the trailing edge, has a low effect on the airflow. On the other hand, actuator #3, placed closed to the separation point, allows to fully reattach the airflow. Therefore, it seems that it is necessary to act near the separation point to obtain a significant effect.

In order to confirm this result, we have determined what is the minimum electrical power consumption of each actuator, by adjusting the applied high voltage, to fully reattach the airflow, for α equal to 14 and 15 degrees (Figure 4b). For instance, at $\alpha = 15°$, one needs about 148 W to fully reattach the airflow with actuator #5 when only 78 W are necessary to induce the same effect with actuator #3. Moreover, it is not possible to reattach the airflow with actuators #6 and #7, because the actuation is placed too far downstream of the separation. This demonstrates that one must act close to the separation point (just upstream or downstream) to be the most efficient.

3.2 Effects of the Unsteady Actuation

Unsteady actuation consists in cycling the AC high voltage switch off and switch on with a period T, which is highly greater than the period of the AC high voltage [2]. The unsteady period T is based according to the Strouhal frequency on the flow. The forcing frequency for the unsteady disturbances is usually optimum when the Strouhal number is near unity. With the present aerodynamic conditions, the frequency based on $S_t = 1$ is equal to 6 Hz.

Here, we present a typical example. The airfoil incidence is equal to 16° and actuator #2 is employed. Figure 5a shows the effect of unsteady actuation on the airflow at 6 Hz. In the absence of actuation, the airflow is naturally detached above the airfoil. When the discharge is switch on using the steady actuation, the airflow is fully reattached with an electrical power consumption of 78 W. When applying the unsteady actuation, the airflow is also reattached. In this case, the discharge is first acting with a 50% duty cycle. Therefore the consumption of the plasma actuator is divided by two (39 W). Moreover, at 25%, corresponding to a consumption of 19.5 W, the airflow is still reattached. However, at 10%, the injected energy is not sufficient to reattach the airflow. This shows that the power consumption may be significantly reduced when the action is non-stationary.

Figure 5b presents velocity profiles at different frequencies, for a duty cycle of 50%. Because the injected energy is sufficient in all cases, one cannot see the frequency effect. This point will have to be more accurately studied in the future.

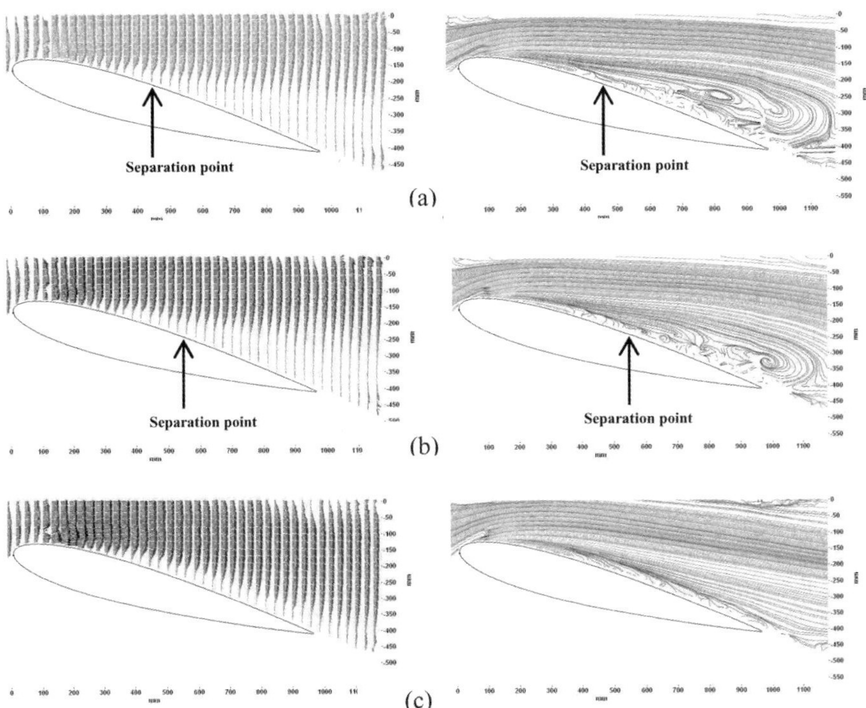

Fig. 3 Velocity vector fields and streamlines of the flow without (a) and with actuation at positions #6 (b) and #3 (c).

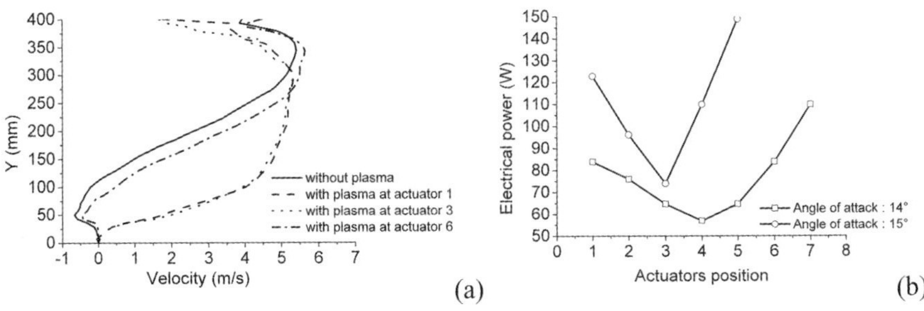

Fig. 4 Velocity profiles for different actuator positions (a) and minimum electric power consumption versus actuator position to fully reattach the airflow (b).

4 Conclusion

This study showed the ability of a dielectric barrier discharge plasma actuator to control an airflow above an NACA 0015 airfoil, at low velocity ($Re = 0.4 \times 10^6$). The results presented in this paper showed that this type of action is able to fully

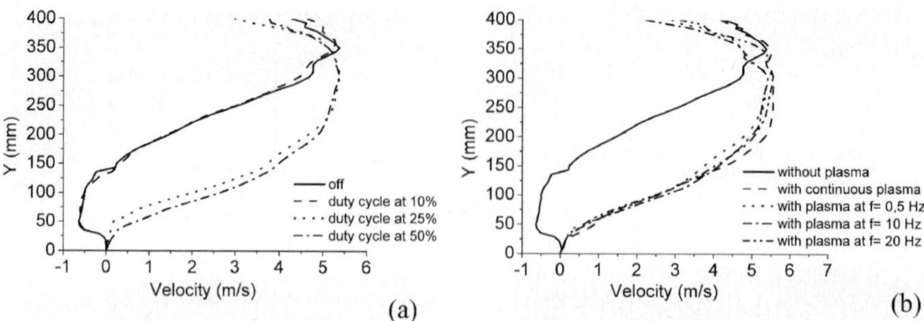

Fig. 5 Velocity profiles at the trailing edge for unsteady actuation at different duty cycles (a), and at different frequencies (b).

reattach an airflow naturally separated, for angles of incidence up to 16 degrees. Moreover, it was clearly demonstrated that the electrical power consumption may be greatly reduced by acting close to the separation point, and by using an non-stationary actuation with a minimum duty cycle.

Other experiments have also shown that the plasma actuator was able to detach a naturally attached airflow, or to move the separation point upstream, for angles between 9 and 14°, with a counter-flow ionic wind (the ionic wind and the free stream are in opposite directions).

Then, experiments showed that it was more difficult to displace the separation point towards the leading edge (upstream) than towards the trailing edge (downstream).

Acknowledgment

The authors acknowledge the support of AIRBUS (contract No. D05028043), under the scientific direction of Dr. Stephen Rolston, AIRBUS UK.

References

1. J.R. Roth, D.M. Sherman and S.P. Wilkinson, Boundary layer flow control with a one atmosphere uniform glow discharge surface plasma, AIAA Paper 98-0328, 1998.
2. T.C. Corke and C. He, Plasma flaps and slats: An application of weakly-ionized plasma actuator, AIAA Paper 2004-2127, 2004.
3. J.R. Roth, Aerodynamic flow acceleration using paraelectric and peristaltic electrohydrodynamic effect of a one atmosphere uniform glow discharge plasma, *Phys. Plasmas* **10**, 2003, 2117–2226.
4. R. Sosa, E. Moreau, G. Touchard and G. Artana, Stall control at high angle of attack with periodically excited EHD actuators, AIAA Paper 2004-2738, Portland, USA, 2004.

5. R. Sosa, G. Artana, E. Moreau and G. Touchard, Stall control at high angle of attack with plasma sheet actuators, *Exp. in Fluids* **42**, 2007, 143–167.
6. D.F. Opaits, D.V. Roupassov, S.M. Starikovskaia, A.Yu. Starikovski, I.N. Zavialov and S.G. Saddoughi, Plasma control of boundary layer using low-temperature non-equilibrium plasma of gas discharge, AIAA Paper 2005-1180, 2005.
7. E. Moreau, Airflow control by non thermal plasma actuators, *J. Phys. D: Appl. Phys.* **40**(3), 2007, 605–636.
8. J. Pons, E. Moreau and G. Touchard, Electrical and aerodynamic characteristic of atmospheric pressure barrier discharge in ambient air, in *Proc. ISNTPT2004*, Florida, USA, May 2004, pp. 307–310.
9. M. Forte, J. Jolibois, F. Baudoin, E. Moreau, G. Touchard and M. Cazalens, Optimization of a Dielectric Barrier Discharge actuator and non-stationary measurements of the induced flow velocity – Application to airflow control, AIAA Paper 2006-2863, 2006.

Dynamic Surface Pressure Based Estimation for Flow Control

Lawrence Ukeiley[1], Nathan Murray[2], Qi Song[3] and Louis Cattafesta[3]

[1] *Department of the Mechanical and Aerospace Engineering, Research and Engineering Education Facility, University of Florida, Shalimar, FL 32579, U.S.A.; E-mail: ukeiley@ufl.edu*

[2] *Department of Engineering and Physics, Harding University, Searcy, AR 72143, U.S.A.*

[3] *Department of Mechanical and Aerospace Engineering, University of Florida, Gainesville, FL 32611-6550, U.S.A.*

Abstract. The need for adaptive control methodologies which involve realistically obtainable information is driving the direction of active flow control research. In this work, time resolved estimates of the velocity field using a mean-square estimation procedure with the unsteady surface pressure as the condition are presented which highlight the relationships between these quantities. Understanding the causal relationships between the velocity field and the surface pressure, an adaptive control strategy solely based on surface pressure can be developed. To this end dynamic stochastic estimation and system identification approaches are shown to accurately predict future surface pressures based on their time histories which can then be used to form the basis of a closed loop control strategy.

Key words: Stochastic estimation, system identification, cavity flows.

1 Introduction

The need for surface measurement based controllers is readily apparent when one considers the type of information that is necessary for a realistic flow control application. Recently, much research has been directed towards the ability to predict or estimate off-body properties from quantities on the surface with the goal to develop control methodologies that only require surface information. In cavities this has been primarily through fluctuating surface pressure. In this study, highlights of an ongoing program are presented where surface pressure has been used to estimate the time-resolved velocity field and is now being transitioned so that it can be used in a dynamic fashion for adaptive control schemes.

The present results are from flow over a two-dimensional (length-to-depth ratio 6) cavity with a free stream Mach number of nominally 0.6. The velocity data was acquired through a particle image velocimetry technique which was sampled synchronously with 14 pressure transducers mounted throughout the cavity. Figure 1

J.F. Morrison et al. (eds.), IUTAM Symposium on Flow Control and MEMS, 183–189.
© 2008 *Springer. Printed in the Netherlands.*

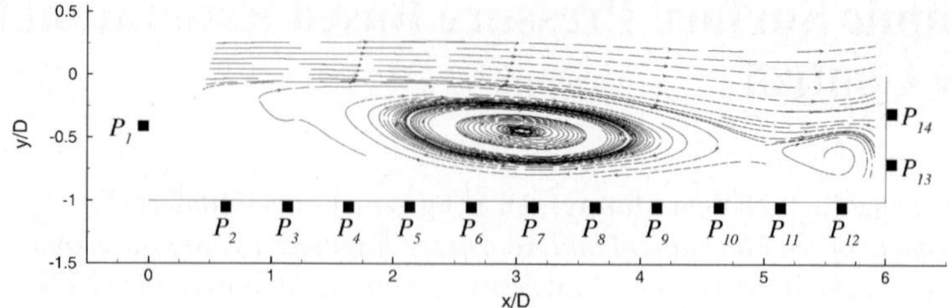

Fig. 1 Mean flow streamlines and surface pressure measurement locations.

shows the mean flow streamlines and the locations of the 14 pressure sensors. For details of the experiments the reader is referred to [7, 8].

2 Surface Pressure-Based Velocity Estimation

Over the past several years there have been efforts from many research groups to estimate flow field quantities from surface pressure measurements. In general this work can be lumped into two categories; those that are designed to investigate the flow physics and those designed for developing models suitable for flow control. The work of Murray and Ukeiley [8] and Naguib et al. [9] have been among those that have concentrated on the former, while those discussed in [2, 11] have concentrated on the latter.

The effort discussed here concentrates on applying the Modified Quadratic Stochastic Estimation (mQSE) to subsonic flow over a two-dimensional cavity. The term modified is included here as this techniques uses the proper orthogonal decomposition [5] in combination with stochastic estimation [1]. The time resolved fluctuating surface pressure is used to estimate the velocity POD expansion coefficients (a^n). This process can be represented by a series expansion as

$$\tilde{a}(t) = \mathbf{L}P_i(t) + \mathbf{Q}P_i(t)P_j(t), \tag{1}$$

where a tilde indicates an estimated quantity. Minimizing the squared error between the estimated and true coefficients via the solution of a matrix equation involving correlations between the pressure measurements and the POD coefficients provides the linear L and quadratic Q coefficients [8]. The need to include the quadratic term in Equation (1) was demonstrated in [6] where numerical simulations were used to validate the estimation procedure. The estimated POD expansion coefficients are used with previously known POD eigenfunctions to obtain the estimate of the velocity field,

$$\tilde{u}_i^n(\mathbf{x},t) = \sum_n \tilde{a}^n(t)\phi_i^n(\mathbf{x}). \tag{2}$$

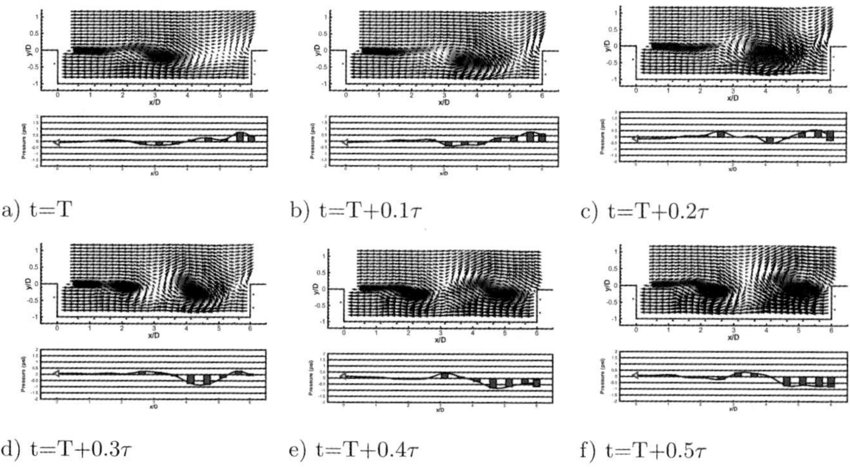

a) t=T   b) t=T+0.1τ c) t=T+0.2τ

d) t=T+0.3τ e) t=T+0.4τ f) t=T+0.5τ

Fig. 2 Estimated velocity fields MQSE application to Mach 0.6 cavity flow.

Figure 2 displays the velocity field estimated using this approach. In each of the sub-plots the top figure displays the estimated velocity vectors in a convective frame of reference with contours of the spanwise vorticity in the background, while the bottom plots display the measured pressures. In order to reconstruct the estimated velocity field, the series in Equation (2) was truncated so nearly 80% of the kinetic energy was utilized. The six plots represent points throughout the evolution of one cycle observed in the cavity and are spaced by $0.1 * \tau$ where $\tau = 0.57L/U$. The sequence represented here starts with a vortex near the center of the cavity and is just after one has been ejected over the aft wall. The free stream flow rushes over the vortex in the aft part of the cavity and impacts the aft wall. This high streamwise velocity fluid is then turned back upstream, further driving the vortex. The pressure rise associated with this action is what pushes the vortex out over the aft deck of the cavity as the cycle goes from (f) back to (a). Then, with this vorticity ejected, the mean flow dips into the cavity bringing along with it the next vortex. Although not included here there are other interesting features observed in the time resolved estimated velocity fields. One particularly noteworthy feature is what appears to be a switching between different dominant Rossiter modes [4]. In these estimated velocity snapshots one can observe the transition between two vortical structures across the cavity to three. Other interesting features observed in the estimated velocity field involve the interaction of the turbulent structures with the aft wall. All of the scenarios of how the vortical structures interact with the aft wall described in [10] were observed. The exact details of the surface pressures for these various events are still under study and will be reported in forthcoming articles.

The work discussed above, and the cited references, have elucidated the relationships between the fluctuating surface pressure and the velocity field, however this, estimating the velocity field, is not necessary for active control applications [11,12].

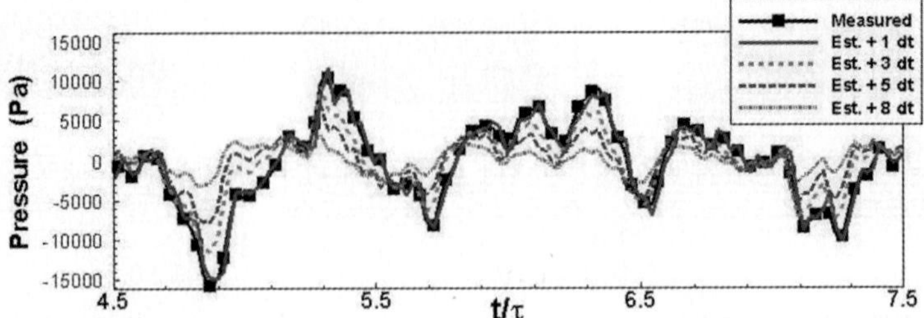

Fig. 3 Dynamic linear stochastic estimation of aft-wall presure.

Essentially, all that is required is to demonstrate the causal relationship between the surface quantity and the flow field, as has been shown above, then one can use the surface pressure alone.

3 Dynamic Stochastic Estimation

Here we demonstrate an application where the stochastic estimation is formulated so that it can be used to predict future pressures based on previous values. This formulation comes from adapting the work of Adrian [1] to formulate stochastic estimation in time instead of space. Through minimizing the squared error of the estimate and expanding in a power series the estimated pressure can be represented by

$$\tilde{P}(t) = A_1 P(t - 1 - S) + A_2 P(t - 2 - S) + \cdots + A_n P(t - n - S). \tag{3}$$

Here the A's are the linear stochastic estimate (LSE) coefficients and are found by inverting a matrix of time-lagged pressure correlations. Equation (3) also has a shift of S which has been inserted to allow for the estimation further out in time (i.e., the prediction horizon). In this formulation, the coefficients are determined *a priori* from previous measurements. It should be noted that a linear estimate is now being used as opposed to the quadratic previously discussed. This is based on the fact that in the application here there is no need to account for the velocity-pressure relationships. This procedure was performed for the aft-wall sensor (P_{14}) in the cavity. Figure 3 displays the measured surface pressure for increasing values of S which range from the smallest possible $1dt$ to approximately one temporal integral scale $8dt$, where the temporal integral scale was estimated from the first zero crossing in the auto-correlation. From this plot it is clear that the frequency of the oscillations are represented quite well even with a prediction horizon approaching the temporal integral scale. However it is clear that the amplitude of the fluctuations is greatly reduced with increasing prediction horizons. The error in the standard deviation of the surface pressure was approximately 2% for the $1dt$ prediction and increased to 72% for the $8dt$ prediction.

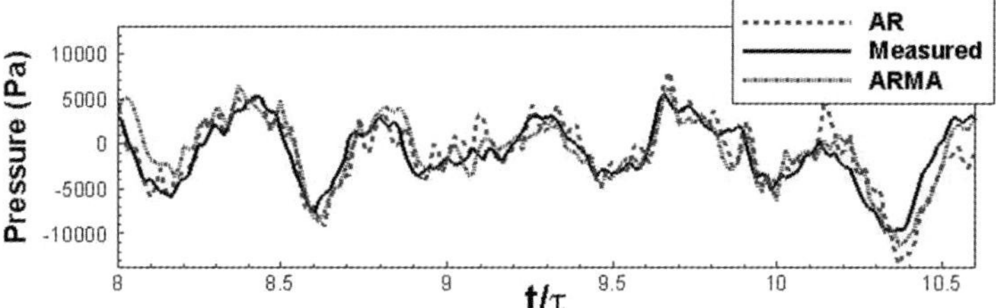

Fig. 4 System ID prediction of aft-wall pressure.

4 ARMA Model

The use of system identification approaches with Auto-Regressive Moving Average (ARMA) filters have long been used in linear control schemes [3]. In the present application, it will be used in a manner similar to the LSE discussed in the previous section to predict future surface pressure measurements based on past surface pressure time histories at one or more locations.

The form of the ARMA model that will be used here can be represented as;

$$\tilde{P}(t) = \alpha_1 P(t-1) + \alpha_2 P(t-2) + \cdots + \alpha_n P(t-n)$$
$$+ \beta_0 \mathbf{u}(t) + \beta_1 \mathbf{u}(t-1) \cdots + \beta_n \mathbf{u}(t-n). \tag{4}$$

Here $\tilde{P}(t)$ is the estimated surface pressure, α and β are auto and moving average estimation coefficients, respectively. $\mathbf{u}$ are other inputs, i.e., neighboring pressure measurements, which are used to improve the quality of the estimation. The estimation coefficients can be determined in one of two methods. First, in a manner analogous to that described above for the dynamic LSE, an off-line least-squares approach. Second, a gradient-descent based algorithm, such as least-mean-squares [13], is used to update the coefficients each time step.

The goal here is predict the surface pressure at future times, hence we will drop the β_0 term from the moving average part of the equations and insert a time lag S. This results in the auto regressive part of the model having an analogous form to Equation (3). The differences arise from coefficients for the LSE approach were calculated using a matrix inversion and in the ARMA model the coefficients are recursively updated. This is important in terms of adaptive flow control applications where it will be important to update these coefficients as the flow state evoles. Results from an application of the ARMA for the prediction of the aft-wall sensor (P_{14}) using a predictive horizon of $8dt$, the same as the dynamic LSE is presented in Figure 4. This figure includes the time history for the measured pressure, the estimated pressure using only information from its own time history (dashed line) and the estimated pressure using its own time history plus that of P_5, P_{10} and P_{13} (dot-dashed line). The AR case, where only the information from P_{14} is used tracks both the amplitude and low frequency quite well even for this long predictive horizon.

This scenario resulted in a 22% difference in the standard deviation and directly compares to the LSE application discussed above, demonstrating that recursively updating the estimation coefficients significantly improves the quality of the estimation. Examination of the time history from the ARMA application shows that using the additional information increases the accuracy of the amplitude of the estimate to approximately 4% error in the standard deviation.

5 Summary/Conclusions

Steps in developing a surface pressure based estimation procedure for adaptive control of fluctuating surface pressure levels were presented. This first step involved developing an understanding of how features of the velocity field relate to the surface pressure. This was done by using the mQSE procedure. This has shown the source of the largest amplitude pressures on the aft wall to be associated with the events just after the large scale structures have been ejected out of the cavity, and the outer flow impacts the aft wall. With the understanding of relationships between the velocity and surface pressure, closed loop flow control strategies for altering the surface pressure fluctuations could be developed. To this end we have presented two different methodologies for predicting the surface pressure based on its previous time histories. It was demonstrated that under certain constraints the more general ARMA model has the same form as the LSE, although its estimation coefficients are determined through a recursively updating process. Both procedures showed that, using predictive horizons on the order of a temporal integral scale, the low frequency oscillations were accurately estimated. However, the ARMA model was better at reproducing the amplitude of the pressure fluctuations.

References

1. Adrian, R., Conditional eddies in isotropic turbulence, *Phys. Fluids* **22**, 1979, 2065–2070.
2. Glauser, M., Higuchi, H., Ausseur, J. and Pinier, J., Feedback control of separated flows, AIAA2004-2521, 2004.
3. Juang, J., *Applied System Identification*, Prentice-Hall, 1994.
4. Kegerise, M., Spina, E., Garg, S. and Cattafesta, C, Mode-switching and nonlinear effects in compressible flow over a cavity, *Phys. Fluids* **16**(3), 2004.
5. Lumley, J.L., The structure of inhomogeneous turbulent flow, in *Atmospheric Turbulence and Radio Wave Propatation*, A.M. Yaglom and V.I. Tatarsky (Eds.), Nauka, Moscow, 1967.
6. Murray, N. and Ukeiley, L., Estimation of the flow field from surface pressure measurements in an open cavity, *AIAA J.* **41**(5), 2003.
7. Murray, N. and Ukeiley, L., An application of gappy POD to PIV data, *Exp. Fluids* **42**(1), 2007, 79–91.
8. Murray, N., Flow field dynamics in subsonic cavity flows, Ph.D. Thesis, University of Mississippi, 2006.
9. Naguib, A.M., Wark, C.E., and Juckenhfel, O., Stochastic estimation and flow sources associated with surface pressure events in a turbulent boundary layer, *Phys. Fluids* **13**(9), 2001.

10. Rockwell, D. and Knisley, C., The organized nature of flow impingement upon a corner, *J. Fluid Mech.* **93**(3), 1979.

11. Rowley C. and Williams, D., Dynamics and control of high Reynolds number flow over open cavities, *Ann. Rev. Fluid Mech.*, 2006.

12. Tian, Y., Song, S., and Cattafesta, L., Adaptive feedback control of flow separation, AIAA2006-3016, 2006.

13. Widrow, B. and Stearns, S., *Adaptive Signal Processing*, Prentice-Hall, 1985.

The Control of Laminar Separation Bubbles Using High- and Low-Amplitude Forcing

Mark Phil Simens[1] and Javier Jiménez[1,2]

*School of Aeronautics, Universidad Politécnica de Madrid, 28040 Madrid, Spain;
E-mail: mark@torroja.dmt.upm.es*
[2]*Center for Turbulence Research, Stanford University, Stanford, CA 94305-3035,
U.S.A.*

Abstract. Two-dimensional simulations are used to demonstrate the existence of two different amplitude regimes to control laminar separation bubbles with periodic zero-mass-flux wall jets. One is based primarily on a shear-layer instability found using low-amplitude forcing. The minimum bubble length is obtained for a Strouhal number approximately equal to 0.018, based on a properly defined momentum thickness. Higher forcing is found to create large vortices, which are responsible for very effective control. A relation is presented between the forcing parameters and the size of the vortices. These estimates are then used to explain the range of effective frequencies to control the separation bubble.

Key words: Strong forcing, Kelvin–Helmholtz, instability, vortex shedding, separated flow, boundary layers, control.

1 Introduction

Separation is an important problem in aviation and in turbo-machinery, which can in part be remedied using passive or active control. A lot of work has been done on active control of separation bubbles and a good overview can be found in [8]. Two other works should be highlighted. It is postulated in [5] that the forcing excites the shear layer formed by the separation bubble in such a way that big shed vortices are formed. It is found in [2] that, at high forcing amplitudes and at certain frequencies, the bubble is absent.

Here we describe the active control of laminar two-dimensional separation bubbles using periodic-zero-mass-flux wall jets. Results for a wide range of pressure gradients, forcing frequencies and forcing amplitudes will be discussed. Finally, a simple model is proposed to explain the results for strong forcing (10% of the free stream velocity).

J.F. Morrison et al. (eds.), IUTAM Symposium on Flow Control and MEMS, 191–197.
© 2008 Springer. Printed in the Netherlands.

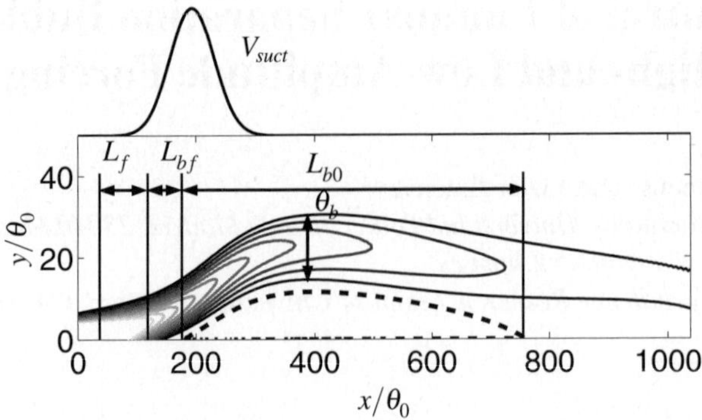

Fig. 1 The numerical domain used throughout the study. $----$: initial separation bubble. ——— : ω_z contours given by $-0.12:0.03:-0.03\ U_\infty/\theta_0$.

1.1 Numerical Techniques and Unperturbed Flows

The Navier–Stokes equations are discretized using second-order central finite difference schemes on a staggered grid. Time integration is done using a low-storage third-order Runge–Kutta scheme [7] for the convective terms, while the viscous terms are treated implicitly. A fractional-step method based on [9] is employed to assure the conservation of mass, and to obtain the pressure in an efficient manner.

The numerical simulations have been done in two dimensions over a flat plate (Figure 1). At the inlet, the streamwise component, u, of the Blasius profile is imposed, while the component perpendicular to the wall, v, is imposed to be zero. No-slip boundary conditions are imposed at the bottom wall, while impermeability is assumed for v along most of the bottom wall. The exception is a small segment in which a periodic zero-mass-flux forcing is imposed, given by $v(x,y = 0,t) = V_f \sin(2\pi f t)$. The u velocity at the upper wall is modeled using a vorticity-free condition $\partial_y u(x,y = L_y,t) = \partial_x v(x,y = L_y,t)$. Adverse pressure gradients are obtained by imposing a suction profile $V_{suct}(x,y = L_y,t) = a_s \exp(-b_s(x - c_s)^2)$, similar to that used in [1].

In the case of arbitrary mass fluxes over the boundaries, a modification of the standard procedure was necessary to guarantee global mass conservation. It consists in expressing the total solution for the correction of the pressure $\phi_{tot} = \phi_1 + C(x^2 - y^2)$ as the sum of the solution ϕ_1 of the Poisson equation, and a function satisfying the Laplace equation $C(x^2 - y^2)$. The Poisson equation for ϕ_1, singular due to Neumann boundary conditions, is solved with the aid of cosine Fourier decomposition. A solution can only be obtained when the zero mode $k = 0$, $\widehat{\phi}_1(k = 0, L_y) = 0$ is imposed. However, if the total mass flow through the boundaries is not zero, the derivative $\partial \widehat{\phi}(k = 0, L_y)/\partial n \neq 0$. The constant C is used to force $\partial \widehat{\phi}(k = 0, L_y)/\partial n = 0$, creating a pressure gradient that assures the correct mass flow, allowing $\partial \widehat{\phi}/\partial n \neq 0$ at the exit. The outflow boundary conditions then become $\partial_t \mathbf{u}_{x=L_x,y,t} + U_\infty \partial_x \mathbf{u}_{x=L_x,y,t} + \partial_x \phi_{tot}|_{x=L_x,y,t} = 0$.

The Reynolds number of most simulations, based on the inlet momentum thickness θ_0, is $Re_{\theta_0} = 30$, while at separation it is $Re_{\theta_s} \approx 110$. For comparison a few other cases were run at higher Reynolds numbers. The length of the numerical domain, adimensionalized with θ_0, is $L_x \times L_y = 1047 \times 133$, and is discretized using $N_x \times N_y = 256 \times 128$ points. The time step is determined using a constant CFL $= 0.6$. At $Re_{\theta_0} = 30$, all the separation bubbles were stable and reattached due to viscous diffusion.

2 Low-Amplitude Forcing

The low-amplitude forcing is expected to trigger the instability of the shear layer formed by separation. This necessarily would imply that the most unstable frequency lies around $St_\theta = f\theta_b/U_\infty = 0.018$ [11], which has been obtained for separated flow in [6]. Here the momentum thickness θ_b has been measured at the position of maximum negative velocity inside the bubble. It was proposed in [10] that triggering shear layer instabilities could be useful in reducing the separation bubble, but they did not scale their results with a suitable momentum thickness. Instead, they used the length of the initial separation bubble, L_b, if the flow reattaches before the trailing edge; in the case where the flow stalls, the length of the airfoil is used. This results in a most unstable frequency of around $St_F = fL_b/U_\infty \approx 1$.

In Figure 2, the most effective Strouhal number as a function of the minimum bubble length are shown. The two different scalings based on θ_b (Figure 2a), and L_b (Figure 2b) have been used. One observes that the scaling based on θ_b gives much better collapse than the one based on the initial length of the separation bubble. The results, when scaled with θ_b, tend to an asymptotic value for long bubbles. The deviation from 0.018 is presumably due to the presence of the wall, and to the shear layer not being parallel.

The physical reasoning behind $St_F \approx 1$ was discussed in [8], still in terms of the shear layer instability. Recently however, the same group [3] related $St_F \approx 1$ to stalled flow in which the important length scales are the wake width and the wavelength of the periodic vortex roll-up. This is different from the flow under consideration here. To explain the different scaling laws obtained by both groups, we propose to make a distinction between stalled flows, and closed separation bubbles. In the latter, the most effective frequency would be $St_\theta \approx 0.018$, and in the former $St_F \approx 1$ may be appropriate.

3 High-Amplitude Forcing

High-amplitude forcing creates an instantaneous separation bubble that rolls-up to form a large vortex. This vortex is very effective in controlling the separation, because it forms upstream of the bubble. The roll-up is difficult to describe in detail,

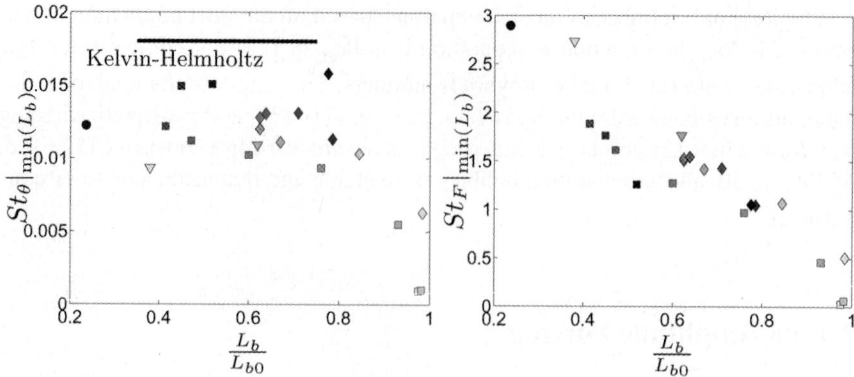

Fig. 2 (a) Scaling of the optimum frequency with θ_b and (b) with L_{b0}. With $c_s/L_x \approx 0.19$: ■ : $Re_{\theta_0} = 30$; ▼ : $Re_{\theta_0} = 100$, with $L_x/\theta_0 \times L_y/\theta_0 \approx 785 \times 84$ and $N_x = 512$, $N_y = 128$; ● : $Re_{\theta_0} = 21$, with $L_x/\theta_0 \times L_y/\theta_0 \approx 2094 \times 133$ and $N_x = 512$, $N_y = 128$ ◆ : $Re_{\theta_0} = 30$, $c_s/L_x \approx 0.38$. In all cases, $V_f/U_\infty = 0.001$. Gray indicates short bubbles, black long bubbles.

but the principal idea is that the vortex size is determined by the mass entrained. The characteristic thickness at forcing is approximated by θ_b. The area of the vortex is given by $\pi r^2 = t_s U_\infty \theta_b$, where $t_s U_\infty \theta_b$ is the fluid the vortex entrains during a time t_s. A second length scale D, defined to first order as the shear layer lift-up V_f/f, determines the shedding frequency, $f_s D/U_\infty \sim 0.2$, assuming that the process is similar to the vortex shedding behind an obstacle. Substituting $t_s = 1/f_s$ in r gives

$$\frac{r}{\theta_b} \sim \sqrt{5\frac{V_f}{\pi f \theta_b}}. \tag{1}$$

A different estimation for the radius was derived in [4] by assuming that the vortex entrains all the mass added during the blowing phase of the forcing, but we now believe that the present estimate is more accurate.

The radius in Equation (1) is used to approximate the circulation $\Gamma = \int \omega dA$, with constant $\omega = U_\infty/\theta_b$ and $A = \pi r^2$. This, at time t_s, results in

$$\frac{\Gamma}{U_\infty \theta_b} = \frac{5V_f}{f\theta_b}. \tag{2}$$

These estimates agree reasonably well with the results from numerical experiments (not shown), especially considering the crude approximations involved.

3.0.1 The Important Parameters for High-Amplitude Control

The correct estimation of the vortices, as a function of the forcing parameters, is only useful if it can be related to the effectiveness in controlling the separation bubble.

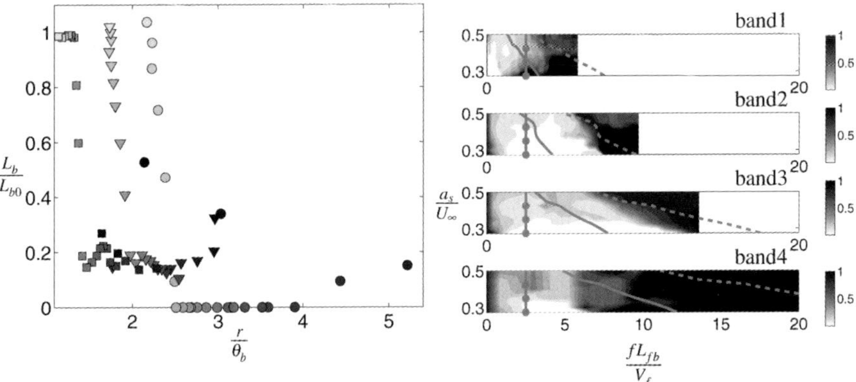

Fig. 3 (a) Relation between measured vortex radius and their effectiveness expressed in bubble reduction L_b/L_{b0}. Gray (high frequency) to black (low frequency) shading is used to distinguish between frequencies. $\circ$: $as/U_\infty = 0.3$, $\square$: $as/U_\infty = 0.375$, $\triangle$: $as/U_\infty = 0.45$. (b) The reduction in bubble length as a function of a measure of the pressure gradient and the adimensional frequency. The bubble length is indicated by shaded contours with levels given by (0:0.1:1). —•— : $fL_{fb}/V_f = 2.5$; - - - - : $fL_{fb}/V_f = 20L_{fb}/25\pi\theta_b$. —— : $0.018\frac{U_\infty L_{fb}}{V_f\theta_b}$. From top to bottom: $L_{fb}/\theta_0 \approx 62, 103, 144, 226$.

We will make two assumptions to obtain this relation. The first one is that the vortex should form before it reaches the separation bubble to be effective. The time that the forming vortex needs to get to the separation bubble is L_{fb}/U_c, where $U_c \approx U_\infty/2$ is the convective velocity of the vortex. This should be larger than the time t_s that the vortex needs to form, giving

$$\frac{L_{fb}}{U_c} \geq \frac{5V_f}{fU_\infty}, \quad \text{or} \quad \frac{fL_{fb}}{V_f} \geq 5\frac{U_c}{U_\infty} \sim 2.5. \tag{3}$$

The second assumption is that the radius of the vortex r/θ_b is larger than some number, which is fitted empirically to 2.5 from Figure 3a. Only then it can effectively mix fluid from the separation zone with the free stream. Using the estimation (1), this gives

$$\frac{f\theta_b}{V_f} \leq \frac{25\pi}{20}, \quad \text{and after multiplying by} \quad \frac{L_{fb}}{\theta_b}, \quad \text{gives} \quad \frac{fL_{fb}}{V_f} \leq \frac{20L_{fb}}{25\pi\theta_b}. \tag{4}$$

The assumption leading to Equation (4) is tested in Figure 3a, where it is seen that $2.5 \leq r/\theta_b \leq 3.5$ is required for maximum efficiency.

Figure 3b tests the limits in Equations (3) and (4), which are given by the gray circles and the dashed lines. In general the agreement is reasonable considering the approximations made. The limits effectively mark the most efficient forcing frequencies.

The failure for higher reduced frequencies, especially in band 4, is probably due to viscous effects. At higher frequencies the radius of the vortices gets smaller, and although they might be strong when formed, they get weaker due to viscous diffusion before they reach the separation bubble. The positive results obtained for band one completely outside the limits of (3) and (4) is probably related to the positive effect of suction.

4 Conclusions

Two ways to force a separated laminar boundary layer are examined. For weak intensities, the effective forcing has been shown to depend on the instability of the shear layer. The most unstable frequency scales well when adimensionalized with the appropriate momentum thickness. Adimensionalization based on the initial length of the separation bubble is less successful.

The second possibility is to create large vortices by strong forcing at a position upstream of the separation bubble. The large vortex enhances mixing along the whole APG region. The instability invoked by the weak forcing experiments causes vortices to be formed downstream of the separation point, and is therefore less effective. A rough model for the process of vortex formation gave results in satisfactory agreement with the numerical experiments. They relate the forcing parameters with the vortex size, and this in turn gives the most effective frequency range. The range also includes the Kelvin–Helmholtz instability, but this seems fortuitous, as the model is not related to an instability mechanism of the initial unperturbed separation bubble. Large vortices were also observed in [2,5], but our explanation of their formation is different from theirs. Four forcing regimes are proposed in [5], and the one related to the creation of large vortices is a shedding instability of the separation bubble, analogous with Kármán vortex shedding. Here, we also invoked a shedding type instability, but related to the instantaneous separation bubble created by the forcing.

This forcing, although more effective than triggering the instability, is more energy demanding, as higher values for the amplitude are required.

References

1. Alam, M., Sandham, N.D.: Direct numerical simulation of short laminar separation bubbles with turbulent reattachment. *J. Fluid Mech.* **410** (2000) 1–28.
2. Kiya, M., Shimizu, M., Mochizuki, O.: Sinusoidal forcing of a turbulent separation bubble. *J. Fluid Mech.* **342** (1997) 119–139.
3. Darabi, A., Wygnanski, I.J.: Active management of naturally separated flow over a solid surface. Part 1. The forced reattachment process. *J. Fluid Mech.* **510** (2004) 105–129.
4. Simens, M.P., Jiménez, J.: Alternatives to Kelvin–Helmholtz instabilities to control separation bubbles. ASME Paper GT2006-90670 (2006).

5. Sigurdson, L.W.: The structure and control of a turbulent reattaching flow. *J. Fluid Mech.* **298** (1995) 139–165.

6. Huppertz, A., Fernholz, H-H.: Active control of the turbulent flow over a swept fence. *Eur. J. of Mech. B/Fluids* **21** (2002) 429–446.

7. Spalart, P.R., Moser, R.D., Rogers, M.M.: Spectral methods for the Navier–Stokes equations with one infinite and two periodic directions. *J. Comp. Phys.* **96** (1991) 297–324.

8. Greenblatt, D., Wygnanski, I.J.: The control of flow separation by periodic excitation. *Progr. Aerospace Sci.* **36** (2000) 487–545.

9. Orlandi, P.: *Fluid Flow Phenomena*, Kluwer Academic Publisher, Dordrecht (2000).

10. Seifert, A., Pack, L.G.: Active flow separation control on wall-mounted hum at high Reynolds numbers. *AIAA J.* **40** (2002) 1363–1372.

11. Michalke, A.: On spatially growing disturbances in an inviscid shear layer. *J. Fluid Mech.* **23** (1965) 521–544.

Control of Subsonic Flows with High Voltage Discharges

Pierre Magnier[1], BinJie Dong[2], Dunpin Hong[2], Annie Leroy-Chesneau[1] and Jacques Hureau[1]

[1]*Laboratoire de Mécanique et d'Energetique, 8 rue Leonard de Vinci, 45072 Orleans Cedex 02, France; E-mail: pierre.magnier@univ-orleans.fr*
[2]*Groupe de Recherche sur l'Energetique des Milieux Ionises, UMR 6606, CNRS/University of Orleans, 14 rue d'Issoudun, BP 6744, 45072 Orleans Cedex 02, France*

Abstract. In this paper, a DC surface corona discharge established on a rounded edge of a dielectric material was studied in atmospheric air. The flow induced by this actuator was measured and experiments on a NACA 0015 were performed in a subsonic wind tunnel. These measurements showed that this discharge modified the fully detached flow on the airfoil up to $Re = 267\,000$ and $17.5°$ of angle of attack.

Key words: Flow control, separation, corona discharge, EAD actuator, ionic wind.

1 Introduction

High voltage discharges established between two electrodes will ionize neutrally-charged gasses. Ions and electrons are generated by collisions in the plasma, and the ion migration results in the movement of neutral molecules by transfer of momentum. This induced flow is generally called "ionic wind".

The addition of energy and mass injection have been used for decades to modify subsonic flows, since flow control perspectives are very important in industrial applications. Electroaerodynamic (EAD) actuators have been studied in this way since the mid-1990s by using the flow induced by the plasma. The electrical energy is directly converted into mechanical energy, without additional moving mechanical parts. The ability to control a flow with electric discharges has been shown by many experiments on a cylinder [1] and on wing profiles. Two kinds of EAD actuators have mainly been developped for this application: the AC surface dielectric barrier discharge type [2] and the DC surface corona discharge type [3].

The work presented here investigated the application of a DC surface corona discharge actuator on the leading edge of a NACA 0015 wing profile (the electrical characteristics of this discharge were presented by Dong et al. [4]). First the induced flow was measured, then this airfoil was studied in a subsonic wind tunnel.

J.F. Morrison et al. (eds.), IUTAM Symposium on Flow Control and MEMS, 199–202.
© 2008 *Springer. Printed in the Netherlands.*

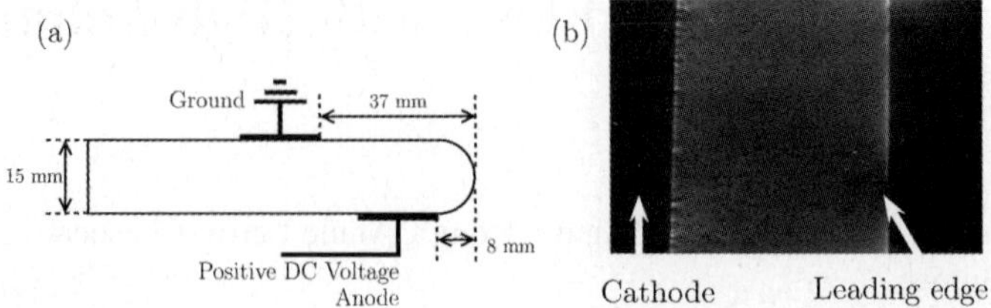

Fig. 1 (a) Schematic side view of a DC surface corona discharge actuator and (b) above photography of the plasma generated on the flat plate.

2 Experimental Setup

The EAD actuator investigated consists of a DC surface corona discharge generated between two flat copper electrodes ($170 \times 25 \times 0.035$ mm) mounted on both sides of the circular leading edge of a flat plate $200 \times 250 \times 15$ mm thick, made of polyvinyl chloride, as shown in Figure 1a. The cathode and anode are placed 37 mm and 8 mm downstream the leading edge, respectively.

The anode was connected to a positive high voltage power supply SPELLMAN SL300 (0–60 kV, 5 mA), and the cathode was connected to the ground. A stable and efficient discharge was obtained with a DC of voltage +44 kV applied to the anode. The mean discharge current measured was 1.2 mA/m. In Figure 1b, the DC surface corona discharge is evident by the presence of homogeneous light in the interelectrode space along the span.

3 Induced Flow Measurements

First, experiments were performed with the flat plate placed in a closed box without external flow. The only flow is that which is generated by the electric discharge by transfer of momentum to the neutral molecules.

The flow topology can be observed in Figure 2, where the important features are: (1) a flow separation at the anode (with a maximum mean velocity of about 1 m/s); (2) a recirculation zone on the top of the flat plate, at the leading edge level; and (3) a flow re-attachment and an acceleration at the cathode to about 0.6 m/s. The flow is parallel to the plate surface downstream the cathode.

The higher the discharge current, the closer the recirculation zone was to the flat plate leading edge. Moreover, the mean velocity magnitude increased with the discharge current. Indeed more particles are ionized with a stronger electric field, and the greater the acceleration between electrodes. Thus this stronger ion movement involves more neutral molecules, therefore the mechanical energy increases with the discharge current (until arc regime).

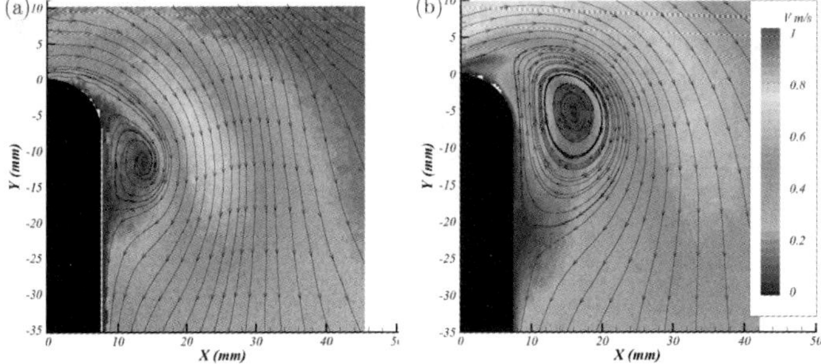

Fig. 2 Velocity fields of the induced flow (a) 0.6 mA/m and (b) 1.2 mA/m.

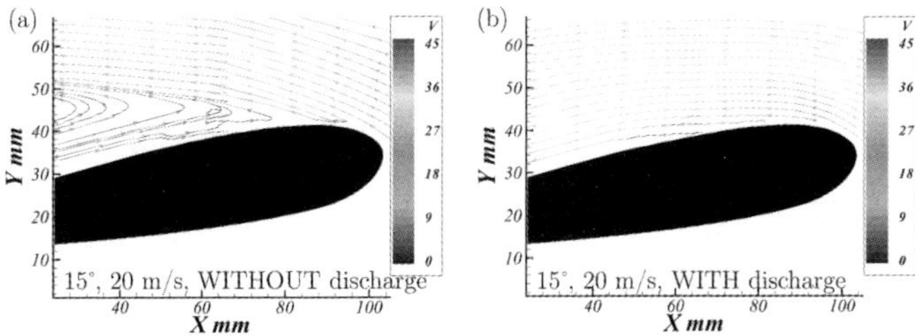

Fig. 3 Comparison of velocity streamlines with and without high voltage discharge (current of 0.8 mA/m) around the leading edge of the NACA 0015 wing profile in the wind tunnel, for an angle of attack of 15 and 20 m/s.

4 Effect of the DC Corona Discharge on the Airflow

Experiments with external subsonic airflow on a NACA 0015 wing profile were performed in a subsonic wind tunnel square test section of 50×50 cm). The effects of the high voltage discharge were observed with a PIV system, for various angles of attack α from 12.5 to 17.5°, and various inlet velocities V from 5 to 25 m/s.

The EAD actuator enables the flow separation to be reduced and can make it disappear in some cases, for example at $\alpha = 15°$ and $V = 20$ m/s (Figure 3). The actuator efficiency depends on the flow regime around the airfoil: moderate fully-separated flows were fully re-attached by the action of the electric discharge (for $\alpha = 15°$ at Re between 67 000 and 267 000) but strongly fully-separated flows were only reduced (for $\alpha = 17.5°$ at Re beetween 67 000 and 267 000).

Without the EAD actuator, the boundary layer is laminar on the leading edge, but is quite unstable and soon detaches. By action of the actuator, an induced flow is added to the boundary layer in the vicinity of the surface and leads to its transition to a turbulent boundary layer. Therefore the boundary layer is thinner, the flow is nearer to the surface and cannot separate. This effect of the electric discharge (and the induced flow) becomes more marked by a Reynolds effect: the more the Reynolds number increases, the more turbulent the reattached boundary layer and the nearer it is to the surface. The re-attached boundary layer parameters were cal-

Table 1 Re-attached boundary layer parameters for various Reynolds numbers: boundary layer thickness δ_{99}, displacement thickness δ_1, momentum thickness δ_2, $H = \delta_1/\delta_2$.

Re	67 000	200 000	267 000
δ_{99} (mm)	13.48	5.97	2.74
δ_1 (mm)	0.21	0.36	0.41
δ_2 (mm)	0.11	0.26	0.33
H	1.83	1.41	1.24

culated at 30% chord and are presented in Table 1. Results indicate that H (the ratio between the displacement thickness δ_1, momentum thickness δ_2) and the boundary layer thickness decreased as the Reynolds number increased; the boundary layer was much nearer to the surface.

5 Conclusion

The present paper describes an experimental study of an EAD actuator constisting of a DC surface corona discharge. The ionic wind induced by the discharge resulted in a detachment, a recirculation zone and a re-attachment. Although the flow velocity induced by this electric discharge was only about 1 m/s, the detached flows on a NACA 0015 wing profile were significantly modified for Reynolds numbers up to 267 000 and for angles of attack up to 17.5°. The efficiency of the electric discharge depends on the flow regime and the degree of detachment of the flow.

In future studies, the EAD actuator will be tested for higher Reynolds numbers. Moreover, further experiments are needed to gain a better understanding of the interactions between the ionic wind induced by the electric discharge and the boundary layer.

References

1. Artana G., DiPrimio G., Desimone G., Moreau E., Touchard G.: Electrohydrodynamic actuators on a subsonic air flow around a circular cylinder. In *Proceedings of 32nd AIAA Plasmadynamics and Lasers Conference and 4th Weakly Ionized Gases Workshop*, Anaheim, CA, June 11–14 (2001) AIAA Paper 20013056.
2. Roth J.R., Sherman D.M., Wilkinson S.P.: Boundary Layer Flow Control with a One atmosphere uniform glow discharge surface plasma. In *Proceedings of 36th Aerospace Sciences Meeting and Exhibit*, Reno, NV, January 12–15 (1998) AIAA Paper 98-0328.
3. Moreau E., Leger L., Touchard G.: Effect of a DC surface-corona discharge on a flat plate boundary layer for air flow velocity up to 25 m/s. *J. Electrostat.* **64**(3/4) (2006) 215–225.
4. Hong D., Magnier P., Bauchire J.M., Leroy-Chesneau A., Pouvesle J.M.: Preliminary study of electric discharges for airflow controls. In *Proceedings of 8th International Symposium on Fluid Control, Measurement and Visualization*, CD Rom (2005) Paper 319, 1–4.

Control of Flow Separation on a Wing Profile Using PIV Measurements and POD Analysis

Julien Favier, Azeddine Kourta and Gillian Leplat
IMFT, French National Institute of Mechanics of Toulouse, 31400 Toulouse, France; E-mail: favier@imft.fr

Abstract. The purpose of this paper is to study experimentally the active control of separation on an ONERA D airfoil using micro-actuators. The configuration is a massively separated flow on the airfoil corresponding to a stalled case at $R_e = 0.5 \times 10^6$ and $\alpha = 16^o$ and the control is performed using continuous blowing microjets. Using PIV measurements and a post-processing based on POD, the main characteristics of the control are highlighted, leading to the developement of a MEMS prototype based on synthetic blowing.

Key words: PIV, POD, separation, flow control, blowing actuator.

1 Introduction

Controlling boundary layer separation can greatly improve the performances of aerodynamic vehicles and means of postponing or suppressing the separation process have received great attention in the last decade. Indeed, enhancing lift and reducing noise or drag constitute valuable technological and economical issues. The diversity of experimental means to control separation is huge, including direct or pulsed blowing, wall movement, plasma-based actuators and MEMS. In order to develop an energy-efficient control procedure, it is critical to highlight the primary characteristics of the separation process and to design an actuation system to affect these key phenomena. Following this idea, the present paper describes a generic blowing actuator in order to study the main effects of a control aimed at suppressing separation.

J.F. Morrison et al. (eds.), IUTAM Symposium on Flow Control and MEMS, 203–207.
© 2008 *Springer. Printed in the Netherlands*

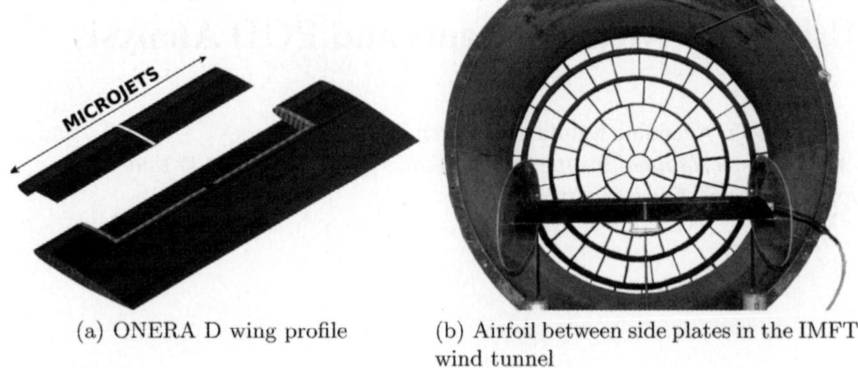

(a) ONERA D wing profile   (b) Airfoil between side plates in the IMFT wind tunnel

Fig. 1 Experimental facilities.

2 Experiments on a Generic Blowing Actuator

The actuator used to control separation consists if 82 microjets blowing continuously and normal to the wall. They are located close to the separation point, i.e. close to the leading edge of the airfoil, and along 40% of the span (Figure 1a). The blowing is achieved using two chambers providing pressurized air (2.5 bars) and generating continuous blowing through the microjets. The velocity of each jet is sonic, as their diameter is tiny (0.4 mm). The actuation system is embedded in the ONERA D wing, which is designed to be easily equipped with multiple actuators (Figure 1a). Experiments are performed in the S1 open loop wind tunnel of IMFT which enables the large degree of optical access needed to realize PIV measurements (Figure 1b).

The principal means of investigation are particle image velocimetry (PIV) to measure velocity fields at midspan all along the chord in both controlled and uncontrolled cases, and a post-processing of PIV velocity fields based on proper orthogonal decomposition (POD) to detect the main energetic features of the flow. Analysis of the PIV velocity fields show that control using microjets is very efficient in suppressing flow separation (Figure 2) and consequently, the global aerodynamic performances of the airfoil are enhanced as shown by measurements of C_l and C_d by an aerodynamic balance. The gain in lift is about 20% for a control zone equal to 40% of span and the performances are increased above the initial stall angle. As shown in Figure 3, the increase of lift coefficient is directly related to the suppression of the recirculating zone, especially at high angles of attack when the flow is stalled. Moreover, the incidence of stall is delayed from 14° to 16°. Control experiments are also performed on a tripped boundary layer in order to force a turbulent behaviour of the near-wall flow and the benefits of control are similar.

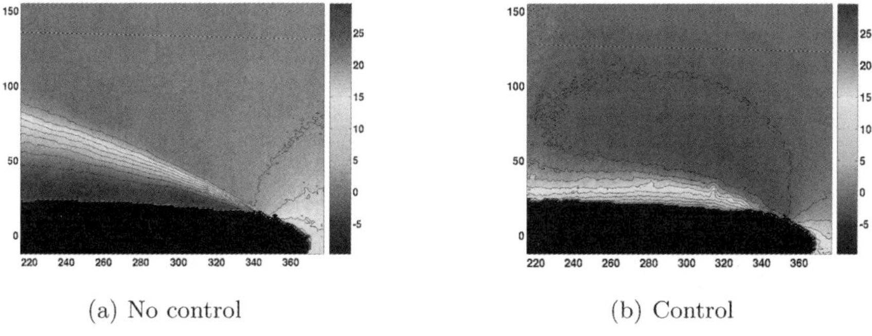

(a) No control　　　　　　　　　　　　(b) Control

Fig. 2 PIV velocity fields for $R_e = 1.5 \times 10^5$ and $\alpha = 16°$ – stalled case.

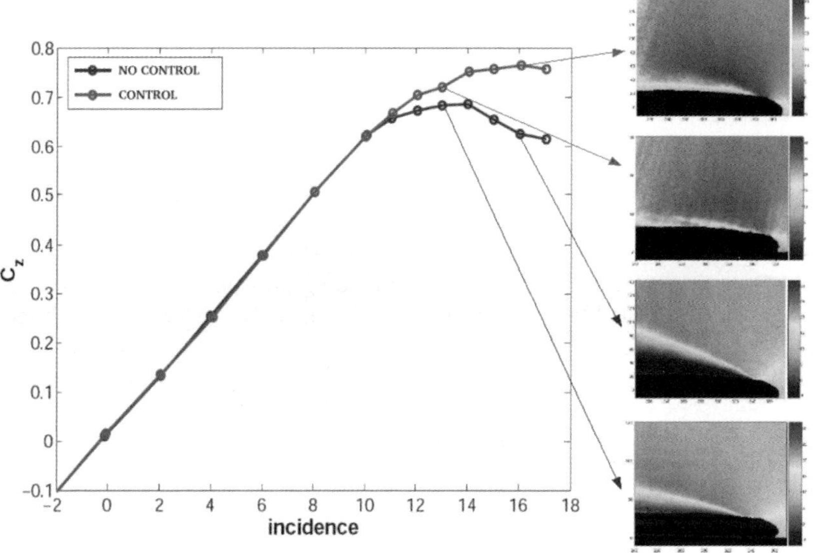

Fig. 3 Loads measurements and PIV velocity fields related to the incidence.

3 POD Analysis and Close Perspectives

The extraction of POD modes is made using the "snapshots method" [1] on the measured velocity fields. Using this methodology, the POD modes are solutions of an eigenvalue problem based on temporal correlations and the velocity field is expressed as an expansion of spatial modes $\Phi_i(X)$ and temporal coefficients $a_i(t)$:

$$u(X,t) = \sum_{i=1}^{N} a_i(t)\Phi_i(X).$$

(1)

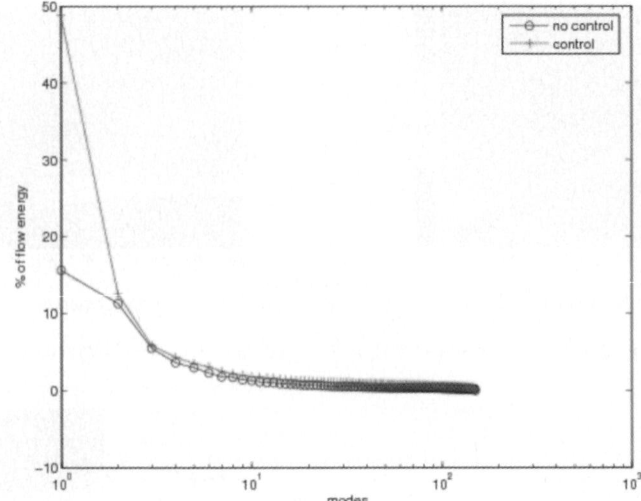

Fig. 4 POD eigenspectrum.

Fig. 5 Synthetic Piezo microjets in the ONERA D airfoil.

Each eigenvalue is related to the energy of its associated POD mode and the eigenspectrum presented in Figure 4 shows that the mechanism of control is mainly related to the most energetic first POD mode, which corresponds to a large scale structure of the order of magnitude of the separated boundary layer thickness. An overview of the first POD modes obtained in this configuration can be found in [2]. Indeed, the action of microjets is to generate streamwise counter-rotating vortices that stabilize boundary layer by mixing the low energetic fluid near the wall with the high energetic fluid of the shear flow.

Within the scope of exploiting this mechanism in terms of vortex shedding frequencies, a prototype of a MEMS actuator is developed to generate synthetic microjets, following the concept of the continuous blowing actuator which has proved its efficiency. Before being miniaturized using the MEMS technology, the synthetic microjets actuator is first being developed using large scale piezoelectric diaphragms. The next step will be to find an appropriate material for the MEMS actuator capable of high performance in terms of blowing/suction velocities and frequencies.

References

1. Sirovich, L.: Turbulence and the dynamics of coherent structures. Part 1: Coherent structures. *Quarterly of Applied Mathematics* **XLV**(3), 1987, 561–571.
2. Favier, J. and Kourta, A.: Étude du contrôle du décollement sur un profil d'aile par mesures PIV et analyse POD. *C.R. Acad. Sci. Paris* **334**, 2006, 272–278.

Control of the Shear-Layer in the Wake of an Axisymmetrical Airfoil Using a DBD Plasma Actuator

Maxime Forte[1], Jérôme Jolibois[1], Eric Moreau[1], Gérard Touchard[1] and Michel Cazalens[2]

[1]*Laboratoire d'Etudes Aérodynamiques (CNRS), Téléport 2,
Bd. Marie & Pierre Curie, BP 30179, 86962 Futuroscope Cedex, France;
E-mail: maxime.forte@lea.univ-poitiers.fr*
[2]*SNECMA (SAFRAN Group), 77550 Moissy Cramayel, France*

Abstract. Several studies have shown that a surface Dielectric Barrier Discharge (DBD) may be used as an ElectroHydroDynamic (EHD) actuator. This actuator adds momentum inside the boundary layer close to the wall and could be used for airflow control. In this paper, the actuator has been set up on a small axisymmetrical airfoil and the discharge is used to modify the characteristics of the shear-layer in its wake. Results show that the plasma actuator modifies strongly the airflow around the airfoil for velocities up to 20 m/s.

Key words: ElectroHydroDynamic (EHD), flow control, plasma actuator, dielectric barrier discharge, shear layer.

1 Introduction

Several studies have shown that discharges in air at atmospheric pressure may be used as ElectroHydroDynamic (EHD) actuators for airflow control along profiles. Applications like drag reduction, flow reattachment or detachment and instability control are particularly concerned. Up to now, two kinds of discharges have succeeded in modifying actively flow characteristics: DC corona discharges and AC dielectric barrier discharges (DBD). Both actuators generate a cold plasma on the surface of the profile and work by creating ionic wind and by injecting kinetic energy inside the boundary layer. This ionic wind is due to the collision between ions and the neutral particles of the gas.

In the present paper, the DBD actuator presented in Figure 1 has been set up on a small axisymmetrical airfoil. This actuator consists of two metallic electrodes asymmetrically mounted on both sides of a dielectric layer. The active electrode is supplied with a high AC voltage, which is needed to maintain the discharge and to evacuate charge build-up on the dielectric surface. The other electrode is grounded. Various studies have shown the ability of this actuator to modify the properties of

J.F. Morrison et al. (eds.), IUTAM Symposium on Flow Control and MEMS, 209–215.
© 2008 *Springer. Printed in the Netherlands.*

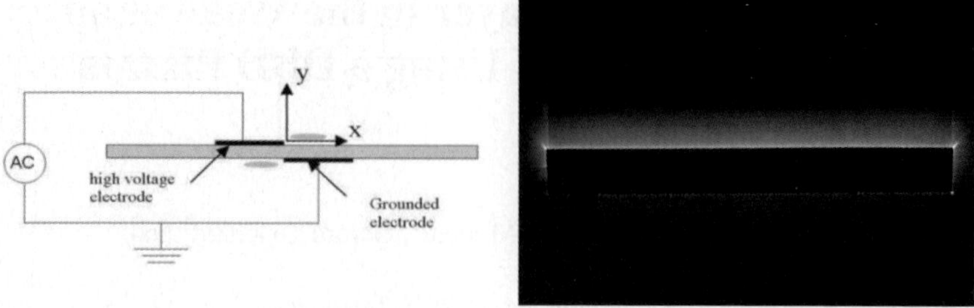

Fig. 1 Sketch (side view) and photograph (top view) of the DBD actuator when it is turned on.

a boundary layer over a flat plate [1, 2] or on turbine blades [3, 4]. Some experimental works [5, 6] and numerical studies [7] with NACA airfoils have shown lift improvement in stall configurations. In order to improve the abilities of this device, several studies have focused on the discharge without free airflow. These works were mainly observations and parametric studies [8–12]. They showed that two different discharge regimes occur during both half cycles. The negative-going portion has been found to produce a more uniform discharge than the positive-going one. Finally, several authors tried to understand how the actuator adds momentum to the air by means of numerical simulation [13–16]. Their results have shown that the nature of the discharge regime seems to play an important role in the momentum transfer to the neutral fluid. Moreover, they have shown that the average induced force is not the same from one half-cycle to the other. This last result has been pointed out recently by an experimental study with Laser Doppler Velocimetry measurements [17]. In this paper, the plasma actuator has been set up on a small axisymmetrical airfoil and the discharge is used to modify the characteristics of the shear-layer in its wake.

2 Experimental Arrangements

2.1 Airfoil & Plasma Actuator

An axisymmetrical airfoil made in PMMA (Plexiglas) has been equipped with a Dielectric Barrier Discharge actuator as shown in Figure 2. The results of a previous study [11] have shown that the ionic wind velocity increases, in an asymptotic way, with the applied voltage amplitude, the frequency and the grounded electrode width. Thus, optimized geometry and electrical parameters have been chosen to induce the fastest velocity. The active electrode is 5 mm wide and is mounted all around the outer side of the airfoil. The grounded electrode is 15 mm wide and is mounted on the inner side of the airfoil. Both electrodes are made of aluminium tape which is about 0.1 mm thick. The active electrode is connected to a power amplifier TREK Model 30/40 (±30 kV, 40 mA peak) applying a sine high voltage in a range of

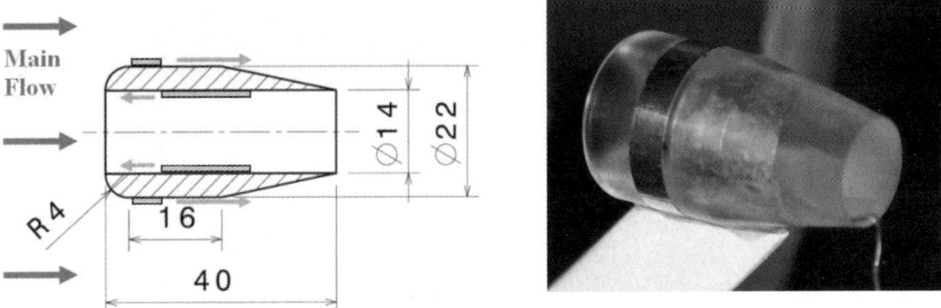

Fig. 2 Cross-section and photograph of the axisymmetrical airfoil equipped with a dielectric barrier discharge actuator.

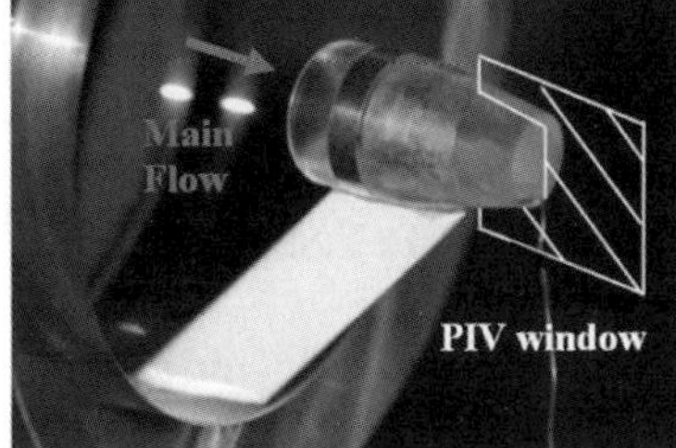

Fig. 3 CAD view and photograph of the airfoil placed in a free jet airflow. The right picture shows the window where PIV measurements have been made.

10–25 kV in magnitude and with frequencies from 1 to 2 kHz. Plasma appears on both sides from an electrode towards the direction of the opposite electrode, with declining luminosity. This geometry allows us to obtain an ionic wind in the main flow direction on the outer side of the airfoil (green in Figure 2) and an opposite ionic wind in the inner side (red in Figure 2).

The plasma is constituted of microdischarges distributed uniformly in time and space along the electrode length. The maximum velocity is reached at the end of the plasma area ($x \sim 10$ mm here) and at $y = 0.5$ mm from the wall. Although the plasma seems to be quite uniform, it has been shown previously that the DBD actuator seems to induce a pulsed airflow at the same frequency than the applied high voltage [2, 17].

2.2 Aerodynamic Set-up & Velocity Measurement Device

The experimental set-up is presented in Figure 3. The airfoil is placed in the potential core of a free jet airflow whose velocity can be adjusted from 5 to 30 m/s.

Velocity measurements have been performed with Particle Image Velocimetry (PIV). Images have been recorded with a CCD camera (LAVISION®, Model

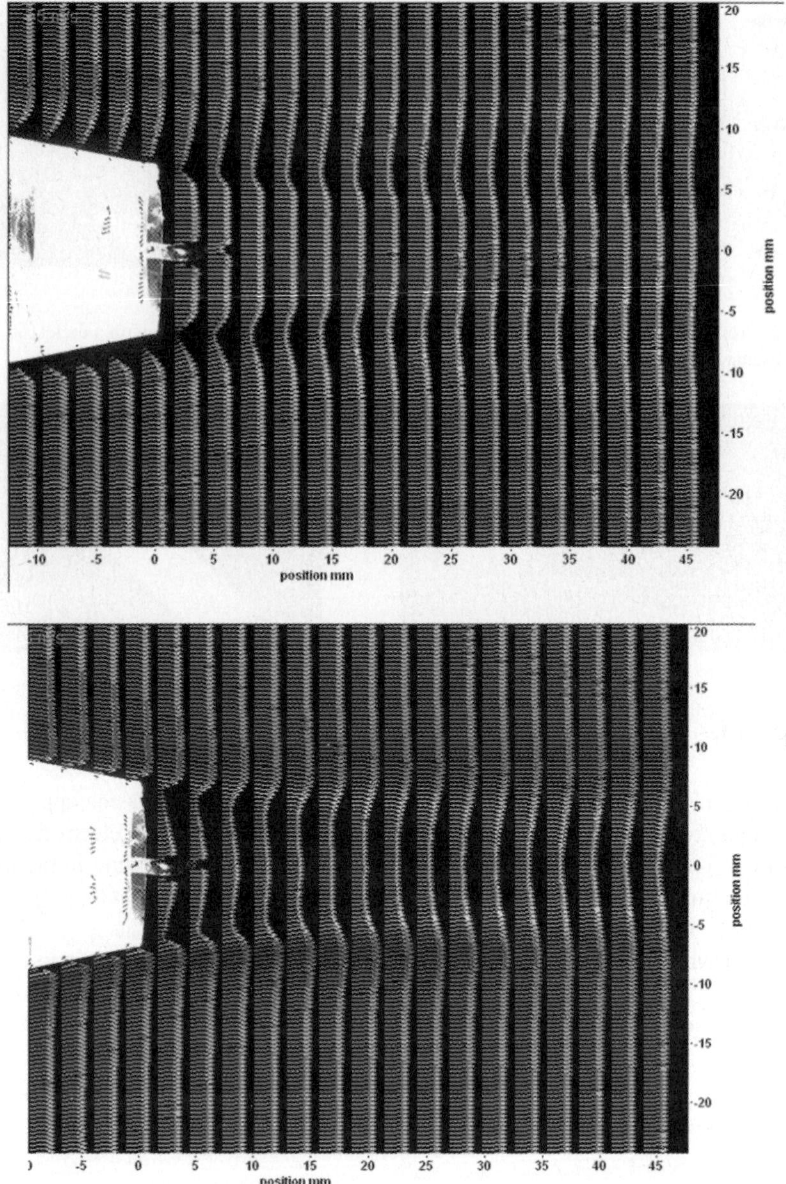

Fig. 4 Time-averaged velocity fields on the wake of the axisymmetrical airfoil with DBD turned off (left) and DBD turned on (right) for a jet velocity of 5 m/s.

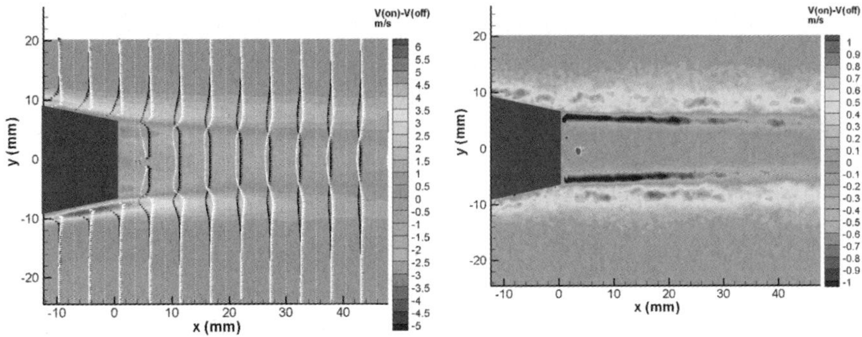

Fig. 5 Norm of the two velocity fields substraction: $V_{(\text{with plasma})} - V_{(\text{without plasma})}$, for $V_{\text{jet}} = 5$ m/s (left) and $V_{\text{jet}} = 20$ m/s (right).

Flowmaster 3, 1280×1024 pixels, 12 bits) using a YAG double impulsion laser (QUANTEL®, Model Twins Ultra, 30 mJ). For each run, 150 image couples have been stored in order to obtain reliable time-averaged flow fields. The images are processed with the cross-correlation algorithm of LAVISION® and the final window size was 16×16 pixels with a 50% overlap.

3 Results

For instance, Figure 4 presents typical time-averaged velocity fields in the wake of the airfoil, at $V_{\text{jet}} = 5$ m/s, with and without actuation. The plasma is ignited with a sine high voltage of 25 kV having a frequency of 1.5 kHz. These pictures show that the actuator modifies strongly the airflow inside the wake. In order to better investigate the exact location of the modifications due to actuation, we decided to compute the vector substraction of the two velocity fields (i.e. $V_{\text{on}} - V_{\text{off}}$ for each vector). Both pictures in Figure 5 show the norm of this substraction, for two different velocities $V_{\text{jet}} = 5$ m/s and $V_{\text{jet}} = 20$ m/s. The left picture even shows velocity profiles without plasma (black curve) and with plasma (white curve). As it was expected to do, the actuator increases the velocity in the outer boundary layer and reduces the velocity in the inner one. For $V_{\text{jet}} = 5$ m/s, the discharge increases the velocity of the outside airflow of about 5 m/s close to the wall, while it slows down the inner flow by about 4 m/s.

The effect of the plasma is clearly noticeable even at $V_{\text{jet}} = 20$ m/s, but obviously, the actuator is less efficient when the main velocity increases. Note that for $V_{\text{jet}} = 20$ m/s, the actuator increases the outside airflow velocity by only 1 m/s and slows down the inner flow by 1 m/s as well. This phenomenon modifies the characteristics of the shear layer (its thickness for example). Data is being processed in order to quantify exactly those changes.

4 Conclusion

In this study, an axisymmetrical airfoil has been fitted with a dielectric barrier discharge actuator, in order to modify the characteristics of the shear-layer. PIV measurements have been performed in the wake of the body so as to investigate the effect of the plasma actuation. Results show that two ionic winds are generated: one which increases the velocity in the outer boundary layer and one which slows down the flow on the inner side of the airfoil. This phenomenon seems to change the parameters of the shear layer, such as its thickness. Data is being processed in order to quantify exactly those changes. The effect of the plasma is clearly noticeable even at $V_{jet} = 20$ m/s, but obviously, the actuator is less efficient when the main velocity increases.

Acknowledgment

The authors gratefully acknowledge the technical and financial support of SNECMA (SAFRAN Group) (Contract No. 920 430961).

References

1. J.R. Roth, D.M. Sherman, and S.P. Wilkinson, Boundary layer flow control with a one atmosphere uniform glow discharge surface plasma, in *Proceedings 36th Aerospace Sciences Meeting & Exhibit*, Reno, NV, 1998, AIAA Paper 98-0328.
2. M. Forte, L. Léger, J. Pons, E. Moreau, and G. Touchard, Plasma actuators for airflow control: Measurement of the non-stationary induced flow velocity, *Journal of Electrostatics* **63**(6/10), 2005, 929–936.
3. R. Rivir, Lt.A. White, C. Carter, B. Ganguly, A. Forelines, and J. Crafton, Turbine flow control, plasma flows, in *Proceedings 41st Aerospace Sciences Meeting & Exhibit*, Reno, NV, 2003, AIAA-Paper 2003-6055.
4. L. List, A.R. Byerley, T.E. McLaughlin, and R.D. Van Dyken, Using a plasma actuator to control laminar separation on a linear cascade turbine blade, in *Proceedings 41st Aerospace Sciences Meeting & Exhibit*, Reno, NV, 2003, AIAA-Paper 2003-1026.
5. M. Post and T.C. Corke, Separation control on high angle of attack airfoil using plasma actuators, in *Proceedings 41st Aerospace Sciences Meeting & Exhibit*, Reno, NV, 2003, AIAA-Paper 2003-1024.
6. T.C. Corke and M.L. Post, Overview of plasma flow control: Concepts, optimization and applications, in *Proceedings 43rd Aerospace Sciences Meeting & Exhibit*, Reno, NV, 2005, AIAA Paper 2005-563.
7. D.V. Gaitonde, M.R. Visbal, and S. Roy, A coupled approach for plasma-based flow control simulations of wing sections, in *Proceedings 44th Aerospace Sciences Meeting & Exhibit*, Reno, NV, 2006, AIAA Paper 2006-1205.
8. C.L. Enloe, T.E. McLaughlin, R.D. VanDyken, K.D. Kachner, E.J. Jumper, and T.C. Corke, Mechanisms and responses of a single dielectric barrier plasma actuator: Plasma morphology, *AIAA Journal* **42**(3), 2004, 589–594.

9. C.L. Enloe, T.E. McLaughlin, R.D. VanDyken, K.D. Kachner, E.J. Jumper, and T.C. Corke, Mechanisms and responses of a single dielectric barrier plasma actuator: Geometric effects, *AIAA Journal* **42**(3), 2004, 595–605.

10. J. Pons, E. Moreau, and G. Touchard, Asymmetric surface dielectric barrier discharge in air at atmospheric pressure: Electrical properties and induced airflow characteristics, *Journal of Physics D: Applied Physics* **38**, 2005, 3635–3642.

11. M. Forte, J. Jolibois, E. Moreau, and G. Touchard, Optimization of a dielectric barrier discharge actuator by stationary and non-stationary measurements of the induced flow velocity – Application to airflow control, In *Proceedings 3rd AIAA Flow Control Conference*, San Francisco, CA, 2006, AIAA Paper 2006-2863.

12. J.R. Roth, and X. Dai, Optimization of the aerodynamic plasma actuator as an ElectroHydro-Dynamic (EHD) electrical device, *Proceedings 44th Aerospace Sciences Meeting & Exhibit*, Reno, NV, 2006, AIAA Paper 2006-1203.

13. G.I. Font, Boundary layer control with atmospheric plasma discharges, in *Proceedings 40th AIAA/ASME/SAE/ASEE Joint Conference and Exhibit*, Fort Lauderdale, FL, 2004, AIAA Paper 2004-3574.

14. D.M. Orlov and T.C. Corke, Numerical simulation of aerodynamic plasma actuator effects, in *Proceedings 43rd Aerospace Sciences Meeting & Exhibit*, Reno, NV, 2005, AIAA Paper 2005-1083.

15. J.P. Boeuf and L.C. Pitchford, Electrohydrodynamic force and aerodynamic flow acceleration in surface dielectric barrier discharge, *Journal of Applied Physics* **97**, Article No. 103307, 2005.

16. A.V. Likhanskii, M.N. Shneider, S.O. Macheret, and R.B. Miles, Modeling of interaction between weakly ionized near-surface plasmas and gas flow, in *Proceedings 44th Aerospace Sciences Meeting & Exhibit*, Reno, NV, 2006, AIAA Paper 2006-1204.

17. M. Forte, F. Baudoin, E. Moreau, and G. Touchard, Non stationary measurements of the induced flow velocity of a single dielectric barrier discharge actuator – Application to airflow control, in *Proceeding of the International Symposium on HydroElectroDynamics (ISHED 2006)*, Buenos Aires, Argentina, 2006.

DRAG REDUCTION AND MIXING

Models for Adaptive Feedforward
Control of Turbulence

Kenneth Breuer and Kevin Wu
Division of Engineering, Brown University, Providence, RI 02912, U.S.A.;
E-mail: kbreuer@brown.edu

Abstract. We present numerical results from an idealized model simulation which implements the Least Mean Square (LMS) and the Filtered-X-Least-Mean-Square (FXLMS) control algorithms as applied to adaptive feedforward control of wall-bounded turbulent shear flows. The FXLMS system is found to work extremely well, effectively controlling the model system which includes phase delays, multiple sensor inputs, high levels of noise and nonlinearities in the forward system.

Key words: Turbulence control, adaptive feedforward control, linear filter, finite impulse response.

1 Introduction

The challenge for practical control of turbulent flows is to find a method that records inputs, applies control signals and monitors the performance of the control algorithm (Figure 1). Numerically, approaches based on this concept have been demonstrated in a variety of ways [1–5]. However, a consistent issue with numerical and theoretical approaches is that they tend to rely on unrealistically large arrays of sensors and actuators and require excessive numerical computations at each time step and are practically unfeasible. For this reason, physical experiments on turbulent boundary layer control have tended to operate at the other extreme – very simple control approaches based on one or two sensors and actuators [6–8]. A modest attempt to integrate formal control in a real experiment was demonstrated some years ago in our own group [9–11] in which a feedforward control algorithm was used and demonstrated to achieve 35% reduction in turbulent fluctuation intensity using only three sensors and three actuators. The control algorithm was based on the concept that turbulence-producing coherent structures were detected by a series of upstream shear sensors and acted upon by an array of synthetic jet actuators. The system was monitored and optimized by a series of downstream shear sensors whose function

J.F. Morrison et al. (eds.), IUTAM Symposium on Flow Control and MEMS, 219–227.
© 2008 *Springer. Printed in the Netherlands.*

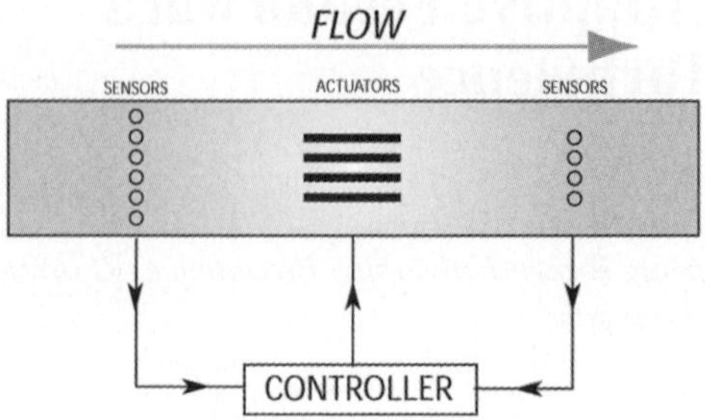

Fig. 1 Schematic of a feedforward control system, including upstream sensors, actuators and downstream control points.

was to close the control loop using an adaptive feedforward system. In that implementation, the feedforward algorithm was trained using system identification, but the loop was not formally closed, rendering the system sensitive to small changes in freestream velocity and to drift in sensor or actuator performance. The present paper explores an attractive method for closing the loop, by using varients of the commonly-used Least-Mean-Square (LMS) algorithm. For the purposes of rapid algorithm design and evaluation, we have used a model system – a simple numerical model that captures the essential statistical features of a turbulent flow.

1.1 Finite Impulse Response (FIR) Filters

The control system uses data from wall-based sensors to determine variations in the shear stress and then attempts to minimize fluctuations by sending signals to the actuators. Because the controller begins operating with little or no prior knowledge of the system, it also relies on feedback from the sensors located downstream to find how best to operate the actuators. Generally speaking, controllers rely on filters to extract information from noisy data. For Finite Impulse Response (FIR systems), the output at any time, $y(n)$, is defined as a weighted average of a sequence of input data:

$$y(n) = \sum_{k=0}^{N-1} w_k x(n-k), \tag{1}$$

where $x(n-k)$ is the sensor input at previous times ($k > 0$), and w_k are the corresponding weights. Infinite impulse response (IIR) filters include input data from future times ($k \leq 0$), and can be written in recursive form. However, for these studies, FIR filters were used exclusively as they have a linear phase response, and unlike IIR filters, are guaranteed to be stable.

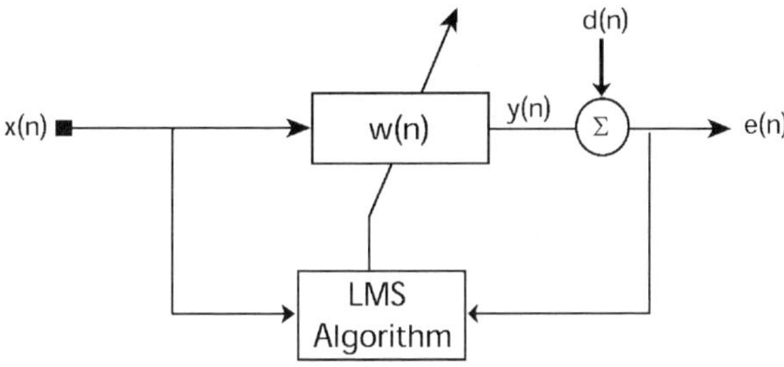

Fig. 2 Schematic of the LMS feedfoward system.

1.2 Least-Means Squared (LMS) Algorithm

The LMS algorithm [12] is a linear adaptive filtering technique that relies on two principal processes, a filtering process and an adaptive process. The filtering process is an FIR filter that relies on the adaptive process to generate its transfer function coefficients. The adaptive process takes downstream data of the system and adjusts the filtering processing coefficients until a specified performance goal is reached. This layout defines a feedback loop [13]. As illustrated schematically in Figure 2, the input signal to the system is $x(n)$, while the output of the system is $y(n)$, $d(n)$ is the desired signal, and $e(n)$ is the error. For our application, $x(n)$ is the signal from the sensors located upstream of the actuators. As in Equation (2), the output, $y(n)$, is the output of an FIR filter, defined as:

$$y(n) = \mathbf{w}^{\mathbf{T}}(n) \cdot \mathbf{x}(n). \tag{2}$$

The error signal, $e(n)$, is defined as:

$$e(n) = d(n) - y(n). \tag{3}$$

Thus, the objective of the control system is to cancel out the signal observed at the upstream sensor. We can see that the controller will continue to adjust the taps until $e(n)$ is zero. The adaptive portion of the algorithm is defined in the coefficient update:

$$\mathbf{w}(n+1) = \mathbf{w}(n) + \mu x(n)e(n). \tag{4}$$

The gain, μ, dictates the magnitude of the changes to the coefficients at each time step. If μ is large, then the influence of the existing transfer function is lessened when compared to the current measurement. Conversely, if μ is small, then changes to the coefficients are less dramatic. In this way, μ must be chosen such that it is small enough that the control system converges accurately and without undue instability, but large enough that the system converges reasonably quickly.

1.3 Implementation

For the current study, a simple model problem was chosen to evaluate the capabilities of adaptive LMS architectures in the control of turbulent shear flows. MATLAB was used to test the performance of the control system. The filters were implemented using MATLAB's Simulink environment. To simulate the dynamics of the true turbulent system, we model a highly idealized system – that of a harmonic signal, representing the large scale coherent structures, corrupted by Gaussian noise (representing the small scale fluctuations). Clearly this is an extreme oversimplification of any true turbulent shear flow. However, it does have relevance, and can be justified by:

1. True turbulent flows in the near wall region are comprised of large scale coherent structures obscured by small scale random fluctuations. These structures are intermittent, but contribute a large fraction of the complete turbulence production [14], and thus a successful control system should focus on the control of these structures. The noisy sine wave provides the first approximation to a physically relevant situation.
2. The premise of this control system, and one justified by recent experiments and computations, is that linear dynamics are quite capable in describing the short-term dynamics of turbulent flows for the short times necessary to execute control. That being so, a sine wave is the natural Fourier decomposition of any linear process in the frequency domain, and thus appropriate as a first step.
3. The key issues that challenge control systems and are treated in this paper are the issues of phase lag between sensors and actuators, as well as spatial coherence of large-scale turbulent structures and lastly the influence of nonlinearities on the performance of a linear control system. As will be described, the simple noisy sine wave system does allow for the testing of the control system's ability to deal with all of these issues.

2 Results and Discussion

Figure 3a shows the performance of the LMS filter operating on a signal of amplitude $2 V_{p-p}$, corrupted by Gaussian noise of variance $1 V$ (both with a mean of 0 V). The RMS of the error signal, the difference between the input signal and LMS filter output, is shown for a 64-tap filter operating at eight varying values of μ. The RMS is calculated using a moving 500-sample window. We use the RMS of the error signal as a gauge for the filter's success at removing fluctuations from the input. We see that the LMS filter, operating with the smallest adjustment factor, $\mu = 0.01$, slowly reduces the variance of the input signal. As μ increases the algorithm adapts faster, and when $\mu = 0.5$, the filter converges the fastest. As μ increases past this value, the rate at which the output converges decreases until $\mu = 2.1$, at which point, the RMS increases dramatically, indicating control system failure. The behavior of

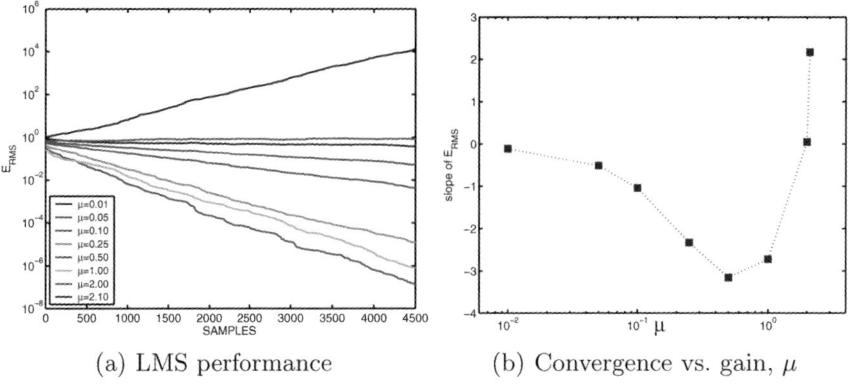

(a) LMS performance   (b) Convergence vs. gain, μ

Fig. 3 The performance of the LMS simulation versus time, and the slope of the convergence (or divergence) as a function of the gain, μ.

the error can be expressed as:

$$E_{\mathrm{RMS}} = Ae^{-\lambda t}, \tag{5}$$

where λ is the rate of convergence, ie. the slope of the line in Figure 3a, and A is a scaling factor. Figure 3b shows the effect of the gain, μ, on the rate of convergence, λ from which we see that as μ increases from zero, the control convergence improves until a critical point at which it rapidly diverges and the control system is no longer stable. This system was also demonstrated to work effectively on arbitrary linear transformations of $x(n)$. For example, using x as the input signal, the LMS system was capable of controlling $\partial x / \partial t$, Ax, etc.

2.1 Filtered-X LMS (FXLMS) Algorithm

While the performance of the LMS system is dependent on μ, it is extremely vulnerable to small delays in the error path (Figure 4). This is particularly severe at the more desirable (e.g. higher) values of μ. This observation is not new, and the feedback loop of the LMS algorithm, and the choice of the gain, μ, is known to be a source of instability [13, 15], especially in applications that experience delays in the error path. The Filtered-X LMS algorithm [16] solves this problem by introducing an additional filter to the reference signal. Illustrated in Figure 5, two filter blocks have been added to the LMS system. The filter with transfer function **C**, represents the dynamics of the shear flow and can include the time delay indicative of fluid propagation from the actuators to the control point, or feedback, sensors. This filter might also represent nonlinearities in the flow, or measurement noise. In order to compensate in the control algorithm, a second filter, with transfer

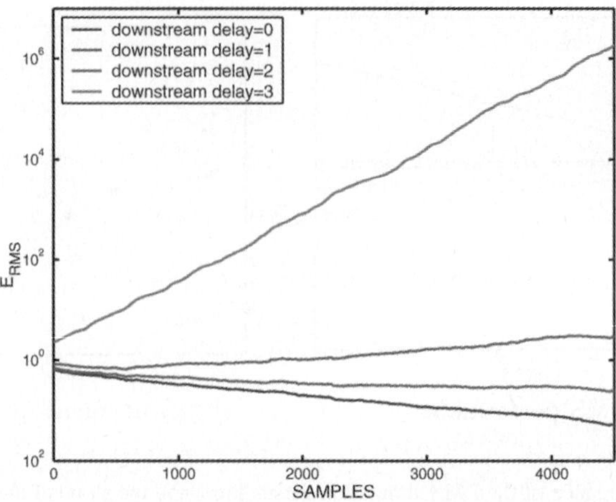

Fig. 4 Time history of plain LMS system, illustrating the LMS sensitivity to delays in the error path. $\mu = 0.1$ $N_{\mathrm{taps}} = 64$.

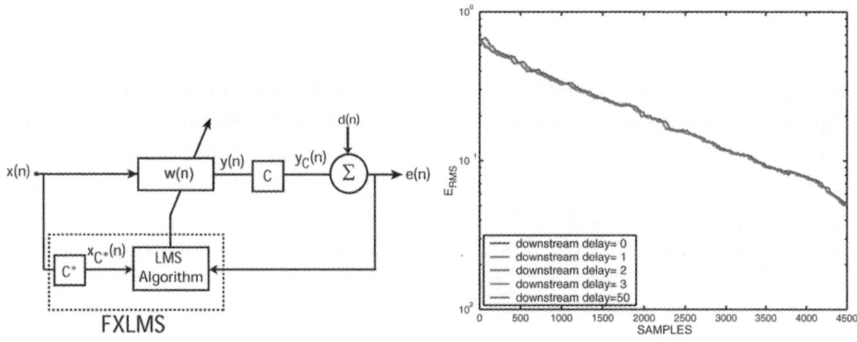

Fig. 5 Schematic of the FXLMS control algorithm which includes an additional filter, $\mathbf{C}^*$, to account for the system dynamics, represented by $\mathbf{C}$. The performance of the FXLMS, subject to delays of 0,1, 2, 3, and 50 samples is shown, clearly indicating the FXLMS's success in adapting to these cases.

function, $\mathbf{C}^*$, is introduced into the *forward* path of the control system. The transfer function $\mathbf{C}$ is defined by the system and filter $\mathbf{C}^*$ serves as an estimate of the true dynamics. Ideally, the estimate is determined by finding the impulse response of the system, but success has been demonstrated by simply implementing a proper phase delay [13, 15, 16].

While Equation (2) still applies, the error signal to the filter becomes:

$$e(n) = d(n) - y_{\mathbf{C}}(n), \qquad (6)$$

where the output $y_C(n)$ is defined as:

$$y_C(n) = \mathbf{C} \cdot \mathbf{y}(n), \tag{7}$$

where $\mathbf{y}(n)$ is now a data sequence and the new reference signal, $x_{C^*}(n)$ is defined as:

$$x_{C^*}(n) = \mathbf{C}^* \cdot \mathbf{x}(n), \tag{8}$$

and the adaptation routine becomes:

$$\mathbf{w}(n+1) = \mathbf{w}(n) + \mu x_{C^*}(n) \cdot e(n). \tag{9}$$

Figure 5 shows the FXLMS system successfully filtering the input signal when operating with the same error path delays as Figure 4. The source signal and the filter parameters are identical, but the reference signal is delayed by the same length as the downstream effect. Additionally, the figure shows a case where the downstream delay is 50 samples and the filter is observed to adapt equally well.

2.2 System Robustness

Another aspect of the true turbulent flow, and one utilized "intuitively" by Rathnasingham and Breuer [11], is the recognition that coherent turbulent structures have a spatial extent, and that one can use multiple sensors to identify coherent structures that span multiple input sensors, and to reject small-scale incoherent noise. Furthermore, nonlinearities in the flow between the upstream sensor and the downstream control point will also serve to degrade the control system's performance. Both of these were tested using the FXLMS architecture. An FXLMS system utilizing one, two and four upstream sensors and a single downstream sensor was simulated. The system is effectively a collection of parallel processes, each receiving input signals from different sensors each of which has the same sine wave (representing a single coherent structure), but different noise sources (representing independent small scale structure). Each filter has its own series of weights and performs separate FIR calculations (desirable for practical implementation), however all of the systems contribute to a single actuator signal (by averaging the output signals) and they receive a single error signal from the lone control point sensor downstream of the actuators.

To further gauge the flexibility of the FXLMS routine and better model a turbulent flow, the simulation was further modified. A delay was added to the *upstream* path of the system in order to simulate the delay due to the propagation from the input sensor to the actuator's location. Lastly, a nonlinearity, N^n, computed from the square of the time derivative:

$$N^n = \left(\frac{u^n - u^{n-1}}{\Delta t} \right)^2 \tag{10}$$

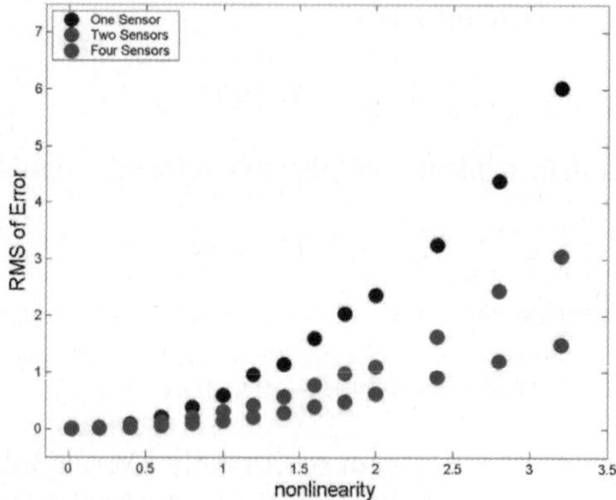

Fig. 6 The effect of simulated system nonlinearities on the RMS of the error output. The system grows more unstable as nonlinearity increases. However, as the number of channels is increased, the variance of the system error (output) is decreased.

was added to the signal (with varying amplitude), simulating the convective nonlinearity of a physical flow.

The inclusion of the additional upstream delay did not significantly affect the performance of the system as long as the FIR filters had enough capacity to accommodate this lag. This simply meant having enough taps (weights) that converged to a value of zero to absorb the upstream delay. Clearly this is not efficient, since we are using filter taps with zero value to accommodate the delay, and each tap requires an add-multiply operation during the filter evaluation. A cleverer implementation would be to incorporate our knowledge of the upstream delay in the same way we have incorporated the downstream delay in the error path, and to reserve the filter taps for productive usage.

However, the effect of multiple sensors and of nonlinearity is more interesting, and the results are shown in Figure 6. As the relative magnitude of the nonlinearity increases, the final RMS of the error signal increases (i.e the control performance degrades). However, in the case of multiple input channels, fluctuations in the error decrease as the number of channels increases. This is simply the effect that we are looking for - that different sensors receiving noisy versions of the same coherent structure will have more capacity to effectively control that structure, and this observation is in good qualitative agreement with our previous experimental results. Although we have not quantitatively evaluated this, one would expect that the performance will increase like the square root of the number of input sensors. However, physically, these sensors will not be able to sample the same coherent structure which does have a finite extent. Thus, from a practical perspective, two or perhaps four sensors will be the limit for this arrangement.

3 Conclusions

Despite the qualitative nature of these results and the fact that the simulations have been conducted on a very idealized system - a noisy sine wave, we have demonstrated the capability of the Filtered-X LMS algorithm as an adaptive control architecture that is quite attractive as a candidate for practical control of turbulent shear flows exhibiting rapid convergence, the ability to integrate several sensors and to reject both noise and nonlinearity. The system's computational simplicity and ability to be parallelized also makes the architecture attractive for practical realizations of flow control.

Acknowledgement

This work was supported by AFOSR(USA), monitored by Dr. T. Beutner.

References

1. Choi, H., Moin, P., Kim, J.: Active turbulence control for drag reduction in wall-bounded flows. *J. Fluid Mech.* **262** (1994) 75–110.
2. Schoppa, W., Hussain, F.: A large-scale control strategy for drag reduction in turbulent boundary layers. *Phys. Fluids* **10**(5) (1998) 1049–1051.
3. Bewley, T.: Flow control: New challenges for a new renaissance. *Progr. Aerospace Sci.* **37** (2001) 21–58.
4. Bewley, T., Moin, P., Temam, R.: DNS-based predictive control of turbulence: an optimal benchmark for feedback algorithms. *J. Fluid Mech.* **447** (2001) 179–225.
5. Lee, C., Kim, J., Babcock, D., Goodman, R.: Application of neural networks to turbulence control for drag reduction. *Phys. Fluids* **9**(6) (1997) 1740–1747.
6. Wilkinson, S., Balasubramanian, R.: Turbulent burst control through phase-locked traveling surface depression. AIAA Paper 85-0536 (1985).
7. Gad-el Hak, M., Blackwelder, R.F.: Selective suction for controlling bursting events in a boundary layer. *AIAA J.* **27** (1989) 308–314.
8. Jacobson, S.A., Reynolds, W.C.: Active control of streamwise vortices and streaks in boundary layers. *J. Fluid Mech.* **360** (1998) 179–211.
9. Rathnasingham, R., Breuer, K.: System identification and active control of a turbulent boundary layer. *Phys. Fluids* **9**(7) (1997) 1867–1869.
10. Amonlirdviman, K., Breuer, K.: Linear predictive filtering in a numerically simulated turbulent flow. *Phys. Fluids* **12**(12) (Dec 2000) 3221–3229.
11. Rathnasingham, R., Breuer, K.S.: Active control of turbulent boundary layers. *J. Fluid Mech.* **495** (2003) 209–233.
12. Widrow, B., Hoff, M.: Adaptive switching techniques. *IRE WESCON Conv. Rec.* **4** (1960) 96–104.
13. Haykin, S.: *Adaptive Filter Theory*, 4th edn. Information and System Sciences Series, Prentice Hall (2002).
14. Johansson, A.V., Alfredsson, P.H., Kim, J.: Evolution and dynamics of shear layer structure in near wall turbulence. *J. Fluid Mech.* **224** (1991) 579–599.
15. Nelson, P., Elliot, S.: *Active Control of Sound*. Academic Press (1992).
16. Hakansson, L.: The filtered-X LMS algorithm, Lecture Notes, University of Karlskrona, Ronneby.

Minimum Sustainable Drag for Constant Volume-Flux Pipe Flows

Ivan Marusic, D.D. Joseph and Krishnan Mahesh
*Department of Aerospace Engineering and Mechanics, University of Minnesota,
Minneapolis, MN 55455, U.S.A.; E-mail: marusic@aem.umn.edu*

Abstract. Comparisons are made between laminar and turbulent flows in pipes with
and without flow control, and a formula is derived that shows just how much the
discrepancy between the volume flux of laminar and turbulent flow at the same
pressure gradient increases as the pressure gradient is increased. Related to this, we
investigate the lowest bound for skin-friction drag in pipes for flow control schemes
that use surface blowing and suction with zero-net volume-flux addition.

Key words: Drag reduction, pipe flow, wall turbulence.

1 Introduction

Recently, an analysis was presented in [1] (hereafter referred to as MJM) that considered laminar and turbulent comparisons in channel flow, with and without flow
control. The study focused on the control strategies that use zero-net volume-flux
blowing/suction at the no-slip walls, and presented a criterion for achieving sub-
laminar drag conditions. The criterion was used to gain insight into why the control
strategy of Min *et al.* [2] was successful in achieving sub-laminar conditions, and
how other improved strategies could perhaps be designed.

Here we extend the analysis by MJM to pipe flows, and consider implications at
high Reynolds number.

2 Equations for Pipe Flow

For fully developed pipe flow of an incompressible fluid, driven by a constant pressure gradient, the Navier–Stokes and continuity equations may be written as

$$\frac{\partial \mathbf{V}}{\partial t} + \mathbf{V} \cdot \nabla \mathbf{V} = -\nabla p + \mathbf{e_x} P + \nabla^2 \mathbf{V}, \tag{1}$$

J.F. Morrison et al. (eds.), IUTAM Symposium on Flow Control and MEMS, 229–235.
© 2008 *Springer. Printed in the Netherlands.*

$$\nabla \cdot \mathbf{V} = 0. \tag{2}$$

Here we use cylindrical polar co-ordinates (r, θ, x), where the x axis is at the pipe center and r is the radial distance from the center of the pipe. Unless indicated, all terms have been non-dimensionalized using v, the kinematic viscosity of the fluid, and a, the radius of the pipe. The domain occupied by the fluid is

$$-\infty < x < \infty; \quad 0 \leq \theta < 2\pi; \quad 0 \leq r \leq 1.$$

Here $P > 0$ is the constant pressure gradient driving the flow, and the total pressure at a point in the fluid is $p(r, \theta, x, t) - Px$, where in all cases pressure has been normalized by ρ, the fluid density. We denote a cylinder average with an overbar:

$$\bar{f}(r,t) = \lim_{L \to \infty} \frac{1}{2L} \int_{-L}^{L} \frac{1}{2\pi} \int_{0}^{2\pi} f(r, \theta, x, t) \, d\theta \, dx,$$

and the over-all average as

$$\langle f \rangle = 2 \int_{0}^{1} \bar{f} r \, dr.$$

Therefore, the Reynolds number $Re_B = \langle \bar{V}_x \rangle$ is that based on the pipe radius and the bulk velocity. We will also decompose the velocity and pressure into mean (cylinder averaged) and fluctuating parts

$$[V_x, V_\theta, V_r, p] = [\bar{V}_x + u, \bar{V}_\theta + v, \bar{V}_r + w, \bar{p} + p'], \tag{3}$$

where the fluctuations have a zero mean: $\bar{u}, \bar{v}, \bar{w}, \bar{p}' = 0$.

The boundary conditions for pipe flow with zero-net-volume flux blowing-suction flow control are

$$V_x = V_\theta = 0; V_r = \phi(x, \theta, t) \text{ at } r = 1.$$

Note that $u = v = 0$, and $w = \phi$ at $r = 1$.

2.1 Energy Equations

In the following we will make use of energy identities. These are derived first by substituting (3) into (1) and using continuity to give

$$\frac{\partial \bar{\mathbf{V}}}{\partial t} + \frac{\partial \mathbf{u}}{\partial t} + \bar{\mathbf{V}} \cdot \nabla \bar{\mathbf{V}} + \bar{\mathbf{V}} \cdot \nabla \mathbf{u} + \mathbf{u} \cdot \nabla \bar{\mathbf{V}} + \mathbf{u} \cdot \nabla \mathbf{u} = -\mathbf{e_r} \frac{\partial \bar{p}}{\partial r} - \nabla p' + \mathbf{e_x} P + \nabla^2 (\bar{\mathbf{V}} + \mathbf{u}). \tag{4}$$

From (2) it follows that $\bar{V}_z = 0$ everywhere. The cylinder average of (4) is

$$\frac{\partial \bar{\mathbf{V}}}{\partial t} + \mathbf{e_r} \left[\overline{\mathbf{u} \cdot \nabla w} - \frac{\bar{V}_\theta^2}{r} - \frac{\overline{v^2}}{r} \right] + \mathbf{e_\theta} \left[\overline{\mathbf{u} \cdot \nabla v} + \frac{\overline{vw}}{r} \right] + \mathbf{e_x} \overline{\mathbf{u} \cdot \nabla u}$$

$$= -\mathbf{e_r}\frac{\partial \bar{p}}{\partial r} + \mathbf{e_x}P + \mathbf{e_\theta}\left[\frac{1}{r}\frac{\partial}{\partial r}\left(r\frac{\partial \bar{V}_\theta}{\partial r}\right) - \frac{\bar{V}_\theta}{r^2}\right] + \frac{\mathbf{e_x}}{r}\frac{\partial}{\partial r}\left(r\frac{\partial \bar{V}_x}{\partial r}\right), \qquad (5)$$

and the difference (4) − (5) is

$$\frac{\partial \mathbf{u}}{\partial t} + \bar{\mathbf{V}}\cdot\nabla\mathbf{u} + \mathbf{u}\cdot\nabla\bar{\mathbf{V}} + \mathbf{u}\cdot\nabla\mathbf{u} - \mathbf{e_r}\left[\overline{\mathbf{u}\cdot\nabla w} - \frac{v^2}{r}\right] - \mathbf{e_\theta}\left[\overline{\mathbf{u}\cdot\nabla v} + \frac{vw}{r}\right]$$

$$- \mathbf{e_x}\,\overline{\mathbf{u}\cdot\nabla u} = -\nabla p' + \nabla^2\mathbf{u}. \qquad (6)$$

Energy identities for the mean and fluctuating components are obtained by forming $\langle\bar{\mathbf{V}}\cdot(5)\rangle = 0$ and $\langle\mathbf{u}\cdot(6)\rangle = 0$, respectively. We also use $\bar{V}_r = 0$ and the following identity

$$\overline{\mathbf{u}\cdot\nabla f} = \overline{\nabla\cdot(\mathbf{u}f)} = \frac{1}{r}\frac{\partial}{\partial r}(r\overline{fw}),$$

which is valid when $\nabla\cdot\mathbf{u} = 0$. Thus,

$$\frac{1}{2}\frac{d}{dt}\langle|\bar{\mathbf{V}}|^2\rangle + \left\langle\frac{\overline{uw}\bar{V}_\theta}{r} + \frac{\bar{V}_\theta}{r}\frac{\partial}{\partial r}(r\overline{vw}) + \frac{\bar{V}_x}{r}\frac{\partial}{\partial r}(r\overline{uw})\right\rangle$$

$$= P\langle\bar{V}_x\rangle - \left\langle\left|\frac{\partial \bar{V}_\theta}{\partial r}\right|^2 + \left|\frac{\partial \bar{V}_x}{\partial r}\right|^2 + \frac{\bar{V}_\theta^2}{r^2}\right\rangle, \qquad (7)$$

and

$$\frac{1}{2}\frac{d}{dt}\langle|\mathbf{u}|^2\rangle - \left\langle\frac{\overline{uw}\bar{V}_\theta}{r} - \overline{vw}\frac{\partial \bar{V}_\theta}{\partial r} - \overline{uw}\frac{\partial \bar{V}_x}{\partial r}\right\rangle + \Gamma = -\langle|\nabla\mathbf{u}|^2\rangle. \qquad (8)$$

Here

$$\Gamma = 2\overline{\phi(p'_w + \phi^2/2)}, \qquad (9)$$

where p'_w is fluctuating pressure at the wall of the pipe. Summing (7) and (8) gives the total energy equation

$$\frac{1}{2}\frac{d}{dt}\langle|\bar{\mathbf{V}}|^2 + |\mathbf{u}|^2\rangle = P\langle\bar{V}_x\rangle - \Gamma - \left\langle|\nabla\mathbf{u}|^2 + \left|\frac{\partial \bar{V}_\theta}{\partial r}\right|^2 + \left|\frac{\partial \bar{V}_x}{\partial r}\right|^2 + \frac{\bar{V}_\theta^2}{r^2}\right\rangle, \qquad (10)$$

where it is noted that the second bracketed terms in (7) and (8) cancel.

3 Volume Flux Comparison between Laminar and Turbulent Flows

First we consider laminar Hagen–Poiseuille flow, for which equation (1) has the solution

$$U_l(r) = 2\langle U_l\rangle(1 - r^2) \qquad (11)$$

for

$$P_l = 8 \langle \bar{U}_l \rangle. \tag{12}$$

In order to evaluate the bulk flow rate for the turbulent flow cases, $\langle \bar{V}_x \rangle$, we specify two properties of statistical stationarity, assuming that a turbulent flow exists. The first is that all cylinder averages ($\bar{}$) are time independent, and second we assume that velocity components have a zero mean value unless a non-zero mean value is forced externally. This latter property implies $\bar{V}_\theta = 0$. Under such conditions the x-component of equation (5) may be written as

$$\frac{d}{dr} \left[r\overline{uw} - r\frac{d\bar{V}_x}{dr} - P\frac{r^2}{2} \right] = 0, \tag{13}$$

and the energy equation (8) becomes

$$\left\langle \overline{uw}\frac{d\bar{V}_x}{dr} \right\rangle + \Gamma = -\langle |\nabla \mathbf{u}|^2 \rangle. \tag{14}$$

We now seek an expression for P by taking the first integral of (13)

$$P\frac{r^2}{2} = r\overline{uw} - r\frac{d\bar{V}_x}{dr}. \tag{15}$$

Forming $\langle (15) \rangle$ gives

$$P = 4 \langle r\overline{uw} \rangle + 8 \langle \bar{V}_x \rangle. \tag{16}$$

A comparison between the volume flux in a pipe for a turbulent flow with control and the base laminar flow can now be made. For flows with the same driving pressure gradient ($P = P_l$), using (16) and (12) the difference between the bulk flow rates between fully developed laminar and turbulent flows is given by

$$\langle \bar{U}_l - \bar{V}_x \rangle = \frac{1}{2} \langle r\overline{uw} \rangle. \tag{17}$$

Therefore a proof that zero-net volume flux blowing/suction control cannot produce a volume flux in excess of laminar flow requires

$$\langle r\overline{uw} \rangle \geq 0. \tag{18}$$

To test this we form $\langle \overline{uw} \cdot (15) \rangle = 0$ and using (14) obtain

$$\frac{1}{2}P \langle r\overline{uw} \rangle = \langle |\nabla \mathbf{u}|^2 \rangle + \langle [\overline{uw}]^2 \rangle - \Gamma. \tag{19}$$

Here $P > 0$ by definition, and therefore, the controlled flow can produce a volume flux in excess of laminar flow if, and only if,

$$\Gamma > \langle |\nabla \mathbf{u}|^2 \rangle + \langle [\overline{uw}]^2 \rangle. \tag{20}$$

The same criterion holds for producing sub-laminar drag as will be discussed in the following section.

3.1 Quantitative Comparisons

For pipe flow without flow control (where $\Gamma = 0$), equations (17) and (19) show that the flow rate in laminar flow is always higher than in turbulent flow at the same pressure gradient. This was first proven by Thomas [3], and while generally well known, we are not aware of any formulas in the literature that quantifies these volume flux differences. Quantitative differences between the bulk flow rates can be obtained using functional forms for the mean velocity profiles (laminar and turbulent). Following MJM, using a law of the wall-wake mean velocity formulation [4] it can be shown that, to a good approximation,

$$\langle \bar{V}_x \rangle = Re_\tau \left(\frac{\bar{V}_1}{U_\tau} - \frac{3}{2\kappa} \right) \tag{21}$$

with

$$\frac{\bar{V}_1}{U_\tau} = \frac{1}{\kappa} \ln(Re_\tau) + 5.16 \tag{22}$$

where $\bar{V}_1$ and U_τ are the non-dimensional center-line and skin friction velocities respectively. (Here $Re_\tau = a\hat{U}_\tau/\nu = U_\tau$.) For laminar flow

$$\langle U_l \rangle = \frac{Re_\tau^2}{4}. \tag{23}$$

Figure 1 shows the resulting comparison between the bulk velocities for laminar and turbulent flows for varying levels of Re_τ compared to the experimental data of McKeon et al. [5] obtained in the Princeton superpipe. It is noted that for a typical practical Reynolds number of $Re_\tau = 10^5$ the ratio $\langle U_l \rangle / \langle \bar{V}_x \rangle$ is seen to be over 1000.

4 Drag Comparisons for Pipe Flow

For a fully developed pipe flow with statistical stationarity, the net skin friction drag is simply obtained from a balance with the pressure gradient forces. That is,

$$\hat{\tau}_0(4\pi a L) = \rho \hat{P}(2\pi a^2 L),$$

where $\hat{\tau}_0$ and $\hat{P}$ are dimensional average wall-shear stress and driving pressure gradient respectively. From this it follows that the bulk skin friction coefficient $C_f = 2\hat{\tau}_0/(\rho\hat{U}_B^2)$, where $\hat{U}_B$ is the dimensional bulk velocity, is related to P by

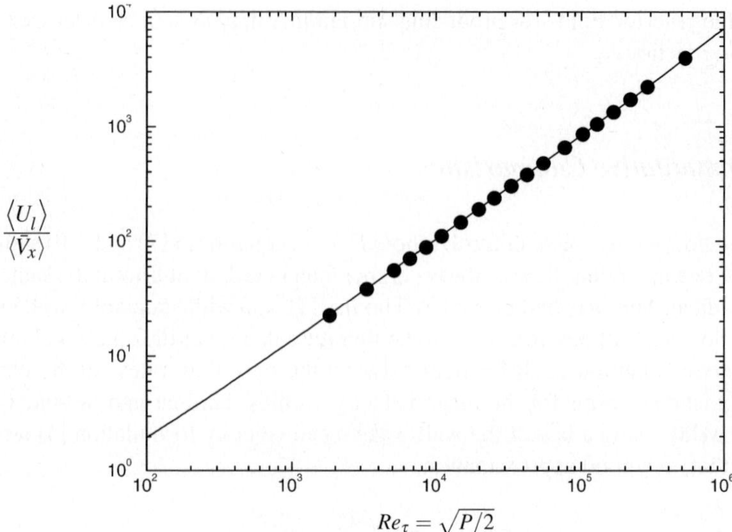

Fig. 1 Ratio of bulk velocities for laminar and turbulent pipe flows as a function of Re_τ. The filled circles are from pipe experiments [5].

$$C_f = \frac{P}{Re_B^2}.\qquad(24)$$

Using (16), (12) and (24) for a given Reynolds number $\langle U_l \rangle = \langle \bar{V}_x \rangle = Re_B$, we obtain

$$C_f - C_{fl} = \frac{4}{Re_B^2}\langle r\overline{uw}\rangle.\qquad(25)$$

Equation (25) is equivalent to the result obtained by Fukagata et al. [6] where they showed that drag reduction is dependent on the weighted integral of Reynolds shear stress.

As expected, the criteria for achieving sub-laminar drag, with control for a fixed volume flux, is equivalent to exceeding the volume flux of laminar flow for a fixed pressure gradient. Both depend on $\langle r\overline{uw}\rangle$. Therefore, for controlled flows to produce sustained sub-laminar skin friction levels requires

$$\Gamma > \langle |\nabla \mathbf{u}|^2 \rangle + \langle [\overline{uw}]^2 \rangle.$$

5 Concluding Remarks

The result in Figure 1 indicates that at very high Re_τ the flow rates in pipes would be phenomenally (and likely aphysically) high for a laminar flow compared to the turbulent case. This would indicate that while sub-laminar conditions require the criterion in equation (20) to hold, other factors likely related to stability need to be considered. Also, the similarity between equation (20) in this paper, and the corresponding expression derived by MJM for a plane channel, suggests that MJM's conclusions about sub-laminar drag control strategies in plane channels, are equally valid for pipe flows.

Acknowledgments

This work was in part supported by the National Science Foundation (IM with CTS-0324898, DDJ with CTS-0302837 and KM with CTS-0133837), and the David and Lucile Packard Foundation.

References

1. I. Marusic, D.D. Joseph, and K. Mahesh. Laminar and turbulent comparisons for channel flow and flow control. *J. Fluid Mech.*, 570:467–477, 2007.
2. T. Min, S.M. Kang, J.L. Speyer, and J. Kim. Sustained sub-laminar drag in a fully developed channel flow. *J. Fluid Mech.*, 558:309–318, 2006.
3. T.Y. Thomas. Qualitative analysis of the flow of fluids in pipes. *Amer. J. Math*, 64:754–767, 1942.
4. A.E. Perry, I. Marusic, and M.B. Jones. On the streamwise evolution of turbulent boundary layers in arbitrary pressure gradients. *J. Fluid Mech.*, 461:61–91, 2002.
5. B.J. McKeon, J.D. Li, W. Jiang, J.F. Morrison, and A.J. Smits. Further observations on the mean velocity in fully-developed pipe flow. *J. Fluid Mech.*, 501:135–147, 2004.
6. K. Fukagata, K. Iwamoto, and N. Kasagi. Contribution of Reynolds stress distribution to the skin friction in wall-bounded flows. *Phys. Fluids*, 14(11):L73–L76, 2002.

Enhancement of Suboptimal Controllability in Wall Turbulence

Olivier Doche, Sedat Tardu and Vincent Kubicki
*Laboratoire des Ecoulements Géophysiques et Industriels, UMR CNRS UJF INPG,
BP 53, 38041 Grenoble Cedex, France;*
E-mail: {olivier.doche, sedat.tardu, vincent.kubicki}@hmg.inpg.fr

Abstract. The wall turbulence is forced to a predictable state through localized time-periodical blowing. It is shown through experiments and direct numerical simulations that the temporal waveform of the localized blowing plays a crucial role in the response of turbulent wall drag. Imposed unsteadiness increases the capacity of the controllability significantly. Results on the optimum periodical temporal waveform of localized blowing are also discussed. We use the characteristics of the cyclostationnarity to achieve this particular goal.

Key words: Near-wall turbulence, active control, localized unsteady blowing.

1 Introduction

Intensive direct numerical simulation investigations conducted during the last decade have clearly shown that the optimal and suboptimal control of the near-wall turbulence are plausible and that appreciable drag reduction of about 30–40% can be achieved through either adaptive or non adaptive schemes. The literature on this topic is vast now and the reader may consult [1, 2] for some recent ideas and developments. The major shortcoming of these methods is the necessity of a dense distribution of sensors (wall shear stress gauges) and actuators (micro blowing-suction jets) with a mesh size roughly equal to the viscous sublayer thickness to achieve significant drag reduction. Increasing the control mesh size decreases the efficiency of the control scheme. This is not always well understood at a first glance. Indeed, the streamwise and spanwise scales of the coherent eddies near the wall are at least an order of magnitude larger than the required control space step. The quasi-streamwise vortices present in the buffer layer are about 300–500 wall units long and are separated by 100 wall units in the spanwise direction. They generate turbulent wall shear by stretching spanwise vorticity zones through ejections and

J.F. Morrison et al. (eds.), IUTAM Symposium on Flow Control and MEMS, 237–242.
© 2008 *Springer. Printed in the Netherlands.*

sweeps. However their regeneration and locations are random in time and space and their capture and subsequent control decision require significantly smaller time and space scales. This poses technical feasibility problems of the sub-optimal strategies, despite the important progress achieved in micro-smart technologies. Suction, on the other hand, is undesired in practical applications. Investigations of somewhat simpler large-scale control methods are therefore still necessary.

One of the ways to remedy to the shortcomings discussed above is to make use of dual-control. The latter consists of exciting a system to increase its predictability and its controllability as a consequence. We applied this strategy by making use of a localized blowing sinusoidal in time with a small severity parameter [5]. We found an unexpectedly strong effect on the near-wall turbulence, especially in the high frequency regime. The local blowing induces a vorticity layer that is of opposite sign to the underlying base flow. The easiest way to explain this phenomenon is to notice that the blowing acts in a manner opposite to suction. The latter suppresses the existing vorticity that is replaced to maintain the non-slip condition. Near the wall the major vorticity component is in the spanwise direction. It is moreover negative at the mean ($\bar{\Omega}_z < 0$) and its instantaneous fluctuating part ω'_z is skewed towards the negative values (the skewnesss of ω'_z is -1 at the wall). Therefore the suction induces a negative spanwise vorticity layer and the local blowing a positive one. The physical mechanism governing the blowing is of course not simply the opposite of suction and more convincing arguments can be found in [5]. The induced vorticity layer advects and diffuses from the wall. In the case of sinusoidal time periodical blowing, however, the diffusion is constrained into a layer of thickness $\delta^+ = 1/\sqrt{f^+}$. Hereafter $+$ denotes the quantities in wall units, non-dimensionalized by the shear velocity and the viscosity. The vorticity layer is negligibly affected by the turbulent mixing, as δ^+ is smaller than the low buffer layer thickness $\delta^+ < 10$; it concentrates and becomes compact under these circumstances. Its first effect is to dilute the prevailing negative vorticity layer near the blowing slot. The flow is consequently partly relaminarized during the acceleration phase of the injection velocity $\langle v_0 \rangle$. The relaminarized phase is unstable and inflexional points appear in the phase-averaged velocity during the deceleration phase. After $x^+ = 50$, however, the reaction of the near-wall turbulence changes somewhat abruptly. Due, on one hand, to the destabilization of the near-wall flow, and on the other, to the constrained diffusion, the induced positive spanwise vorticity rolls up into a coherent vortex. The latter increases the drag in a predictable fashion as it is convected downstream. The system results in a drag penalty for $x^+ = 50$, and can be used in separation rather than in drag control. More details are available in [5].

We have shown through detailed experiments and direct numerical simulations in different recent publications (for example [6]) that local asymmetric blowing with a rapid acceleration followed by a slow deceleration prevents the roll-up of the induced vorticity layer and the further extension of the positive spanwise vorticity downstream resulting in total relaminarization during 70% of the oscillating period. Figure 1 shows, for example, the net suppression of the quasi-streamwise vortices near the slot in the case of asymmetric blowing compared to sinusoidal. These results show how the temporal waveform of the localized perturbation is important in

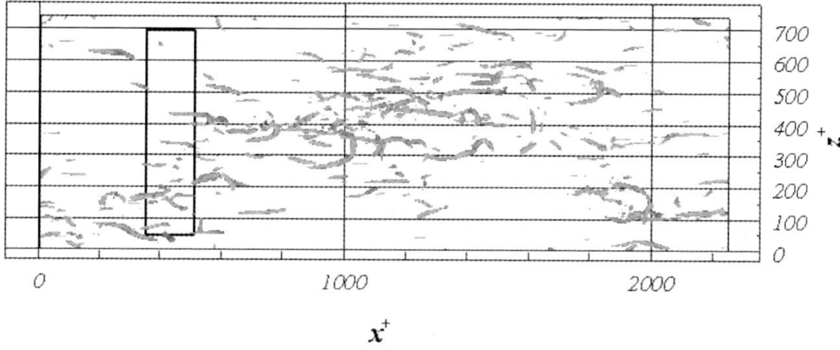

Fig. 1 λ_2 contours in plane (x^+, z^+) for a sinusoidal blowing.

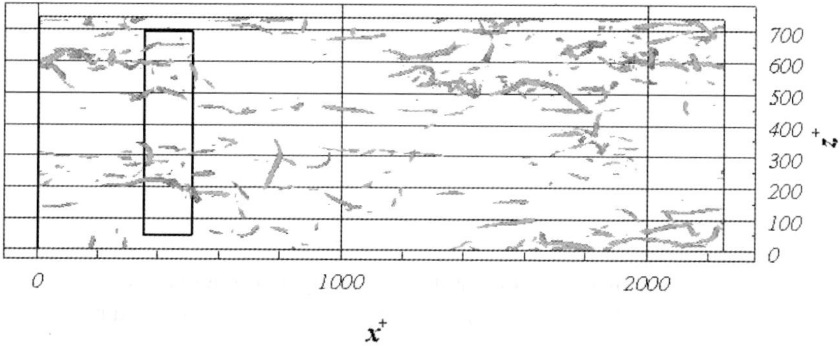

Fig. 2 λ_2 contours in plane (x^+, z^+) for a asymetric blowing.

the drag reduction mechanism, a point that we believe, having clearly investigated and detailed in the past. The question raised in this paper, is however different. We will raise the question of the controllability under the effect of a localized unsteady blowing.

2 Direct Numerical Simulations

The direct numerical simulation (DNS) code developed in [4] has been modified and adapted to investigate the effect of localized blowing of different temporal shapes. The code is of finite difference type combined with a fractional time procedure. The non-linear terms are explicitly resolved by an Adams–Bashforth scheme. Periodical boundary conditions are used in the homogeneous streamwise and spanwise directions. The size of the computational domain is $(4\pi h \times 2h \times 1.33h)$ in respectively the streamwise x, wall normal y and spanwise z directions, h being the channel half width. There are 513, 129, and 129 computational modes in x, y, z, respectively.

Uniform and stretched coordinates are used in the streamwise, spanwise and wall normal directions. The first mesh from the wall is at 0.2 wall units. The mesh sizes in the x and z directions are respectively 4.5 and 5.5 $l_v = v/u_\tau$, where v and u_τ are the viscosity and the shear velocity. The Reynolds number based on the channel height and the centerline velocity is fixed at $Re = U_c \cdot h/v = 4200$ corresponding to $Re = u_\tau \cdot h/v = 180$. The computational time step is $\Delta t^+ = 0.1$.

Contrarily to the optimal control whose aim is to relaminarize the flow in a given time interval, the suboptimal strategy attempts to decrease at each time step the cost function. The latter is:

$$J(\phi) = \frac{k}{2\Gamma} \iint_w \phi^2 dS + \frac{1}{\Gamma} \iint_w \tau dS, \tag{1}$$

where τ is the shear at the wall whose area is denoted by Γ, ϕ is the action at the wall in the form of local blowing/suction distribution and k is a constant. The first integral above is clearly the energy expended to achieve the drag reduction. The control problem consists of determining the optimum ϕ at each time step. The sensitivity of the cost function to the actuation modifications ϕ is measured through Frechet derivatives as in classical non-linear control theory. The variation of a functional $\xi(\phi)$, denoted by $\tilde{\xi}(\phi,\tilde{\phi})$ is given by:

$$\tilde{\xi}(\phi,\tilde{\phi}) = \lim_{\varepsilon \to 0} \frac{\xi(\phi + \varepsilon\tilde{\phi}) - \xi(\phi)}{\varepsilon} = \frac{1}{\Gamma} \iint_w \frac{F\xi(\phi)}{F\phi} \tilde{\phi} dS, \tag{2}$$

where F stands for the Frechet operator. In practice, the Navier–Stokes equation is discretized in time and space, and the resulting operators are transformed through the Frechet operator. Using a Crank–Nicholson scheme for the time discretization results for instance in the decomposition $Q^{n+1} + R^n = 0$ of Navier–Stokes where Q^{n+1} and R^n regroup the terms at the time steps $n+1$ and n. Computing the resulting Frechet transformation of 1 leads to:

$$\frac{DJ}{D\phi} = \frac{k}{\Gamma}\phi - \frac{\beta_1}{\beta_2\Gamma}\pi_w, \tag{3}$$

relating the Frechet variation of the cost function to the adjoint pressure field π_w at the wall. Thus both the Navier–Stokes equation and its related adjoint are resolved in time and space to determine π_w and the suboptimal distribution of blowing/suction actuation at the wall.

The procedure is the same as used [2] with some subtle differences. We noticed for instance that the research of minima algorithm in the cost function at the time step n is particularly efficient when it is based on the gradient $DJ/D\phi$ computed at $n-1$ and not n. Indeed, the wall shear stress, thereby the cost function cannot have proper information on the instantaneous change induced by a sudden variation in the boundary condition. A time lag of about $\Delta t^+ \approx \Delta y^{+2}$ in wall units (related to the shear velocity and viscosity) is necessary in order that the information at the wall diffuses to the first resolved mesh points near the wall. Taking $\Delta y^+ = 0.1$ where the

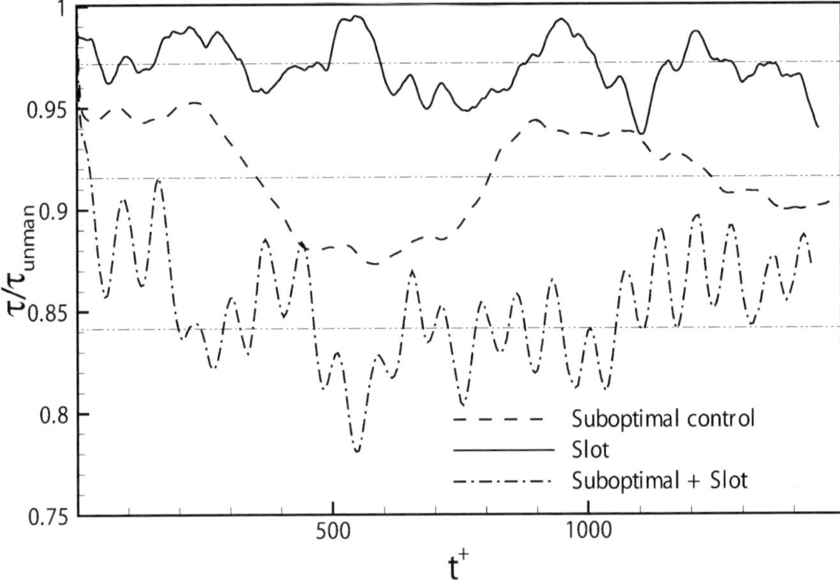

Fig. 3 Drag reduction under suboptimal control combined with localized unsteady blowing.

estimation of the shear is performed results in $\Delta t^+ = 0.01$ which is just the time step of the simulations.

3 Results

The controllability can be easily defined and analyzed in some stability active control problems but is rather difficult to define in the absence of uncoupled modes as in the fully developed near-wall turbulence. We will rather couple here the notion of controllability with that of predictability. A predictable stochastic process should have a discrete spectrum [3]. The time-periodical, localized, blowing allows us to inject the imposed unsteadiness in the spectrum about the median range. Figure 3 shows the drag reduction obtained through suboptimal control schemes with and without localized blowing. It is seen that the imposed unsteadiness increases significantly the capacity of suboptimal control to reduce the drag, as expected. In the case of non-adaptable *ad hoc* schemes, however, the imposed unsteadiness has no significant effect also as expected. One of the main points discussed is the determination of the optimum waveform of temporal localized blowing. This is obtained by first stationnarizing the near-wall turbulence and applying subsequently the suboptimal strategy and phase averaging the resulting blowing velocity. The obtained waveform is similar to the asymmetric one previously discussed.

4 Conclusion

We have shown through well resolved direct numerical simulations that localized imposed unsteadiness increases the suboptimal controllability of the near-wall turbulence. Thus, the drag reduction doubles in the presence of dual control, i.e. when local unsteady blowing through a spanwise slot is coupled with the suboptimal strategy, compared with the suboptimal control alone. More interestingly, the wall shear stress is synchronized with the oscillating injection temporal waveform over a surface of large extent downstream the slot. Synchronization is a fundamental concept in, for instance, the chaos control, and the strategy we propose here may have several practical implications in the management of wall turbulence.

References

1. Bewley T., 2001, Flow control: New challenges for a new renaissance, *Progress in Aerospace Sciences* **37**(1), 21–58.
2. Bewley T., Moin P., Temam R., 2001, DNS-based predictive control of turbulence: An optimal benchmark for feedback algorithms, *J. Fluid Mech.* **447**, 179–225.
3. Papoulis A., 1982, *Probability, Random Variables and Stochastic Processes*, McGraw-Hill, New York.
4. Orlandi P., 2000, *Fluid Flow Phenomena, A Numerical Toolkit*, Kluwer Academic Publishers, Dordrecht, pp. 3–51 and 188–230.
5. Tardu S., 2001, Active control of near wall turbulence by local unsteady blowing, *J. Fluid Mech.* **43**, 631–649.
6. Tardu S., O. Doche, (2005) Optimal active control of turbulent drag by dual strategies, in *Proceedings 4th International Symposium Turbulent Shear Flow Phenomena*, Virginia, June 27–29, 2005.

An Improvement of Opposition Control at High Reynolds Numbers

Mathieu Pamiès[1], Eric Garnier[1], Pierre Sagaut[2] and Alain Merlen[3]

[1]*Applied Aerodynamic Department, ONERA, BP 72, 29 Avenue de la Division Leclerc, F-92322 Châtillon Cédex France; E-mail: mathieu.pamies@onera.fr*

[2]*Institut Jean Le Rond d'Alembert, Université Pierre et Marie Curie, 4 Place Jussieu, 75252 Paris Cédex 05, France*

[3]*Laboratoire de Mécanique de Lille, Université des Sciences et Technologies de Lille, Cité Scientifique, Bâtiment M3, 59655 Villeneuve d'Ascq Cédex, France*

Abstract. Opposition control is a simple feedback control method traditionnally used to attenuate near-wall turbulence and reduce drag in wall-bounded turbulent flows. The idea is to impose blowing and suction at the wall to counteract near-wall quasi-streamwise vortical structures. Unfortunately, the efficiency of this method decreases as the Reynolds number increases. The present study proposes a simple but efficient modification of opposition control (OC) to increase its performance at large Reynolds numbers. We demonstrate a 300% improvement when performing a blowing-only opposition control (BOOC), where OC's suction part has been removed, on a spatially developing turbulent boundary layer at $Re_\tau = 920$. It is shown that BOOC only applies blowing at the location of high skin friction events, which suppresses the latter without altering the "natural" low skin friction events. As a result, BOOC dramatically changes the probability density profile of wall shear stress but does not weaken turbulence intensity near the wall.

Key words: Flow control, drag reduction, turbulent boundary layer, LES.

1 Introduction

The control of wall-bounded turbulence for drag reduction has been extensively studied, particularly in the case of opposition control [1]. This simple feedback control strategy uses blowing and suction at the wall to counteract the velocity induced by near-wall turbulent vortical structures. The velocity at the wall is set to the opposite of the wall-normal velocity taken at a certain height y_{opp}. When the sensing plane ($y = y_{\mathrm{opp}}$) is optimally located, the method succeeds in reducing the wall shear stress (up to 25%) and in attenuating the turbulent intensities in the near-wall region. In this case, the method establishes a virtual wall (at $y = y_{\mathrm{opp}}/2$) where fluctuating quantities are nearly zero [2]. Interaction between the flow and the wall is then considerably reduced. Indeed, physical studies have shown that the instantan-

J.F. Morrison et al. (eds.), IUTAM Symposium on Flow Control and MEMS, 243–249.
© 2008 Springer. Printed in the Netherlands.

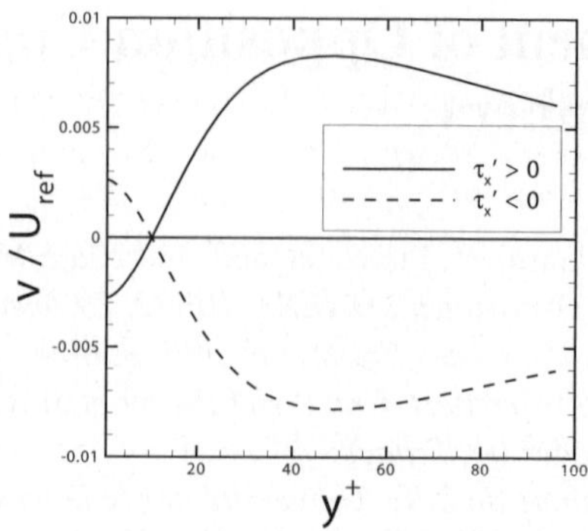

Fig. 1 Conditional average of wall normal velocity. Velocity is averaged when positive (solid line) or negative (dashed line) skin friction fluctuation occurs at the wall.

eous wall shear stress is highly correlated with vortical structures located a small distance above the wall [3] and with turbulence intensities [4]. More precisely, high skin friction zones are correlated with sweep events (downward fluid movement) and low skin fiction zones with ejection events (upward fluid movement), both induced by the near-wall streamwise vortices. Opposition control counteracts both ejection and sweep events, lowering at the same time the intensity of high and low skin friction zones. The control method put forward herein suggests targetting only the sweep events, in order to keep the drag-reducing benefits of the low skin friction zones. In fact, the drag-reducing effect of wall blowing has already been reported, but exclusively for fixed wall transpiration strips. Among others, Park and Choi [5] have shown that localized constant blowing at the wall locally decreases the drag but increases the turbulence intensities. For high blowing rates, this can even lead to a significant increase of skin friction downstream the blowing zone. The exact opposite effect is reported for suction. On the contrary, the blowing and suction zones of opposition control are not stationary: they are convected with the vortical structure which they are trying to counteract. Figure 1 demonstrates that Park and Choi's results also apply to moving, non-uniform, blowing or suction zones, showing the conditionnal average of the wall normal velocity in an opposition-controlled boundary layer. The average is computed in cases of positive and negative skin friction fluctuations. It appears clearly that low (high) skin friction events appear under a wall blowing (suction) condition. Consequently, removing the suction part of the opposition control can have a significant drag-reducing effect.

Such a modification implies the injection of fluid into the boundary layer, which could increase its sensitivity to separation. *A posteriori* verification shows that the amount of fluid introduced into the flow, measured by the integral of the wall velocity averaged over the controlled surface S and the simulation time T

$$\langle V_w \rangle = \frac{1}{ST} \int_T \int_S v \, dS \, dt$$

is less than $0.001U_\infty$. The ratio of momentum flux gain due to the blowing and momentum flux of the incoming boundary layer is defined as $\sigma = V_w L / \theta U_\infty$, where θ is the momentum thickness of the uncontrolled flow at the begining of the control and L the streamwise extent of the control area [6]. For the new strategy proposed here, σ is equal to 0.037, which is a relativelty small value, and is unable to trigger any separation effect. Another modification induced by this new method is the loss of suction's stabilization effect due to the decrease of the turbulent activity above the wall. The method proposed here will *a priori* not be able to achieve drag reduction without an increase in the root-mean-square velocities, which leads to the same result as an uniform blowing strip. Thus, an *a posteriori* verification is needed to assess the method's effectiveness for drag-reduction purpose. This will be verified by a comparison between the new method and a control with a constant velocity imposed at the wall.

2 Numerical Procedure

We investigate the potential of this method (referred to as blowing-only opposition control, or BOOC) using a large eddy simulation of a spatially evolving turbulent boundary layer at Mach number 0.1 and $Re_\theta \approx 3300$. The Reynolds number is based on the momentum thickness, θ, and the free stream velocity, U_∞ ($\approx 25 u_\tau$). The velocity, distance, and time were non-dimensionalized using wall units, where the friction velocity, $u_\tau = (\tau_w / \rho)^{1/2}$, is determined from the average wall shear stress τ_w for the uncontrolled flow. Simulations were performed using a finite-volume solver for the compressible Navier–Stokes equations. LES filtering operation uses the formalism developed by Vreman [7] and subgrid scales are modeled using the selective mixed scale model proposed by Sagaut [8]. Time advancement was accomplished using the second-order accurate backward scheme of Gear with a time step of $0.00126\delta/U_\infty$. The spatial scheme is a modified AUSM+(P) scheme, proposed by Mary and Sagaut [9]. The size of the computational domain is $(20\delta \times 4.8\delta \times 2.7\delta)$ in the longitudinal, wall-normal and transverse directions, respectively. Grid spacings are: $\Delta x^+ = 50$, $\Delta y^+_{\min} = 1$ and $\Delta z^+ = 17$. The computational domain has been split along the streamwise direction into four blocks of equal size: the two first ones are uncontrolled and devoted to the imposition of realistic inflow boundary conditions with the rescaling method of Lund et al. [10]. The control takes place in the third block and the last domain is used to record possible relaxation effects. The transpiration velocity at the wall is the positive part of the one obtained with a classical opposition control: $v_{\text{wall}} = \max(0; -v(x, y_{\text{opp}}, z))$. No slip boundary conditions were applied at the wall for the u and w components. In order to compare the method to classical OC, uniform continuous blowing has been also applied at the entire controlled wall at a rate corresponding to the mean blowing rate of a BOOC simulation.

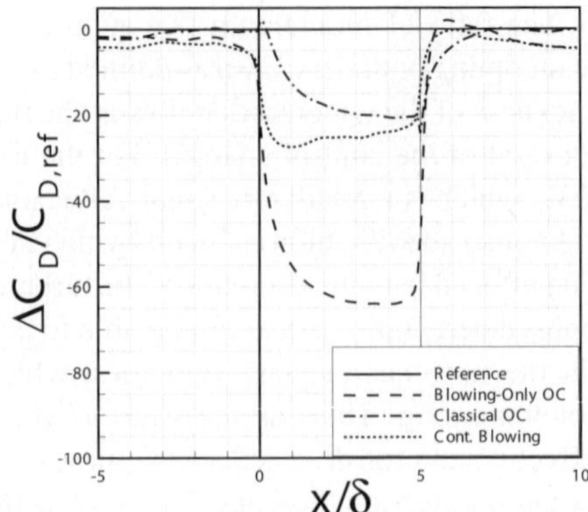

Fig. 2 Spatial evolution of the percentage of relative drag reduction for an opposition control (dash-dotted line), a blowing-only opposition control (long-dashed line) and a continuous blowing one (dashed line). Control starts at $x = 0$ and stops at $x = 5\delta$.

3 Results

Most studies dealing with opposition control have been conducted using direct numerical simulations in channel flows. Although they have demonstrated drag reductions up to 25%, the Reynolds number of the flows were in the range $80 < Re_\tau < 180$. Chang et al. [11] performed large eddy simulations of the same case up to $Re_\tau = 720$ and found that the efficiency of opposition control decreases as the Reynolds number increases. They also observed that the optimal sensing plane location gets closer to the wall. In the present work at $Re_\tau = 920$, we used an extrapolation of their results and chose $y_{\mathrm{opp}}^+ = 11$. For this case, opposition control yields 20% drag reduction, which is in good agreement with an extrapolation of the results of Chang and co-workers. Figure 2 shows that BOOC is much more efficient and reaches a 64% drag redution. The drag-reducing effect stops immediately after the controlled zone, indicating that all methods do not establish the modification in a sustainable way.

Nevertheless, the ability of opposition control to reduce root-mean-square velocities [1, 11] is still observed in our simulations, as shown in Figure 3(a). On the contrary, rms velocities are increased by approximately 20% when BOOC is applied, and by even more when continuous blowing is applied, as expected. This modification is persistant in the relaxation zone only for BOOC, as it can be seen in Figure 3(b). The error from the reference is then close to 10% for all components and is propagating vertically into the logarithmic region while decreasing in the streamwise direction. Consequently, the susbtantial loss in skin friction when using the BOOC cannot be achieved without relatively small but persisting alterations of the turbulence.

Physical insights [2, 12] into opposition control drag-reducing mechanisms have demonstrated the presence of the so-called virtual wall. As illustrated in the inset in Figure 3(a), the establishment of this wall for our opposition control is not fully

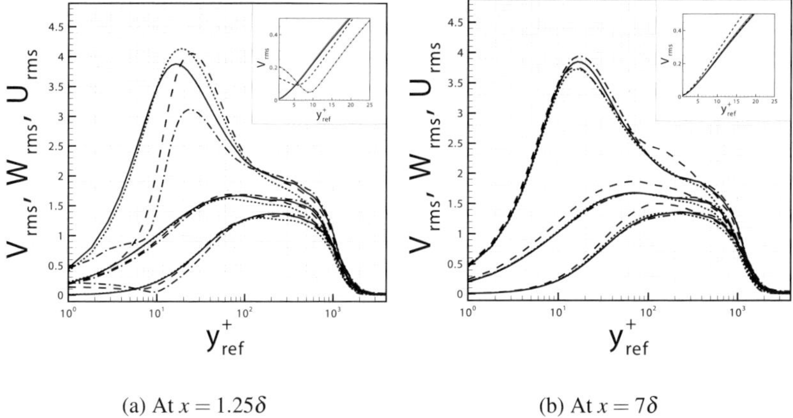

(a) At $x = 1.25\delta$ (b) At $x = 7\delta$

Fig. 3 RMS profiles of the velocity. V_{rms} has been plotted in linear scale for the near-wall region in the upper-right corner of figures. (For legend see previous caption.)

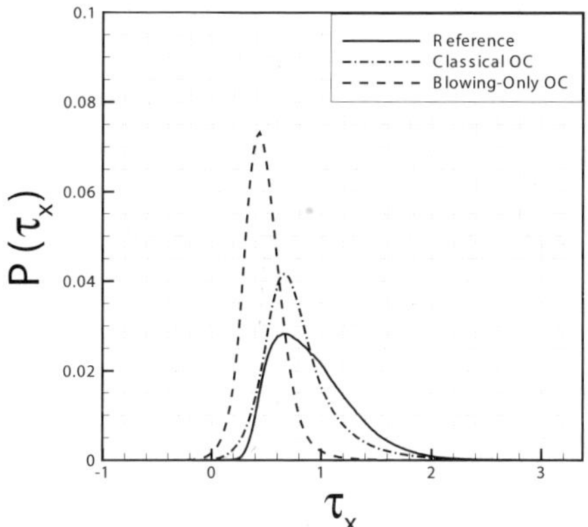

Fig. 4 Probability density function of the streamwise wall shear stress.

efficient because the minimum of v_{rms} occurs at a non-zero value and close to the sensing plane. However, the control succeeds in weakening and pushing away the near-wall vortical structures. The sweeps and ejections induced by them do not reach the real wall, but only the virtual one. BOOC is designed to not counteract the drag-reducing ejection events. This fact probably explains why the minimum of v_{rms} occurs at a value twice as large as the minimum obtained for opposition control. In that sense, BOOC also establishes a virtual wall, but one which is only impermeable to sweeps. As a result, BOOC reduces the turbulent transport towards the wall but somehow increases the turbulent transport away from the wall.

The statistical effect on skin friction can be seen in Figure 4. As already noticed by Hu et al. [13], the probability density function of the uncontrolled flow exhibits a important asymmetry: the maximum occurs at $\tau'_x = -0.7\sigma$ (where σ is the standard deviation) and kurtosis is about 4.14. Events that statistically contribute the most to the low-skin friction are concentrated beneath and near the mean in a hump. On the contrary, events which are responsible for the high skin friction are spread after the mean up to relatively large values. Consequently, an efficient control method should reduce this spreading and concentrate skin friction events around the mean. Figure 4 ensures that both classical and blowing-only opposition controls fulfil this condition, but BOOC is by far more efficient, since standard deviation is reduced by half ($\tau_{x,\mathrm{rms}}$ passes from 0.36 to 0.18) and maximum of PDF occurs closer to the mean ($\tau'_x = -0.19\sigma$).

4 Conclusion

This study has shown that an improvement of opposition control is possible, even at relatively high Reynolds number. Deep modifications on skin friction behaviour can be achieved through a mechanism similar to that of opposition control. Nevertheless, the main disadvantage of BOOC compared to classical OC remains the alteration of the turbulence properties due to the application of the control. But to put things into perspective, the gain in drag reduction may widely compensate the 10% increase in turbulence intensities.

References

1. Choi, H., Moin, P., Kim, J.: Active turbulence control for drag reduction in wall-bounded flows. *J. Fluid Mech.* **262** (1994) 75–110.
2. Hammond, E.P., Bewley, T.R., Moin, P.: Observed mechanisms for turbulence attenuation and enhancement in opposition-controlled wall-bounded flows. *Phys. Fluids* **10** (1998) 2421–2423.
3. Kravchenko, A.G., Choi, H., Moin, P.: On the relation of near-wall streamwise vortices to wall skin friction in turbulent boundary layers. *Phys. Fluids A* **5** (1993) 3307.
4. Fukagata, K., Iwamoto, K., Kasagi, N.: Contribution of Reynolds stress distribution to the skin friction in wall-bounded flows. *Phys. Fluids* **14** (2002) 73.
5. Park, J., Choi, H.: Effects of uniform blowing or suction form a spanwise slot on a turbulent boundary layer flow. *Phys. Fluids* **11** (1999) 3095–3105.
6. Antonia, R.A., Zhu, Y., Sokolov, M.: Effect of concentrated wall suction on a turbulent boundary layer. *Phys. Fluids* **7** (1995) 2465.
7. Vreman, A.W.: Direct and large eddy simulation of the compressible turbulent mixing layer. PhD Thesis. University of Twente, Twente (1995).
8. Sagaut, P.: *Large-Eddy Simulation for Incompressible Flows, An Introduction.* Springer (2002).
9. Mary, I., Sagaut, P.: LES of a flow around an airfoil near stall. *AIAA J.* **40** (2002) 1139–1145.
10. Lund, T.S., Wu, X., Squires, K.D.: Generation of turbulent inflow data for spatially-developing boundary layer simulations. *J. Comput. Phys.* **140** (1998) 233–258.
11. Chang, Y., Collis, S.S., Ramkrishnan, S.: Viscous effects in control of near-wall turbulence. *Phys. Fluids* **14** (2002) 4069–4080.

12. Kang, S., Choi, H.: Active wall motions for skin-friction drag reduction. *Phys. Fluids* **12** (2000) 3301–3304.
13. Hu, Z.W., Morfey, C.L., Sandham, N.D.: Wall Pressure and Shear Stress Spectra from Direct Simulations of Channel Flow. *AIAA J.* **44** (2006) 1541–1549.

Direct Numerical Simulation of Alternated Spanwise Lorentz Forcing

Stéphane Montesino, Jean-Paul Thibault and Sedat Tardu
*Laboratoire des Ecoulements Géophysiques et Industriels, UMR CNRS UJF INPG,
BP 53, 38041 Grenoble Cedex, France;
E-mail: {stephane.montesino, jean-paul.thibault,sedat.tardu}@hmg.inpg.fr*

Abstract. Spanwise electro-magnetic forcing is used to study turbulence control and drag reduction in a numerical channel flow with a constant mass flow rate and low Reynolds number. The originality of this study comes from the computation of the force field from the geometry of the magnet and the electrode. It is shown that the tilt of the wall-normal component of the vorticity in the spanwise direction characterise the drag reduction caused by alternated spanwise forcing.

Key words: Electromagnetic turbulence control, spanwise oscillation.

1 Introduction

Based on the use of "flush mounted wall" electrodes ($\mathbf{j}$, current density) and "subsurface" magnets ($\mathbf{B}$, magnetic induction), electromagnetic (EM) forcing can create local Lorentz body forces ($\mathbf{j} \wedge \mathbf{B}$) within a conducting boundary layer. This direct source of momentum has no moving parts, holes, nor protuberances and there is no mass injection nor suction. Moreover, an unsteady forcing with an arbitrary wave form and a high frequency response can be used.

EM forcing has academic applications like boundary layer manipulation and vortical structures generation or destruction [1]. Furthermore, it has also naval applications like flow separation prevention [2] and drag reduction [3, 4]. In most of the litterature about EM forcing [4, 5], the EM force model is simplified to an exponential function with a given penetration depth. The originality of this study lies in the computation of the EM forces based on the geometry of the electrodes and the magnets.

The present work is focused on the so-called parallel EM actuators, where electrodes and magnets are parallel to each other and generate a quasi unidirectional force field parallel to the wall. When the actuators are oriented perpendicularly to the mean flow, they generate spanwise forces, which can be periodically excited.

J.F. Morrison et al. (eds.), IUTAM Symposium on Flow Control and MEMS, 251–257.
© 2008 *Springer. Printed in the Netherlands.*

Direct Numerical Simulations (DNS) of a turbulent channel flow are performed in order to give a better understanding of the mechanisms involved in drag reduction caused by a spanwise electromagnetic flow control. To do so, we analyze the behaviour of the forced flow through quantitative analysis of the flow statistics as well as qualitative analysis of the flow stuctures evolution.

2 Direct Numerical Simulation

This choice of DNS is motivated by the strongly non-uniform distribution of the applied EM forces which can act very locally in the flow at the scales imposed by the actuator. Prior to simulate any forcing of the turbulent channel flow, the first step is the computation of the fully ordinary (unforced) turbulent flow. The one obtained is fully comparable to that of Kim et al. [7] which is classically taken as a reference. These results are used as initial flow fields for the simulation presented hereafter. The simulations are based on a parallel (Open MP) and slightly modified version of Paolo Orlandi's code [8] using fractional step method and second-order finite central difference scheme, where the mean pressure gradient is adjusted in order to globally conserve the mass flow rate. In a first step, the viscous terms are treated implicitly, the convection one explicitly. A correction step solves a Poisson equation, which drives to a conservative velocity field (local divergence free). It uses Fourier transforms (FFT) in the periodic direction and a tridiagonal solver in the wall-normal direction. The time advancement uses a third-order Runge–Kutta method for its stability and its low storage cost. The computational domain of the simulations is $4\pi h \times 2h \times \frac{4}{3}\pi h$, where h is the half height of the channel. The necessary minimal mesh ($512 \times 129 \times 256$) is larger than that classically used for the simulation of unforced turbulent channel flow. Non-uniform meshes with a *tanh* evolution law (stretching parameter $= 3.8$) in the wall-normal direction ($\mathbf{y}$) and periodic boundary conditions in the streamwise ($\mathbf{x}$) and spanwise ($\mathbf{z}$) directions are used. The Reynolds number of the simulation (based on the shear velocity U_τ and the half height h of the channel) is quite low:

$$Re^* = \frac{U_\tau h}{\nu} = 178.5. \tag{1}$$

The grid spacing is $\Delta x = 4.4^+$, $\Delta z = 2.9^+$, $\Delta y_{\min} = 0.5^+$ at the wall and $\Delta y_{\max} = 5.5^+$ at the center of the channel. The subscript $^+$ denotes normalisation by the viscous length $l_\nu = \nu/U_\tau$, where ν is the kinematical viscosity and U_τ is the wall shear velocity defined as

$$U_\tau = \sqrt{\nu \left.\frac{d\langle u\rangle}{dy}\right|_{\text{wall}}}. \tag{2}$$

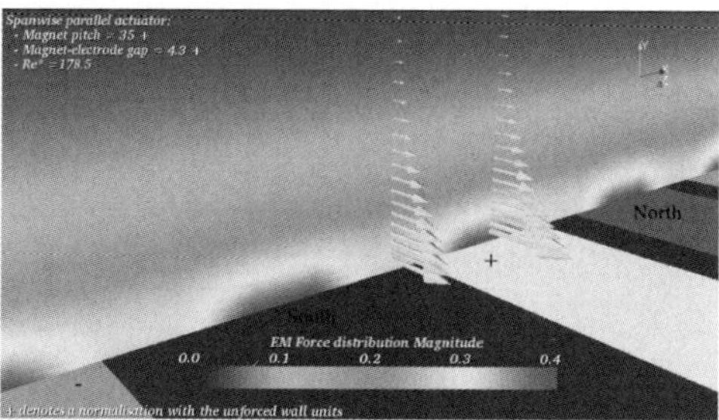

Fig. 1 Forces distribution.

3 EM Force Model

The electromagnetic source-terms $(\mathbf{j} \wedge \mathbf{B})$ included in the Navier–Stokes equations are quasi-independent from the velocity field of the flow, principally because of the poor apparent electrical conductivity of seawater. As a consequence, the EM force field only depends on the actuator geometry and the current power supplied. So, the forcing scales have to be adjusted to the flow scales. The originality of this study comes from the computation of the EM forces from the 2D or 3D geometry of the electrodes and the magnets by an analytical method based on magnetic and electric flux conservation [6]. The current and magnetic density are assumed constant over the surface of the magnets and the diffusion layer of the electrodes. In the case of parallel EM actuatators, it is assumed that electrodes and the magnet are infinitely long, and the 2D model is used to compute the electric and magnetic field in this study. The solution represented in Figure 1 is a cut of the electromagnatic force field distribution over the electrodes and the magnets of the parallel EM actuator. It clearly demonstrates that the force field is far from being uniform and that a maximum arises over the magnets and the electrodes. The force field intensity is directly controlled by the current intensity and its sign is fixed by the polarity of the electrodes. The maximum of the distribution arise at 0.4 from the wall. The magnet width is equal to the electrode width and is 13.1^+. The distance between magnet and electrode is 4.4^+. The magnet height is equal to the half size of the channel $h^+ = 178.5$. A magnet height equal to five times its width guaranties that the magnetic field saturation is reached. The mean EM force field distribution is represented in Figure 2. The maximum of the mean value is around 3.3^+ and 63% of this value is still reached at $y^+ \approx 7$.

4 Simulation of EM Forcing

The EM forcing presented hereafter is performed with a magnet pitch equal to 35^+. The forcing intensity is characterised by a Stuart number based on the shear velocity U_τ and the half size of the channel h,

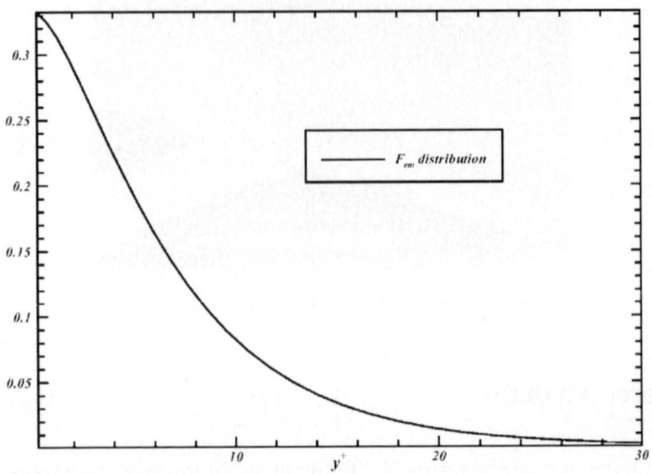

Fig. 2 Mean forces distribution.

$$\text{Stuart}^* = \frac{jBh}{\rho U_\tau^2} = 2400. \tag{3}$$

The Stuart number is the ratio between a convective time scale h/U_τ and an electromagnetic time scale $\rho U_\tau / jB$. Considering the distribution of the EM forces, the mean value of the force at the wall is around 800. With this high intensity, the forcing is able to strongly interact with the velocity fluctuations and its main effect on the channel flow can be underlined. The sign of the forcing is alternated with a temporal period $T = 53^+$ and around 40% of drag reduction is achieved after only 6 periods. The statistics presented hereafter are integrated in the streamwise and spanwise direction during four forcing period after achieving a steady drag.

The root mean square (RMS) of velocity fluctuations are plotted in Figure 3. It shows a net decrease of the streamwise and wall normal velocity fluctuatuations. The RMS value of the spanwise velocity is increased by the alternated spanwise forcing. It achieve a maximum $w'_{RMS} = 6.5^+$ at $y \approx 5^+$ that mostly characterise the Stokes layer generated by the alternated forcing.

The RMS of vorticity fluctuations are shown in Figures 4 and 5. We can notice an enhancement of the RMS streamwise vorticity fluctuations at the wall directly caused by the forcing. It is one order of magnitude more important than for the freely evolving turbulent flow. This recalls that the intensity of the forcing is really strong. We can also notice a local minimun ($\omega'_{x_{RMS}} = 0.6^+$) at $y \approx 5^+$ which characterises the height of the Stokes layer generated by the alternated forcing. The RMS of the spanwise vorticity fluctuations is subject to a net decrease near the wall with a local minimum around $\omega'_{z_{RMS}} = 0.15^+$ under $y \approx 5^+$. This decrease translate into a stabilisation of the wall shear as it can be seen in the detailed flow visualisation we performed (not shown here).

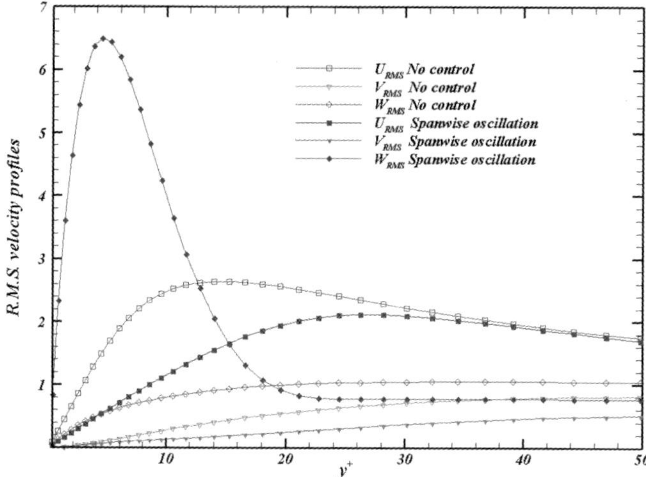

Fig. 3 RMS velocity profiles.

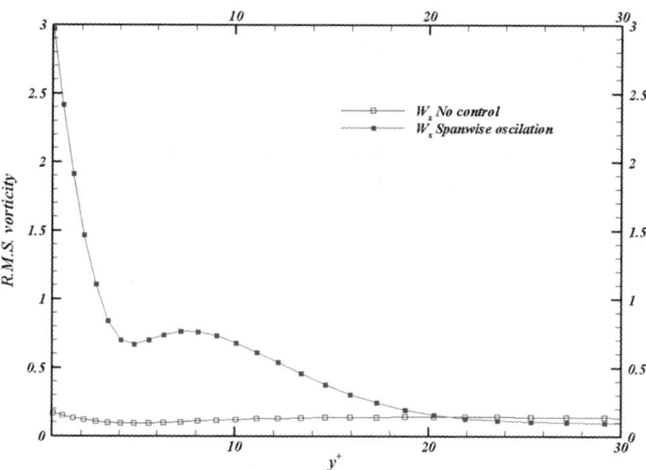

Fig. 4 RMS vorticity (x) profiles.

The RMS of the normal vorticity fluctuations is increased near the wall (under $y^+ \approx 2.5$) and is next subjected to a net decrease. A local maximum arises $\omega'_{y_{RMS}} = 0.07^+$ at $y^+ \approx 1.5$ and a local minimum arises ${\omega'_{y_{RMS}}}^+ = 0.07$ at $y^+ \approx 5$ at the edge of the Stokes layer. We may explain this fact by a tilt of the normal vorticity towards the spanwise direction by the Stokes layer.

Flow visualisations in Figures 6 and 7 show that the spanwise oscillation break down the wall normal vorticity layers which are strongly tilted in the spanwise direction. The consequence of this is the reduction of their tilting in the stream-

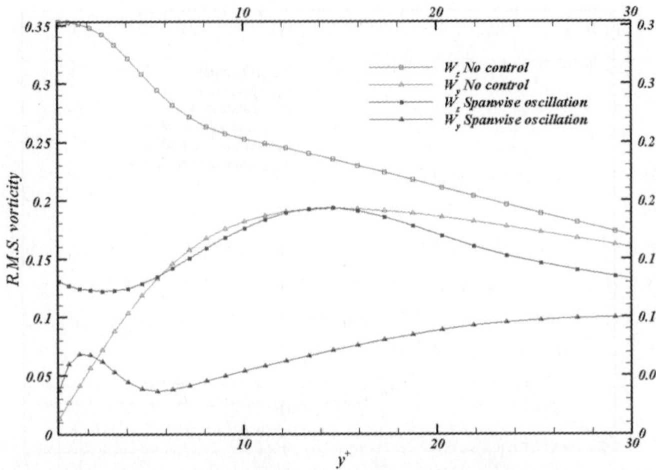

Fig. 5 RMS vorticity (*y-z*) profiles.

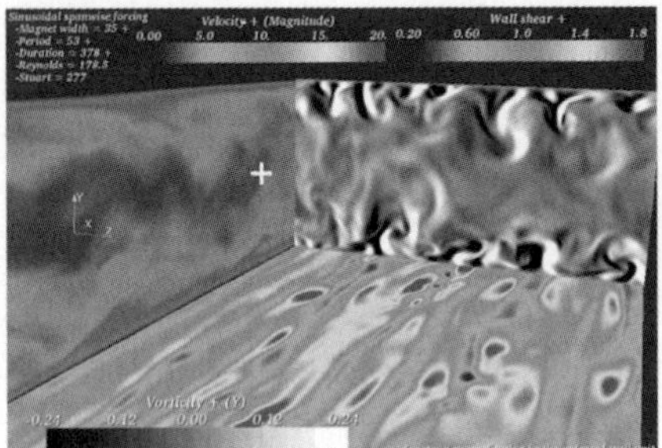

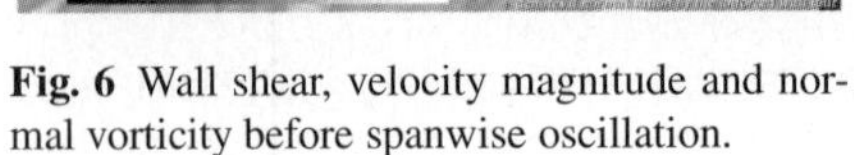

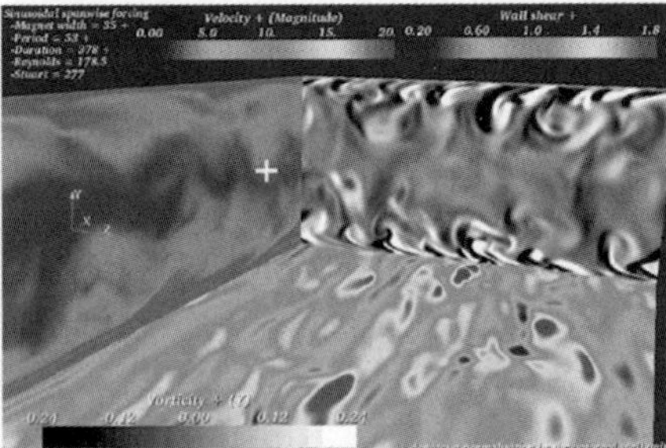

Fig. 6 Wall shear, velocity magnitude and normal vorticity before spanwise oscillation.

Fig. 7 Wall shear, velocity magnitude and normal vorticity at the initial stage of the spanwise oscillation.

wise direction by the mean shear and the formation (regeneration) of new quasi-streamwise vortices. This is also clearly seen in the flow statistics (Figure 5) which show two local maxima in the distribution of the RMS of wall normal vorticity fluctuations $\omega'_{y_{\mathrm{RMS}}}$. The $\omega'_{y_{\mathrm{RMS}}}$ layers are thereby concentrated in the viscous sublayer and strongly stabilised in the buffer layer.

5 Conclusions

In this study, it is shown that the tilt of the wall-normal component of the vorticity in the spanwise direction characterises the drag reduction caused by alternated spanwise forcing. The legs of the hairpin vortex are convected by the forcing whereas the head is not affected. Consequently, the streamwise vortices are tilted in the spanwise direction as in spanwise wall oscillation [9].

6 Perspectives and Applications

The DNS flow solver and the EM forcing pre-solver have demonstrated their capability to simulate the intense EM forcing of a turbulent channel flow. In the presented case, a single configuration in term of penetration length, oscillation period, Stuart and Reynolds number is used to validate the code. This study is a first step towards the achievement of a parametric approach of a spanwise forcing with lower intensities. The expected results of this parametric study is the understanding of the action playing the keyrole in drag reduction and, consequently, to propose an actuation scheme involving a limited power supply to reduce the drag.

References

1. Thibault J-P., Rossi L., 2003, Electromagnetic flow control: Characteristic numbers and flow regimes of a wall-normal actuator, *J. Phys. D: Appl. Phys.* **36**, 2559–2568.
2. Weier T., Fey U., Gerbeth G., Mutschke G., Avilov V., 2000, Boundary layer control by means of electromagnetic forces, *ERCOFTAC Bull.* **44**, 36–40.
3. Nosenchuck D.M., 1996, Boundary layer control using the Lorentz force, in *Proceedings of ASME Fluids Engineering Meeting*, San Diego.
4. Berger T.W., Kim J., Lee C., Lim J., 2000, Turbulent boundary layer control utilizing the Lorentz force, *Phys. Fluids* **12**(3), 631–649.
5. Du Y., Symeonidis V., Karniadakis G.E., 2002, Drag reduction in wall-bounded turbulence via a transverse travelling wave, *J. Fluid Mech.*, **457**, 1–34.
6. Akoun and Yonnet, 1984, 3D analytical calculation of the forces exerted between two cuboidal magnets, *IEEE Transactions on Magnetics* **20**(5), September.
7. Kim J., Moin P. and Moser R., 1987, Turbulence statistics in fully developed channel flow at low Reynolds number, *J. Fluid Mech.* **177**, 133–166.
8. Orlandi P., 2000, *Fluid Flow Phenomena, A Numerical Toolkit*, Kluwer Academic Publishers, Dordrecht, pp. 3–51 and 188–230.
9. Choi K.S., 2002, Near-wall structure of turbulent boundary layer with spanwise-wall oscillation, *Phys. Fluids* **14**(7), 2530–2542.

Boundary Layer Control for Drag Reduction by Lorentz Forcing

Peng Xu and Kwing-So Choi
School of Mechanical, Materials and Manufacture Engineering, University of Nottingham, NG7 2RD, U.K.; E-mail: kwing-so.choi@nottingham.ac.uk

Abstract. A study was carried out with an aim to better understand the drag reducing mechanisms by spanwise oscillation and spanwise travelling wave via Lorentz forcing flow control. A maximum 47% of drag reduction was achieved with $w^+ \approx 12.2$ when the Lorentz forcing spanwise oscillation was applied in a turbulent boundary layer. It was, however, shown that the spanwise travelling wave forcing can reduce or increase the skin friction drag depending on the operating conditions, which offers a flexibility for flow control. A maximum 28.9% of drag reduction and 22.8% of drag increase have been achieved, respectively. Flow visualization indicated that the spanwise displacement of the streaky structures may play an important role in obtaining the drag reduction by spanwise travelling wave actuation.

Key words: Flow control, turbulent boundary layer, Lorentz force.

1 Introduction

Electro-magnetic (Lorentz) force is a body force which is non-intrusive to the flow and allows many different types of control to be introduced such as streamwise force control [8], wall-normal blowing or suction [11], and spanwise oscillation [2, 3, 12]. Due to its benefits, flow control by Lorentz forcing (or called Electro-Magnetic Turbulence Control) has attracted many researchers' attention and has been extensively studied in the past decade.

Recent research showed that the turbulence regeneration cycle is associated with the near-wall region [9], where one of the strategies for drag reduction is to modify or disturb any part of the near-wall activities [10]. The weakening of streamwise vortices [7] was successful in reducing skin friction drag. In the present paper, two flow control schemes were introduced: spanwise oscillation and spanwise travelling wave via Lorentz forcing. The aim was to better understand the drag reducing mechanisms by Lorentz forcing. To achieve this, the velocity profiles and turbulent stat-

J.F. Morrison et al. (eds.), IUTAM Symposium on Flow Control and MEMS, 259–265.
© 2008 *Springer. Printed in the Netherlands.*

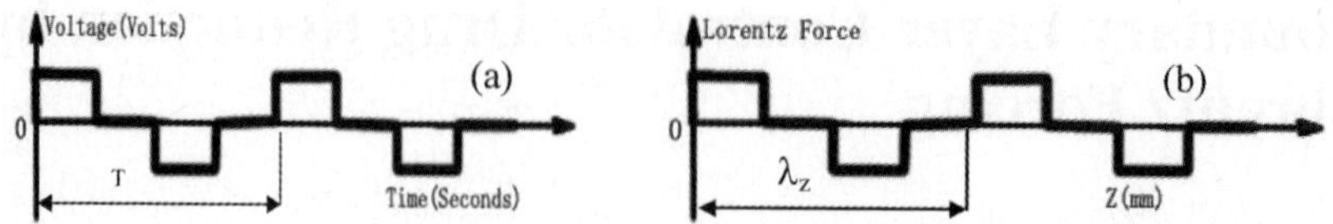

Fig. 1 Actuating signal to each electrode (a) for spanwise travelling wave control associated with Lorentz force (b).

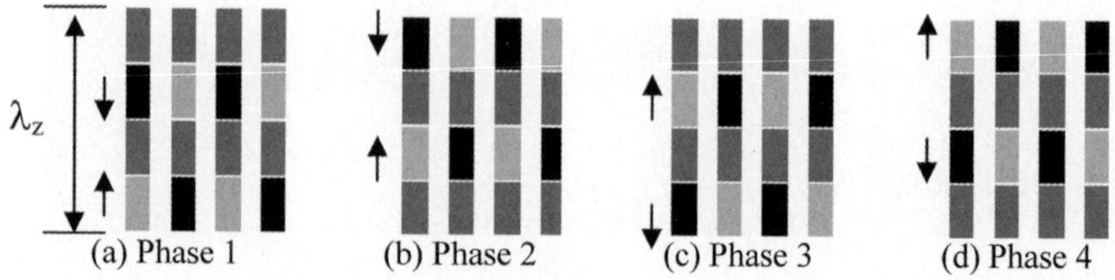

Fig. 2 (a)–(d) Four phases of the travelling wave actuation. Black colour denotes positively activated electrodes; light grey denotes negatively activated and dark grey not activated. Flow is from right to left.

istics with and without Lorentz forcing were obtained using hot-film anemometry. Flow visualizations were also performed.

2 Experiments

The experiments were carried out in a 7.3 m long close-loop open-water channel with a working section of 600 mm by 300 mm where a 4.0 m long test plate was placed horizontally. A trip device was placed just after the leading edge to make sure the flow in test section was fully turbulent. The freestream velocity was 0.1 m/s, and the Reynolds number was 388 based on the friction velocity and boundary layer thickness. Tap water was made conductive by introducing electrolyte solution with a conductivity of 1.0 S/m through a wall slot 24 mm upstream of the electromagnetic (EM) actuator. Copper Sulphate solution was injected for velocity measurements, while Potassium Permanganate solution was used for flow visualization because of its dark red colour. Water temperature was controlled by a heat exchanger in the upstream tank and the temperature change was within 0.1°C throughout the measurements. Velocity signals were measured by the Dantec CTA hot-film anemometry with a single boundary-type sensor. The sensor was mounted in a vertical traverse gear with a step motor of 6 μm resolution. All the turbulent boundary layer profile measurements were made at the 7 mm downstream of the end of the EM actuators, where no electro-magnetic effects on hot-film sensor were detected.

A 0.25 mm thick Mylar sheet with etched copper laminate (0.017 mm thick) was laid flush with the surrounding test plate, below which permanent magnets (1.2 T)

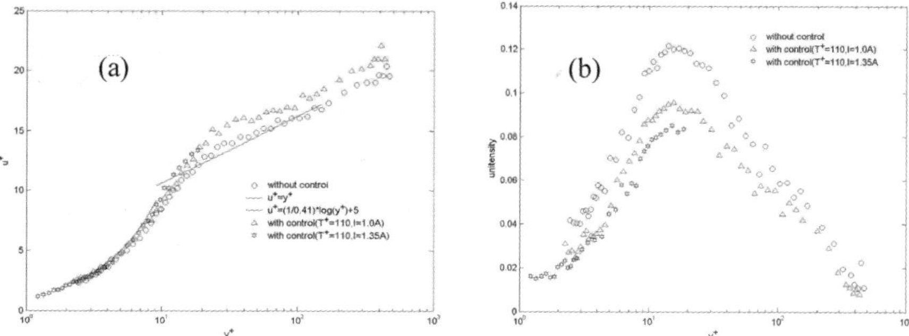

Fig. 3 (a) The logarithmic velocity profiles with and without spanwise oscillation actuation. (b) Turbulence intensity profiles.

were placed. The EM actuator covered the streamwise distance of 300 mm ($x^+ = 1580$) and a spanwise distance of 230 mm ($z^+ = 1214$). The penetration depth, as defined by $\Delta^+ = a^+/\pi$ was 7.6, where a^+ is the width of the electrode in wall units. Figure 1a shows the square wave signal to each electrode, where T denotes an excitation period. A four-phase scheme was introduced here for the travelling wave excitation. The spatial distribution of the spanwise Lorentz force is shown in Figure 1b, where the wave length λ_z was 495 wall units. Figures 2a–2d give an example of how the phase of the travelling wave is shifted.

3 Results

As shown in Figure 3, the logarithmic velocity profiles are shifted upward, indicating that the viscous sublayer is thickened by the spanwise oscillating Lorentz force; similar to [12]. This is also similar to the results for spanwise wall oscillation [4]. The turbulent intensity profiles show significant reduction as well, not only in the near-wall region but also in the logarithmic region of the boundary layer. Meanwhile, both skewness and kurtosis of the streamwise velocity fluctuations were increased in the viscous sublayer and the buffer layer (not shown here). These increases in higher-moments are the consequence of a reduction in turbulence activity, agreeing well with the experiments carried out by Pang and Choi [12] and others on spanwise wall oscillation [1, 5, 13].

Figure 4a shows a visualization picture of the near-wall streaky structures with Lorentz force oscillation. We can clearly see that the low-speed streaks were twisted into the spanwise direction due to spanwise flow oscillation. These are very similar to those observed in spanwise wall oscillation (Figure 4b). Choi et al. [6] proposed that a negative spanwise vorticity is created in turbulent boundary layers due to the spanwise wall oscillation, which seems to apply to the Lorentz force flow oscillation

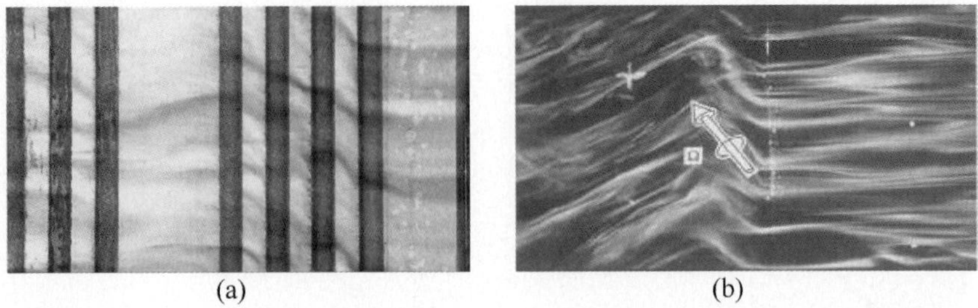

(a)   (b)

Fig. 4 (a) Visualization of near-wall streaky structures with Lorentz force spanwise oscillating actuation. Window area is 475 (x^+) by 382 (z^+) wall units. (b) Visualization of near-wall streaky structures with spanwise wall oscillation [6]. Flow is from right to left.

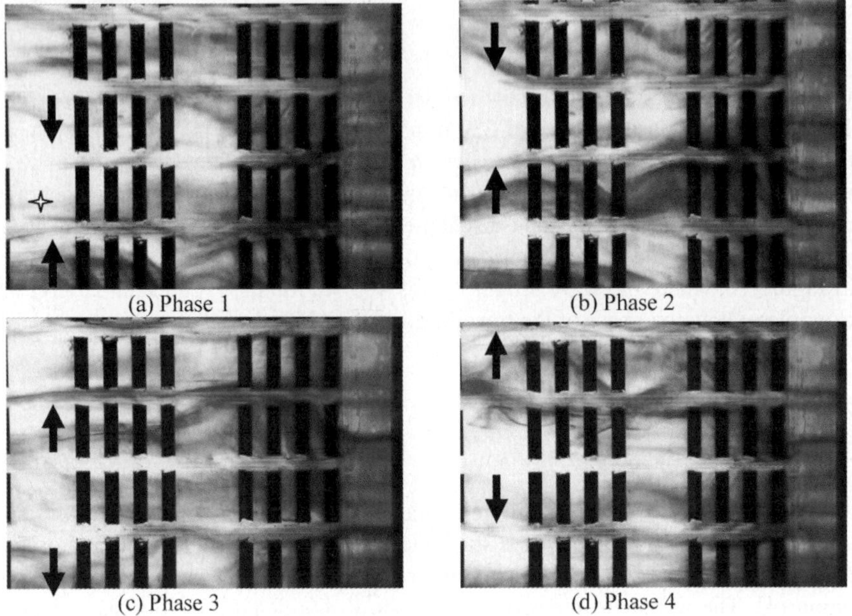

(a) Phase 1   (b) Phase 2

(c) Phase 3   (d) Phase 4

Fig. 5 Visualization of near-wall streaky structure with the Lorentz force travelling wave actuation. Window size is 625 (x^+) by 500 (z^+) wall units. Flow is from right to left. Travelling wave direction is upward. Arrows show the activated rows of electrodes with Lorentz force. The star in (a) shows the spanwise position of z where the velocity measurements were carried out.

equally. In our study, we have achieved a drag reduction up to 47% with a maximum spanwise velocity, w^+, up to 12.2.

When the spanwise travelling wave forcing was applied, the logarithmic velocity profile was shifted upwards similar to that with spanwise flow oscillation. The turbulence intensity also showed a significant reduction across the entire region of the turbulent boundary layer (figures not shown here).

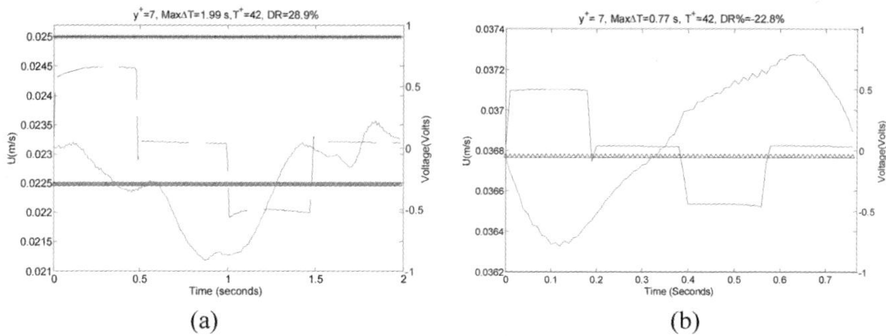

Fig. 6 Phase-averaged streamwise velocity with spanwise travelling wave at $y^+ = 7$ at drag reducing condition (a) and at drag increasing condition (b). Triangles denote the local mean velocity; stars denote the mean velocity of no-control flow; solid lines denote the phase-averaged velocity; dash lines denote the excitation signal.

Figure 5 shows a series of flow visualization pictures, where the near-wall streaky structures were modified to give an appearance of 'ribbons', similar to [7]. As seen in Figure 5, the low-speed streaks were twisted and propelled along the spanwise direction when the positive (upward direction in pictures) Lorentz force was imposed. If the force was strong enough, the streaky structures were able to travel out of the activated force region into the adjoining area. Then, the streaks could not move further and clustered together to form 'ribbons' until the non-activated region became active. This spanwise motion of the streaky structures repeated across the entire section, where the spanwise displacement of the low-speed 'ribbons' can be up to the total width of the test sheet ($z^+ = 1214$ wall units). To achieve drag reductions, we found that the spanwise displacement of low-speed 'ribbons' must be greater than 115 wall units (z^+). This almost compares to the spacing between low-speed streaks ($z^+ \sim 100$). Otherwise, drag increase was observed.

The phase-averaged streamwise velocity profiles at $y^+ = 7$ in one excitation period, as seen in Figure 6, clearly show that the streamwise velocity of the turbulent boundary layer was modulated by the spanwise travelling wave actuation. High-speed region (velocity greater than the local mean velocity) and low-speed region (velocity less than the local mean velocity) appeared alternately in one actuation cycle regardless of the amount of drag reduction. This confirms that the low-speed and high-speed 'ribbons' are not solely responsible for the drag reduction in spanwise travelling wave flow control by Lorentz forces. The phase-averaged velocity profiles also show that the period of the wave-form velocity is consistent with the excitation signal. Moreover, the phase-averaged velocity in Figure 6a confirms the findings of the flow visualization in Figure 5. In Figure 5a (phase 1), some low-speed fluid had been able to move close to the position of the hot-film sensor, which can be seen from Figure 6a where the velocity was decreasing. At phase 2, the velocity had dropped down to the minimum, which supports the view in Figure 5b that the low-speed 'ribbon' was passing by the sensor. After that, Figures 5c and 5d show the high-speed region was approaching the sensor, which was affirmed by the increasing velocity in Figure 6a. Please notice that the velocity measurement point was about 7 mm downstream the actuator, and here, the velocity signal lagged behind the input signal by about 0.3 seconds.

4 Conclusions

In this study, two turbulent flow control schemes were examined: spanwise oscillation and spanwise travelling wave via Lorentz forcing.

The maximum 47% of drag reduction was achieved by spanwise flow oscillation at $w^+ \approx 12.2$. This shows a good agreement with the DNS results [2] and our previous results [12]. The results of flow visualization show the streaky structures were periodically twisted into the spanwise direction by the oscillating Lorentz force, suggesting a similarity in the drag reduction mechanisms between the spanwise wall oscillation and Lorentz force oscillation. The mean velocity was reduced in the near-wall region and turbulent intensity decreased across the boundary layer, and both of the skewness and kurtosis were increased in the near-wall region ($y^+ < 10$).

The spanwise travelling wave actuation produced a drag reduction up to 28.9% and a maximum 22.8% of drag increase. When the drag reduction was achieved, the logarithmic velocity profile was shifted upward, indicating the thickening of the sublayer by the spanwise travelling wave. The turbulent intensity also reduced as described in spanwise oscillation.

Flow visualizations showed that the near-wall streaky structures were modified by the spanwise travelling wave actuation to form low-speed 'ribbons'. Phase-averaged streamwise velocity profiles, however, suggested that the low-speed 'ribbons' are not the sole reason for drag reduction. We found that the spanwise displacement of the low-speed 'ribbons' is also important to achieve drag reduction. In our experiments, drag reduction was achieved only when the low-speed 'ribbons' moved more than 115 wall units in the spanwise direction in one cycle of actuation sequence. When drag reduction was observed, the spanwise travelling wave appeared to congregate the low-speed streaks to form 'ribbons', then propelled them to move more than 115 wall units in the spanwise direction.

References

1. Baron, A. and Quadrio, M., Turbulent drag reduction by spanwise wall oscillations, *Appl. Sci. Res.* **55** (1996) 311–326.
2. Berger, T.W., Kim, J., et al., Turbulent boundary layer control utilizing the Lorentz force, *Phys. Fluids* **12** (2000) 631–649.
3. Breuer, K.-S. Park, J., Henoch C., Actuation and control of a turbulent channel flow using Lorentz forces, *Phys. Fluids* **16** (2004) 897–907.
4. Choi, K.-S., Near-wall structure of turbulent boundary layer with spanwise-wall oscillation, *Phys. Fluids* **14** (2002) 2530–2542.

5. Choi, K.-S. and Clayton, B.R., The mechanism of turbulent drag reduction with wall oscillation, *Int. J. Heat Fluid Flow* **22** (2001) 1–9.

6. Choi, K.-S., DeBisschop, J.-R., et al., Turbulent boundary-layer control by means of spanwise-wall oscillation, *AIAA J.* **36**(7) (1998) 1157–1163.

7. Du, Y., Symeonidis, Y. and Karniadakis, G.E., Drag reduction in wall-bounded turbulence via a transverse travelling wave, *J. Fluid Mech.* **457** (2002) 1–34.

8. Henoch, C. and Stace, J., Experimental investigation of a salt water turbulent boundary layer modified by an applied streamwise magnetohydrodynamic body force, *Phys. Fluids* **7** (1995) 1371–1383.

9. Jimenez, J. and Pinelli, A., The autonomous cycle of near-wall turbulence, *J. Fluid Mech.* **389** (1999) 335–359.

10. Karniadakis, G.E. and Choi, K-S., Mechanisms on transverse motions in turbulent wall flows, *Ann. Rev. Fluid Mech.* **35**, 45–62.

11. Nosenchuck, D.M. and Brown, G.L., Discrete spatial control of wall shear stress in a turbulent boundary layer, in *Near-Wall Turbulent Flows*, R.M. So et al. (Eds.), Elsevier, Amsterdam (1993) pp. 689–698.

12. Pang, J. and Choi, K.-S., Turbulent drag reduction by Lorentz force oscillation, *Phys. Fluids* **16** (2004) 35–38.

13. Quadrio, M. and Sibila, S., Numerical simulation of turbulent flow in a pipe oscillation around its axis, *J. Fluid Mech.* **424** (2000) 217–241.

14. Schoppa, W. and Hussain, F., A large-scale control strategy for drag reduction in turbulent boundary layers, *Phys. Fluids* **10** (1998) 1049–1051.

Multi-Scale Flow Control for Efficient Mixing: Laboratory Generation of Unsteady Multi-Scale Flows Controlled by Multi-Scale Electromagnetic Forces

S. Ferrari[*], P. Kewcharoenwong, L. Rossi and J.C. Vassilicos
Turbulence, Mixing and Flow Control Group, Department of Aeronautics, Imperial College, London SW7 2AZ, U.K.; E-mail: l.rossi@imperial.ac.uk

Key words: Flow control, multi-scale, electromagnetic forcing, experiments, laminar, unsteady.

1 Introduction

Flow control may be used to achieve efficient mixing which is important in many applications including in various combustors and chemical reactors. Mixing and its rate can be measured in terms of the concentration variance, c'^2, and its time dependence. Efficient mixing can be achieved if the energy input required can be minimised for values as low as c'^2 or as high as possible mixing rates.

The idea behind multi-scale flow control, as used here, lies in the generation of fully controlled multi-scale laminar flows. Being laminar, such flows may require relatively little power to be run; being multi-scale, such flows may have turbulent-like fast/effective mixing properties. Control of the flows' unsteadiness in time can further enhance mixing, giving these flows both turbulent-like and chaotic advection type properties (unsteady laminar flows can sometimes display chaotic advection which causes good mixing [6]).

Electromagnetic flow control has been used in various works to this day (e.g. [2, 9, 11]). Here, a new class of unsteady quasi-two-dimensional (Q2D) multi-scale laminar flows is generated in the laboratory. The flows are generated using a shallow layer of brine and controlled by multi-scale electromagnetic forces resulting from the combination of an electric current and a fractal magnetic field created by a fractal permanent magnet distribution (see Figure 1). Our fows are laminar yet turbulent-like in that they have multi-scale streamline topology in the shape of "cat's

---

[*] Permanent address: Dipartimento di Ingegneria del Territorio, Università degli Studi di Cagliari, Italy

J.F. Morrison et al. (eds.), IUTAM Symposium on Flow Control and MEMS, 267–272.
© 2008 *Springer. Printed in the Netherlands.*

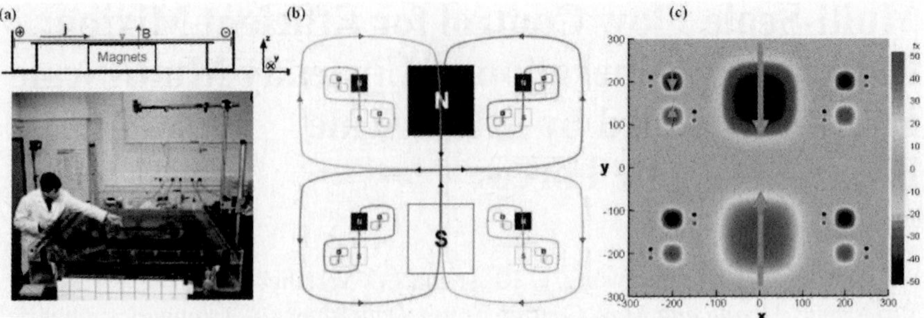

Fig. 1 (a) Schematic for electromagnetic forcing of a shallow brine layer and experimental rig. (b) Schematic of a fractal flow and associated fractal distribution of permanent magnets ($D_s = 0.5$). (c) Electromagnetic forcing distribution computed with $I = 1A$, $B = 1T$; fy in N/m^3.

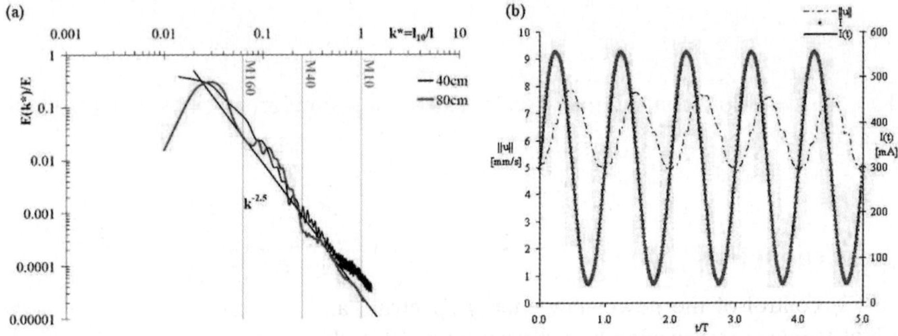

Fig. 2 (a) Average flow energy spectrum for frame 80 cm and 40 cm. The three sizes of forcing scale (M10, M40, M160) are indicated by vertical straight lines. Diagonal straight line gives $E(k) \sim k^{-2.5}$. (b) Preliminary result of mean velocity from PTV of time-dependent flow, $\|\mathbf{u}\|$ in mm/s and I in mA. Electrical currents (measured) are well appointed by forcing $I(t) = 0.298 + 0.258 \sin(2t/T)$ with $T = 12.5$ s.

eyes" within "cat's eyes" or 8 within 8 (Figure 1b), similar to the known schematic streamline structure of two-dimensional turbulence [5].

This multi-scale topology is invariant over a broad range of Reynolds numbers, Re_{2D} from 600 to 9900 [8]. As a result, the flows have a power-law energy spectrum $E(k) \sim k^{-p}$ over a broad range of electromagnetically forced scales $2\pi/L < k < 2\pi/\eta$ (see Figure 2a) where p is controlled by the imposed fractal dimension, D_s of the laminar multi-scale streamline structure's stagnation points. Specifically, p is a decreasing linear function of D_s and D_s can be set by the multi-scale electromagnetic control scheme. It is interesting to note that our multi-scale control strategy allows separate control over L/η and Reynolds number. In particular, our approach is in principle promising for microfluidic mixing where the flow is laminar, but separate control of L/η might nevertheless make it multi-scale with interesting stirring and mixing properties [7].

As mixing efficiency is defined and quantified in terms of the levels of energy input rate and c'^2, two central aspects of flow control for mixing efficiency are con-

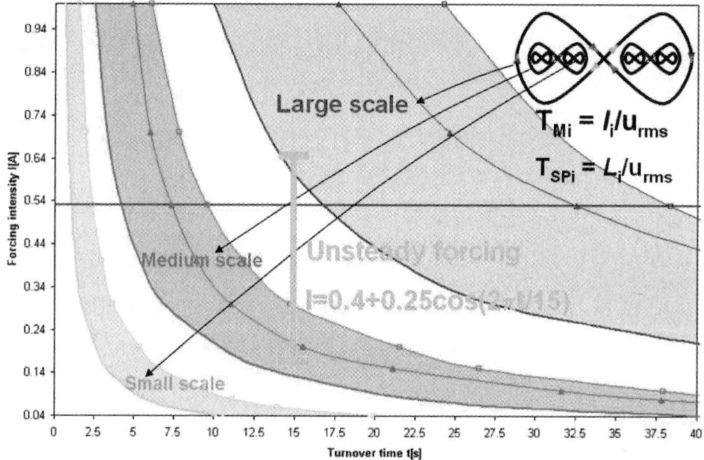

Fig. 3 Approximation of characteristic turnover time at various forcing intensities.

sidered: (i) control of the acceleration field $\mathbf{a}(\mathbf{x},\mathbf{t})$ as $\langle \mathbf{a} \cdot \mathbf{u} \rangle$ (where $\mathbf{u}(\mathbf{x},\mathbf{t})$ is the velocity field and the average $\langle \rangle$ is taken over space) is related to the power input required (see e.g. Figure 6) and as the stagnation point velocity, introduced by [4] to characterise the flow persistence, is proportional to the local acceleration; and (ii) control of the power-law energy spectrum and the spatio-temporal structure of velocity and acceleration stagnation points, which relate to Lagrangian pair statistics [1, 5]; themselves known to determine concentration variances in many different types of flows [10]. We seek to achieve low values of c'^2 and/or high value of $\partial c'^2/\partial t$ by controlling the r.m.s. pair separation $\sqrt{\bar{\Delta}^2(t)}$ and its growth via our multi-scale control of the flow geometry, topology and time dependence. The control of the spatio-temporal structure of the acceleration field influences, where energy and momentum can be most effciently (i.e. economically yet effectively for mixing) injected in the flow.

The acceleration field consists of two terms, which may be independently controllable in the frame of the apparatus, where there is no mean flow. One is the local time derivative $\partial u/\partial t$ and the other is the convective derivative $(u \cdot \nabla) \cdot u$. The former is particularly sensitive to the controlled timed ependence of the EM forcing (via the controlled time dependence of the current). The latter's spatial structure depends strongly on the multi-scale topology of the flow. We obtained spatially and temporally resolved velocity and acceleration Eulerian and Lagrangian fields and their statistics (e.g. Figures 2b and 5, see also [3]).

Unsteadiness is introduced to the flows by means of time-dependent forcing, specifically time-dependent electrical current (see Figures 2b, 3, 5, 6). We investigate various time dependencies of the forcing, e.g. frequency, mean intensity and magnitude, chosen so as to excite different flow scales. For example, we excite the forced flow scales "one by one" and/or all together by adjusting the frequency of the forcing and its amplitude, see Figure 3.

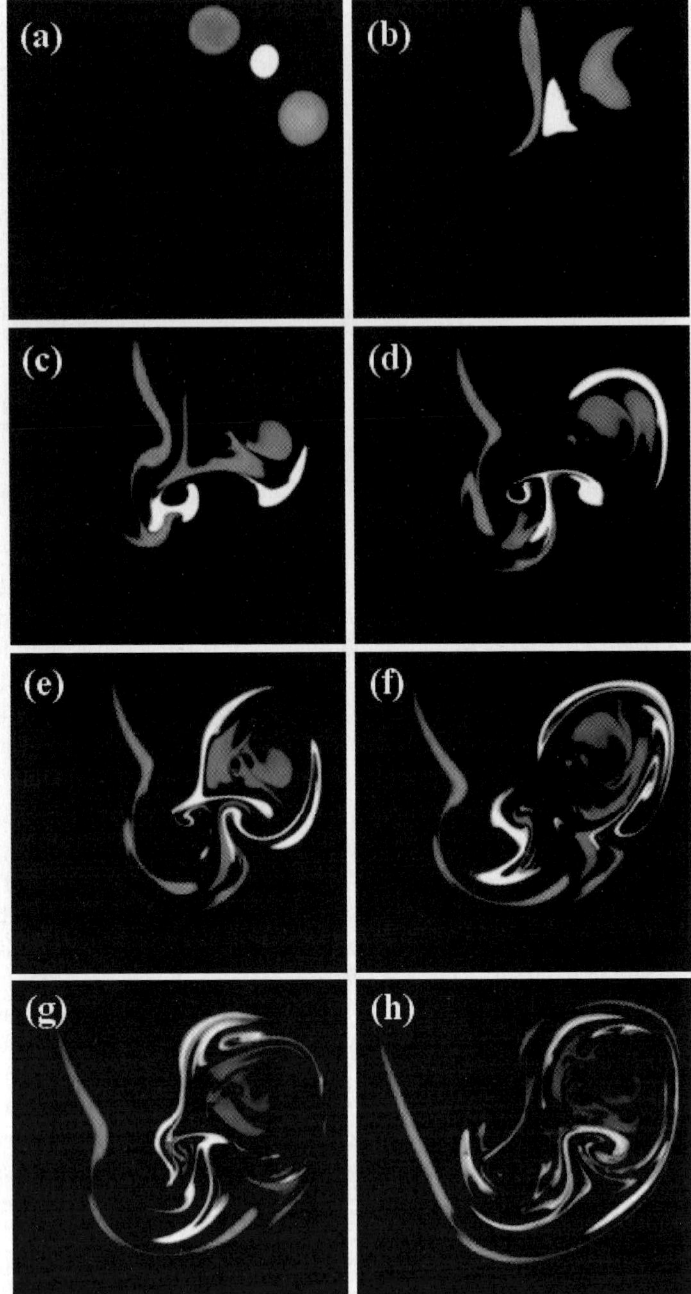

Fig. 4 Dye visualisation of an unsteady multi-scale laminar flow (quarter of flow) with time-dependent forcing: $I(t) = 0.4 + 0.25 \sin(2t/T)$ where $T = 12.5$ s. (a–h) show flow at times $t/T = 0$, 1, 2, 3, 4, 5, 6 and 9, respectively.

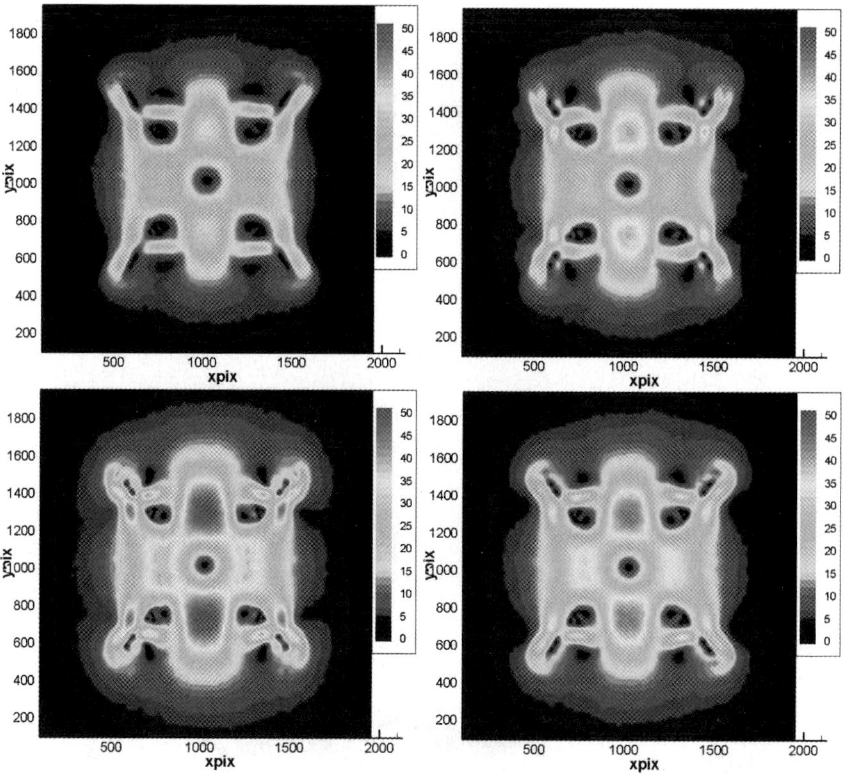

Fig. 5 Preliminary result of Eulerian velocity fields extracted from adaptive PTV measurements [3] with time-dependent forcing of period T. (a–d) show velocity fields with time-increment of $T/4$.

2 Conclusion

We explore the impact of our time-dependent control strategies on various Eulerian and Lagrangian statistics, including wave-number and frequency energy spectra as well as acceleration, pair dispersion and scalar mixing statistics and their time-dependency with a view to develop efficient multi-scale flow control approaches to mixing, which maybe extendable to fully 3D flows as well as microflows.

Acknowledgements

We would like to acknowledge Dr. Yannis Hardalupas and John Laker as well as the funding support of the EPSRC, the Royal Society, the Marie Curie Multi-Partner European Training Site on Environmental Turbulence and the Leverhulme Trust.

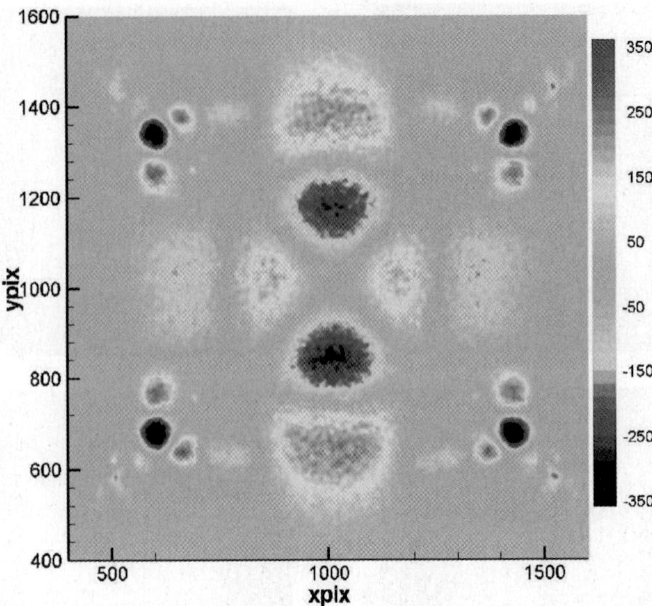

Fig. 6 Example of PTVA measurements of the field $\mathbf{u} \cdot \mathbf{a}$ for the case of steady forcing. See [3] for a description of the PTVA used.

References

1. Davila, J. and Vassilicos, J.C., 2003, Richardson pair diffusion and the stagnation point structure of turbulence, *Phys. Rev. Lett.* **91**(14), 144–501.
2. Du, Y., Symeonidis, V. and Karniadakis, G.E., 2002, Drag reduction in wall-bounded turbulence via a transverse travelling wave, *J. Fluid Mech.* **457**, 1–34.
3. Ferrari, S. and Rossi, L., 2007, Measurements of velocity and acceleration from particle tracking, PTVA and its applications to electromagnetically controlled quasi-two-dimensional multiscale flows, *Exp. Fluids*, to appear.
4. Goto, S., Osborne, D.R., Vassilicos, J.C. and Haigh, J.D., 2005, Acceleration statistics as measures of statistical persistence of streamlines in isotropic turbulence, *Phys. Rev. E* **71**, 015301(R).
5. Goto, S. and Vassilicos, J.C., 2004, Particle pair diffusion and persistent streamline topology in two-dimensional turbulence, *New J. Phys.* **6**(65), 1–5.
6. Ottino, J.M., 1989, *The Kinematics of Mixing: Stretching, Chaos, and Transport*, Cambridge University Press.
7. Rossi, L., Vassilicos J.C. and Hardaluppas, Y., 2006, Multiscale laminar flows with turbulent-like properties, *Phys. Rev. Lett.* **97**, 144–501.
8. Rossi, L., Vassilicos J.C. and Hardaluppas, Y., 2006, Electromagnetically controlled multiscale flows, *J. Fluid Mech.* **558**, 207—242.
9. Rossi, L. and Thibault, J.P., 2002, Investigation of wall normal electromagnetic actuator for seawater flow control, *J. Turbulence* **3**, 1–15.
10. Vassilicos, J.C., 2002, Mixing in vortical, chaotic and turbulent flows, *Philos. Trans. R. Soc. London A* **360**, 2819–2837.
11. Weier, T., Gerbeth, G., Mutschke, G., Lielaulis, O. and Lammers, G., 2003, Control of flow separation using electromagnetic forces, *Flow, Turbulence and Combustion* **71**, 5–17.

Multi-Scale Flow Control for Efficient Mixing: Simulation of Electromagnetically Forced Turbulent-Like Laminar Flows

E. Hascoët, L. Rossi and J.C. Vassilicos
Turbulence, Mixing and Flow Control Group, Department of Aeronautics, Imperial College, London SW7 2AZ, U.K.; E-mail: l.rossi@imperial.ac.uk

Abstract. We perform Direct Numerical Simulations (DNS) of electromagnetically fractal-forced and Rayleigh-damped two-dimensional flows. Our simulations show broad band power law energy spectra. When the fractal dimension of the magnets' distribution is $D_f = 0.5$ then $p \approx 2.5$ in agreement with previous laboratory experiment. Moreover, when the fractal distribution of magnets is changed, p varies linearly with D_f, the fractal dimension of the magnet set up. Hence, fractal control of the energy spectrum is possible.

Key words: Flow control, multi-scale, body forces, 2D DNS, mixing, laminar.

1 Introduction

Mixing of scalar and momentum are of central importance for numerous flow control studies. The control and enhancement of scalar mixing play a key role in e.g. pollutants dispersion, combustion, etc. The control of momentum mixing (completed by the control of pressure field and acceleration fields) have a broad range of applications, e.g. drag reduction, noise reduction, energy transfers.

Multi-scale flow control relies on the ability to force the flow at various scales controlling where and how the energy is injected. This concept is extremely broad and is also supported by theory, e.g. [1–3]. In this first particular contribution, we focus on two-dimensional (2D) flows where the multi-scale distribution of stagnation points [1, 2, 4] (i.e. flow topology) is controlled by the forcing.

Recent experimental work [5] has generated such a controlled multi-scale flow in the laboratory using electromagnetic flow control, which is used in various flow control concepts [6–8]. These experiments have confirmed the theoretical relation $p + D_s = 3$ [1], where p is the exponent of the power law energy spectrum generated by the multi-scale forcing and D_s is the fractal dimension of this forcing. However,

J.F. Morrison et al. (eds.), IUTAM Symposium on Flow Control and MEMS, 273–277.
© 2008 *Springer. Printed in the Netherlands.*

the geometry of the forcing (and thus D_s) was not varied in the experiments and numerical simulations are needed to confirm experimental results, guide future experimental campaigns and indicate how p may vary when D_s in particular and the multi-scale geometry of the forcing in general are varied. Previous numerical works have focused on fractal forcing in Fourier space [3, 9]. Instead, the numerical simulations presented here use the same fractal forcing as in [5], defined in physical space, and we perform Direct Numerical Simulation (DNS) of this multiscale electromagnetically forced flow in regimes including and extending those of [5]. Of particular interest are $p = p(D_s)$ and inter-scale energy transfer.

To model these 2D electromagnetically forced flows, we consider the following modified 2D Navier–Stokes equation:

$$\partial_t \mathbf{u} + (\mathbf{u} \cdot \nabla)\mathbf{u} = -\nabla p - \alpha \mathbf{u} + \nu \Delta \mathbf{u} + f_0 \mathbf{f}(\mathbf{r}),$$

$$\nabla \cdot \mathbf{u} = 0.$$

The boundary conditions are taken periodic on a $2\pi \times 2\pi$ domain and $|\mathbf{f}(\mathbf{r})| = 1$. In order to take into account the dissipation related to the finite thickness of a real fluid, we have added a bottom friction term with friction coefficient α. This friction coefficient is known to scale like $\alpha \propto \nu/h^2$ where ν is the fluid kinematic viscosity and h the fluid layer width. The last term in the equation models the multi-scale Lorentz force, $\mathbf{j} \times \mathbf{B}/\rho$, produced by the combination of a self-similar distribution of magnets, $\mathbf{B}$ (Figure 1d) lying underneath a conducting fluid layer and a constant electric current, $\mathbf{j}$, circulating through the fluid. The forces are directed towards the center of each pair. The overall force intensity, $\mathbf{f}_0$, is fixed by comparison with the experiments [5]. The vector field $\mathbf{f}(\mathbf{r})$ sets the geometrical distribution of the force by means of 2D unit step functions reflecting the magnet's position and size. In order to reproduce the experimental set-up, the size of the large square magnets was fixed to 16/170 times the domain size, the size of the other magnets being obtained by iteratively dividing this size by the scaling factor, R, which was fixed to four in [5] and is varied here.

The modified Navier–Stokes equation has been integrated in spectral space with a pseudo-spectral algorithm using a dealiasing 3/2 rule and a fourth order Runge–Kutta time integration. The number of grid points has been taken equal to 1024^2 in order to accurately resolve the smallest scale given by the smallest magnet size.

As in [5], the flow exhibits a broad band power law energy spectrum with an exponent clearly in-between the exponents $-5/3$ and -3 of 2D turbulence forced at only one, small or large, scale [10]. In Figure 2a we plot three energy spectra for $\alpha = 0.1445$ which corresponds to the conditions of [5]. As in the experiments, the oscillations in the energy spectra are found to decrease with increasing flow intensity (i.e. forcing and Reynolds number).

The exponent p is well defined between $3l_0$ and $2l_2$ (where l_0 and l_2 are the sizes of the largest and smallest magnets, respectively) when the flow topology is similar to that of the experiment and is given in Figure 2b as a function of f/α^2. These values are consistent with the ones found in [5]. In addition, DNS demonstrate that

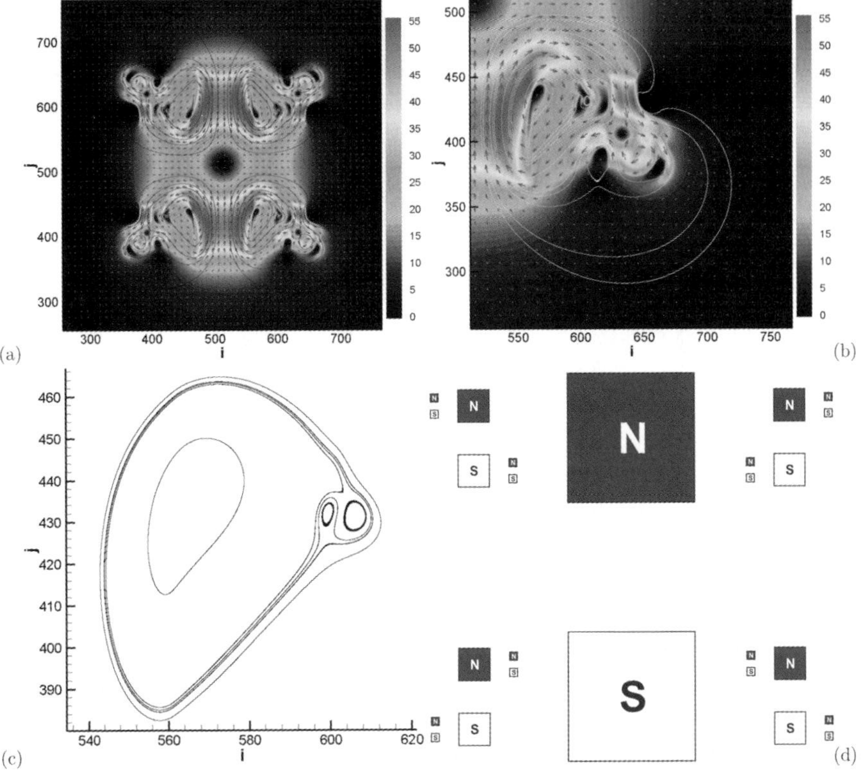

Fig. 1 (a) and (b) Velocity field of the DNS for $\alpha = 0.1445$ s^{-1} and $f = 0.00873$ Nm^{-3}. The grey scale represents the velocity intensity in mm/s; (c) illustration of the streamlines at the small scale; (d) fractal forcing of the flow, N and S refer to the polarities of the magnetic induction and are related to pairs of opposite forces.

flow intensity and flow topology can be controlled independently: the former one by f/α the latter one by f/α^2.

The fractal dimension D_s being proportional to $1/\ln(R)$ the relation $p + D_s = 3$ implies

$$p + \frac{1}{\ln(R)} = cte.$$

We choose to vary R so as to test this linear dependence of the exponent of the power law energy spectrum with the geometry of the forcing. Figure 2c clearly demonstrates the validity of such a dependence and that the multi-scale forcing considered is amenable to a real control of the energy spectrum.

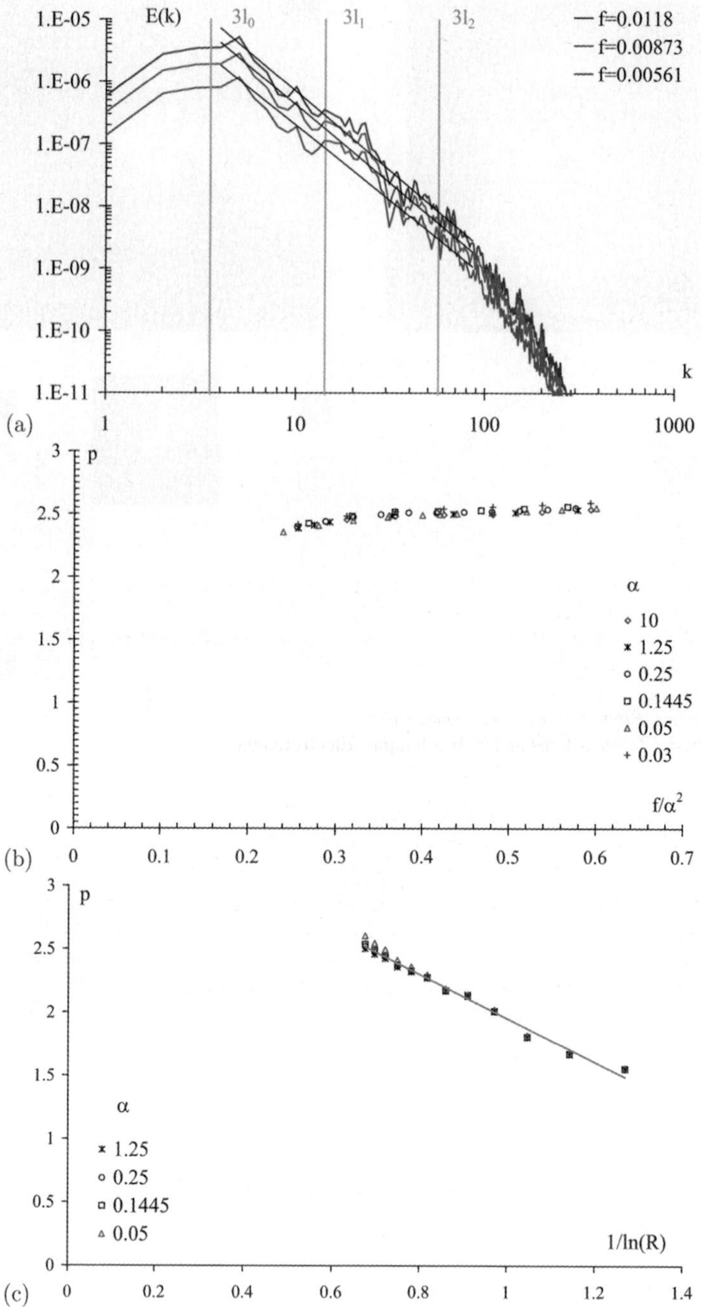

Fig. 2 (a) Illustration of energy spectra for various intensity of the forcing with $\alpha = 0.1445$ s^{-1}; (b) exponent p of the power law energy spectrum ($E(k) \sim k^{-p}$) versus the f/α^2 for various α; (c) dependence of the power law exponent of the energy spectrum, p, with the fractal geometry of the forcing $1/\ln(R)$.

2 Conclusion

These results are extremely encouraging for the design of multi-scale flow control devices supported by and amenable to theory.

Acknowledgment

We would like to acknowledge the EPSRC and The Leverhulme Trust for their support.

References

1. J. Davila and J.C. Vassilicos. Richardson's pair diffusion and the stagnation point structure of turbulence. *Phys. Rev. Lett.* **91**, 2003, 144501.
2. S. Goto and J.C. Vassilicos. Particle pair diffusion and persistent streamline topology in two-dimensional turbulence. *New J. Phys.* **6**, 2004, 1–35.
3. B. Mazzi, F. Okkels and J.C. vassilicos. A shell model approach to fractal induced turbulence. *Europ. Phys. B* **28**(2), 2002, 231–241.
4. L. Rossi, J.C. Vassilicos and Y. Hardalupas. Multi-scale laminar flows with turbulent-like properties. *Phys. Rev. Lett.* **97**, 2006, 144501.
5. L. Rossi, J.C. Vassilicos and Y. Hardalupas. Electromagnetically controlled multi-scale flows. *J. Fluid Mech.* **558**, 2006, 207–242.
6. L. Rossi and J.P. Thibault. Investigation of wall normal electromagnetic actuator for seawater flow control. *J. Turbulence* **3**, 2002, 1–15.
7. Y. Du, V. Symeonidis and G.E. Karniadakis. Drag reduction in wall-bounded turbulence via a transverse travelling wave. *J. Fluid Mech.* **457**, 2002, 1–34.
8. T. Weier, G. Gerbeth, G. Mutschke, O. Lielausis and G. Lammers. Control of flow separation using electromagnetic forces. *Flow, Turbulence and Combustion* **71**, 2003, 5–17.
9. B. Mazzi and J.C. Vassilicos. Fractal generated turbulence. *J. Fluid Mech.* **502**, 2004, 65–87.
10. P. Tabeling. Two-dimensional turbulence: A physicist approach. *Phys. Rep.* **362**, 2002, 1–62.

CLOSED-LOOP CONTROL

Active Control of Laminar Boundary Layer Disturbances

M. Gaster
*Department of Engineering, Queen Mary, University of London, Mile End Road,
London E1 4NS, U.K.; E-mail: m.gaster@qmul.ac.uk*

Abstract. Active suppression of the naturally occurring travelling wave disturbances that amplify in laminar boundary layers and cause the transition from laminar to turbulent flow is considered. Both open-loop and closed-loop schemes are discussed. Numerical predictions, based on linear stability theory, have been used to model the behaviour of the flow disturbances and the controlled waves. Predictions based on these models have shown that the instability waves that occur on a simple flat plate can be stabilized significantly by both types of control. Wind tunnel experiments have so far been used to validate some of the predictions in the open-loop case.

Key words: Control, boundary layer, laminar, transition, instability.

1 Introduction

Drag reduction is becoming an increasingly important aspect of aircraft design. This is not only so that the burden of ever increasing fuel costs can be offset, but also so that noxious exhaust gases can be reduced for ecological reasons. The time will almost certainly come when legislation, possibly through fuel taxation, will be introduced to force the Industry to operate aircraft with greater fuel economy. Airbus is asking the research community to help provide schemes for reducing skin-friction drag by some 50% within 6 years. This target is almost certainly out of reach with current technologies. It is therefore important to consider and research a number of different avenues so far not seriously pursued.

Aircraft surfaces generally have turbulent boundary layers over them and consequently most effort has been directed to the reduction of the resulting turbulent skin friction. Moderate drag reduction can be obtained by passive means using riblets [1] or similar devices, but it seems likely that greater reductions can be achieved by active means. Active techniques use information about the unsteady flow field

J.F. Morrison et al. (eds.), IUTAM Symposium on Flow Control and MEMS, 281–292.
© 2008 *Springer. Printed in the Netherlands.*

to determine what disturbances need to be introduced into the flow so as to modify the turbulent structures and thus reduce skin friction [2]. It appears from numerical experiments that turbulent structures can indeed be modified by suitable control schemes so as to reduce skin-friction drag. In any practical implementation this involves measuring the pressure or skin-friction fluctuations on the surface and then perturbing the surface geometry in some way. To translate these ideas into laboratory demonstrations, let alone onto an aircraft, is so daunting that it is likely to be only a remote possibility far in the future. The reason is partly because the physical length scales involved are small and the frequencies large, making the instrumentation difficult, but also because the control scheme would involve a truly vast number of detectors and actuators to be intelligently controlled. Nevertheless, the topic of active boundary layer control is worth studying because the rewards would be significant if all these problems could be solved. The problem has been tackled both numerically and experimentally by a large number of research groups and are briefly discussed in [3].

My belief is that it is also worth studying the related problem of the active control of laminar boundary layers so as to suppress the occurrence and growth of the travelling waves that cause the laminar flow to breakdown into turbulence. Even though it may turn out that this approach is not applicable to an aircraft situation, the problem is more tractable than the turbulent case and much may be learnt from the study. The laminar problem is, at least in the early stages, linear and therefore amenable to analytical treatment. This enables different control scenarios and control algorithms to be modelled so that a suitable arrangement of detectors and actuators can be found. Once a satisfactory scheme has been developed the whole process can be validated in a wind-tunnel experiment with a reasonable likelihood of success. The scales of the instability waves are relatively large in the laminar problem and this makes a practical demonstration possible without excessive demands on instrumentation. If the laminar problem cannot be solved there would appear to be little chance of achieving control over a turbulent flow with current instrumentation and control ideas.

More than 20 years ago [4] it was shown that a suitable compliant coating panel fitted into a plate in a water towing tank could inhibit the growth of growing Tollmien–Schlichting waves. That work had to be carried out in water because it did not seem possible to find materials of the right density, mass and stiffness that would work in air. Basically the modifications to the growing eigenvalues came about through the mixed boundary condition linking the pressure and vertical displacement of travelling wave disturbances on the surface. The properties of the soft rubbery substrate controlled this quantity in water naturally and resulted in greatly reduced amplification of the travelling waves as predicted by linear theory. It is conceivable that an active scheme, linking arrays of detectors and actuators could, through a suitable control scheme, provide the same functional behaviour in air. Such a scheme could also be configured so that the unwanted interfacial instabilities, such as flutter and divergence that sometimes interfered with the stabilizing effect of the compliant surface could be entirely eliminated by the control algorithm. It appears that relatively few transducers would be needed to mimic an active surface, at least in the two-dimensional example so far considered.

2 Experimental Setup

It is convenient first to explain the experimental arrangements so that modelling requirements can properly represent a situation that can be tested in the laboratory. The experiments were to be carried out in a low-speed wind tunnel of 0.91 m square cross-section on a 1.6 m long plate. The experiments to be discussed here were all concerned with the zero pressure gradient flat plate boundary layer flow. Typical free-stream velocities were 12 m/s and this resulted in displacement thickness Reynolds numbers of around 1000. The wind tunnel had a very low level of background turbulence and without any artificial excitation a hot-wire anemometer could not detect any discernable perturbations. A span-wise row of point source exciters was therefore located upstream at either 200 mm or 400 mm from the leading edge across the plate to provide a controllable disturbance source. These exciters were feed by pre-computed stored digital records so that simulation of a range of on-coming disturbance fields could be created. The fact that any experiment using a prescribed upstream excitation could be repeated as often as required made this arrangement very helpful. Artificial excitations of increasing complexity were the periodic point source, followed by the impulsive point excitation that created a wave packet and finally a full three-dimensional field of random waves generated by the array of computer controlled exciters. Electro-magnetic devices (miniature speakers) were used for both the upstream exciters and for the actuators. The flow velocity was sensed by a traversable hot-wire probe and by hot-wire sensors mounted just off the surface.

3 Analysis

Modelling of flow perturbations was carried out for parameters appropriate to the experimental setup described above as indicated in Figure 1. The evolution of the excited flow field, as well as that resulting from the motion of an actuator, were evaluated by linear stability theory coupled with the usual *parallel mean flow* approximation. The magnitudes of the disturbances during the initial phase of the transition process are sufficiently weak for the linearization approximation to be valid. Treating the mean boundary layer as a parallel flow does introduce some approximation, but the simplification obtained enables the governing partial differential equations to be reduced to the more tractable form of a set of ordinary differential equations in Fourier space. These equations defining the velocities and vorticities could readily be solved and the physical realization obtained by Fourier transforms. The six equa-

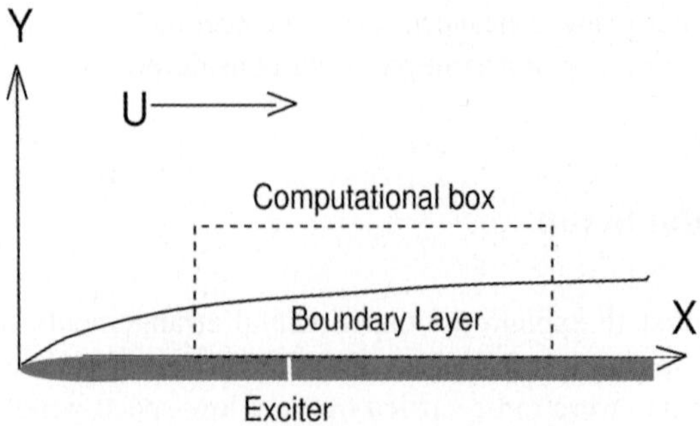

Fig. 1 Flow setup for both experiment and analysis.

tions are homogeneous and excitation of the disturbances arises through boundary values on the wall.

In physical space we can express the variables, such as velocity or pressure, in terms of the Fourier summations of the transformed variables that occur in the ODEs.

$$Q(x,y,z,t) = \iiint \lambda(\alpha,\beta,\omega)\hat{Q}(\alpha,y,\beta,\omega)e^{i(\alpha x+\beta z-\omega t)}\,d\alpha d\beta d\omega, \qquad (1)$$

where

$$\lambda(\alpha,\beta,\omega) \qquad (2)$$

is the spectral content compatible with the boundary values and

$$Q(x,y,z,t) \qquad (3)$$

the velocity or vorticity component. The simplest wall boundary condition is that imposed by a two-dimensional periodic jet issuing from a narrow slit. The integral formulation may be written:

$$u(x,y,t) = \frac{e^{i\omega^* t}}{2\pi} \int \frac{\hat{u}(\alpha,y,\omega^*)}{\hat{v}(\alpha,o,\omega^*)}\,d\alpha. \qquad (4)$$

The zeros of the denominator define the eigenmodes. The above integral consists of two components: the dominant eigenmode that defines the *far-field* and the decaying modes and continuum that control the *near-field*. Another example of a simple excitation is that of a periodic jet at a point. The excitation can be modelled by a spatial delta function of the normal velocity, in both the stream and the span-wise directions. The solution can be constructed by a summation of two-dimensional problems over a range span-wise wavenumbers. The skin-friction distribution on the surface of the plate is shown at one instant of time in Figure 2. This source creates a wedge shaped region downstream consisting of a travelling wave pattern. The near-field foot-print, from the decaying modes and continuum is shown in Figure 3.

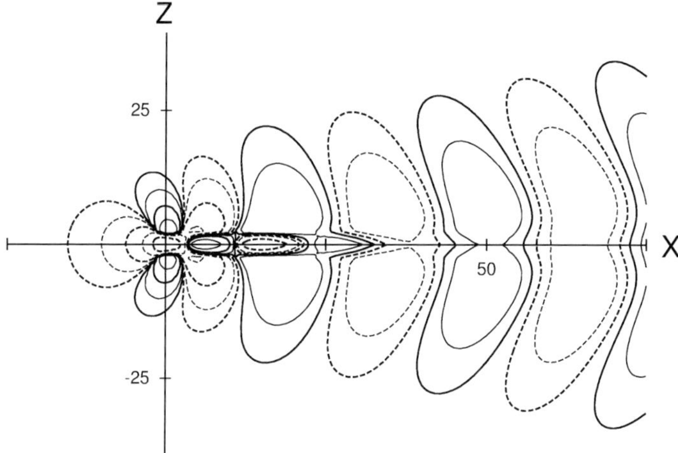

Fig. 2 Contours of skin-friction fluctuation from a periodic point source.

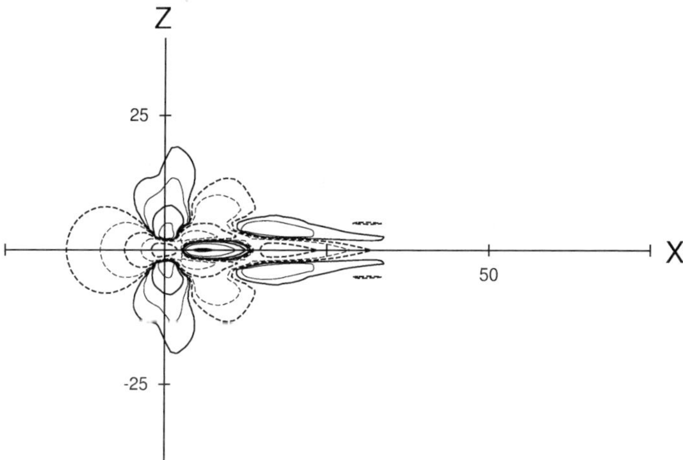

Fig. 3 Contours of the near-field portion of the skin-friction fluctuations.

The impulse response can be derived by further summation of the above solutions over all frequencies. This provides the building block for the solution of the control problem.

Fourier transforms can be applied to predict the on-coming disturbance field approaching the sensors in physical space that were generated by the upstream drivers. The cancelling flow created by the control actuators can also be modelled in the same way.

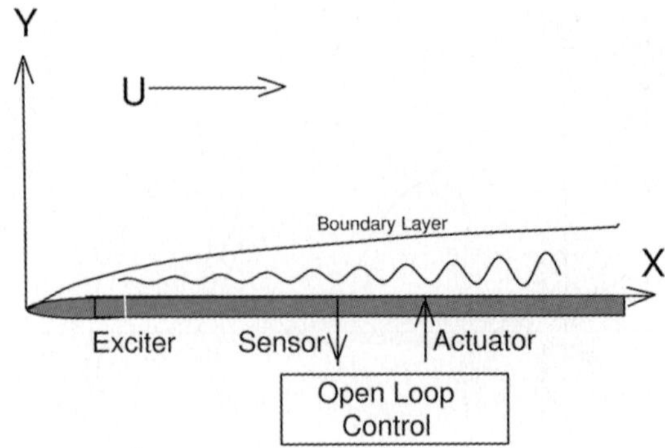

Fig. 4 Schematic of the open-loop control.

4 Open Loop

The simple open-loop control scheme considered here consisted of a span-wise array of detectors followed downstream by a similar array of actuators. A schematic of the setup is drawn in Figure 4. The signals used to drive the actuators were derived from the detectors through a suitable transfer function. Many people have used this type of control to suppress travelling wave disturbances. Mostly the situations where success has been achieved in the past involved disturbances that were largely two-dimensional in character. Then the transfer function was simple and a degree of control could be fairly readily obtained by feeding a filtered version of the detector signal to a downstream actuator, while adjusting the phase and gain so as to achieve good cancellation. This worked well when the wave system was virtually two-dimensional, but a more complex scheme has to be considered when the oncoming wave system was a random fully three-dimensional field. At very large Reynolds numbers and when sweep is used, much greater consideration of the transfer function behaviour will certainly be required.

The transfer function, that links the required actuator driving signals to those detected, is shown in Figure 5. The transfer function only has meaning in the region of positive time so as to conform to causality. This means that even using the complete transfer function, without the portion to the left of the axis, one cannot generate a perfect cancelling signal. The random three-dimensional oncoming disturbance field was generated in the experiment by a random sequence of impulses directed through the acoustic generators mounted in the plate across the span well upstream. This flow field was modelled by a convolution of the impulse response with the driving signal pattern in time and location. The flow over the detectors was then determined as well as that far downstream at some target location. A convolution in span-wise direction and in time with transfer function provided the controlling flow field created by the actuators. This was then compared with the original field at the downstream target location. The correlation between the original field and the cancelling field was found to be roughly 0.97. In this implementation each controlling

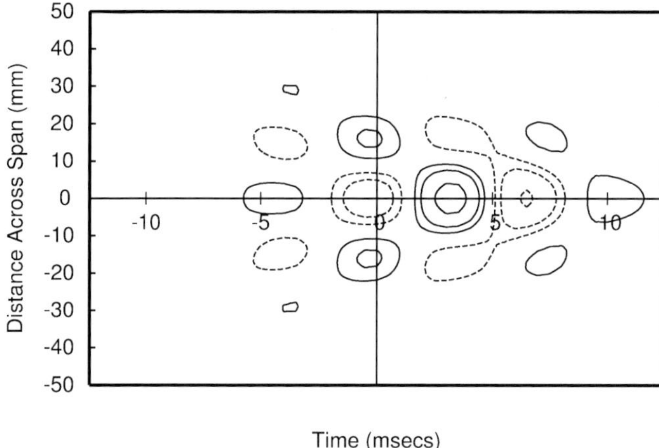

Fig. 5 Transfer function linking the skin friction to actuation.

actuator contained information from all the sensors. This could be accomplished in the simulation model, but is far too complex to consider for an actual real-time control in the wind tunnel. It turned out that control of the actuator was dominated by the sensor directly upstream and only slightly by the neighbouring ones. The transfer function could therefore be approximately represented by a much simpler form consisting of only two delta functions on the centre line at appropriate time delays. Maybe this simplified form was an obvious possibility, but by going through the analysis in some detail the end result could be properly understood. Predictions of the amount of disturbance reduction obtained for different control models were then made for various approximations of the transfer function. For the best choice of time delay and pulse amplitude a correlation coefficient of 0.97 was again achieved.

These predictions were compared with measurements made in the low-turbulence wind tunnel. The oncoming pseudo-random disturbance was created by feeding an upstream span-wise array of exciters with pre-computed pulses. A plan view of this setup is shown in Figure 6.

We did not have the luxury of being able to use large numbers of sensors and controlling transducers to mimic the above ideal control scheme. Instead a single detector-actuator setup was used and the measurements obtained compared with the predicted outcome of such a control scheme. The sensor signal could be recorded for one sweep of the exciter field. Then the controlling signal was evaluated for any specified algorithm and the experiment repeated with the control operating. This enabled *off-line* experiments to be carried out where the computation of the signal feeding the actuator did not have to be done in *real time*. Later a hard-wired digital controller showed that the result from a real-time controller was virtually identical to that from the simpler off-line experiment. The results of these experiments are compared with the model predictions in Figure 7. Details of this work are given in [4].

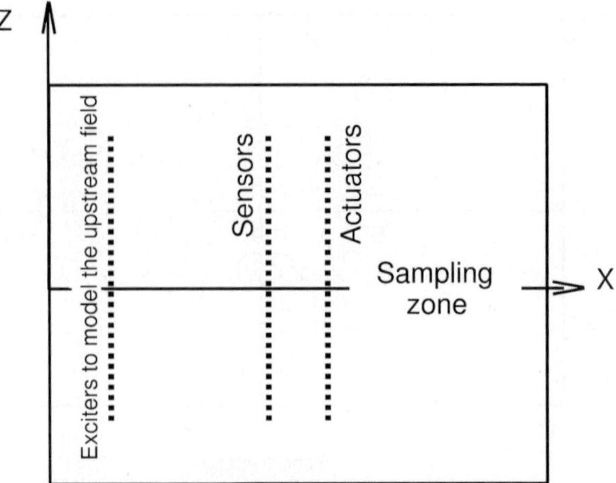

Fig. 6 Plan view of the full control setup.

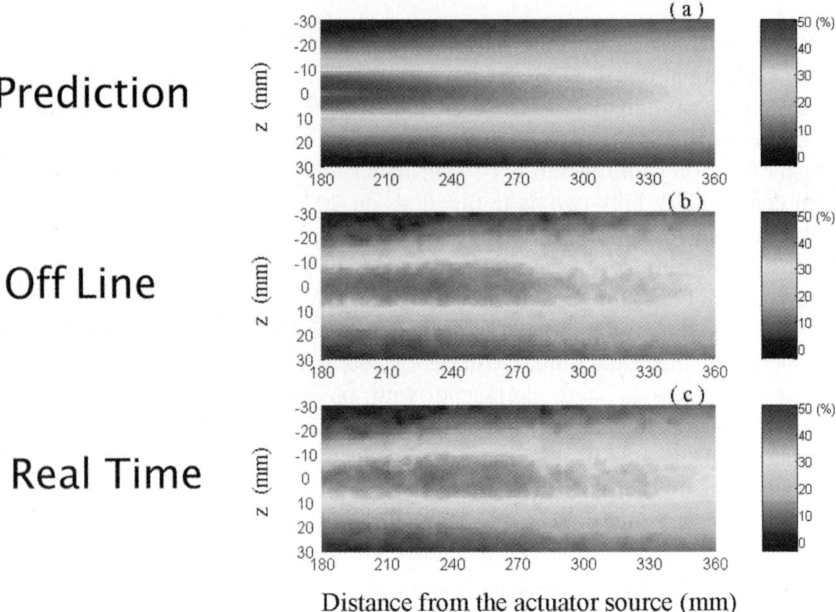

Fig. 7 Cancelling achieved by the open-loop control.

5 Closed Loop

Closing the control loop by a *feed-forward* scheme presents other challenges. Again linear theory has been used to provide the disturbance field on the wall created by the exciters as shown in Figure 8. This information was then used in formulating

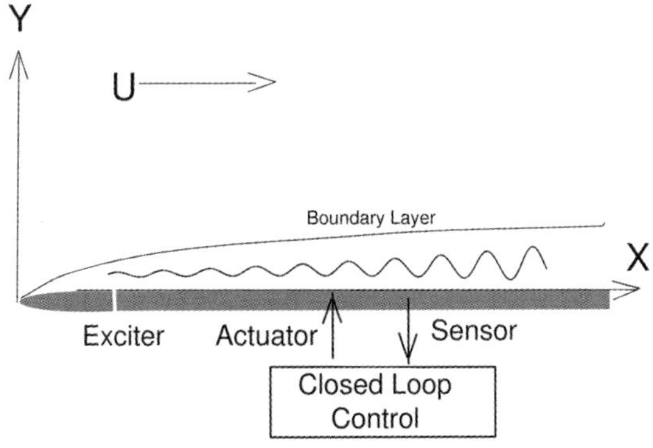

Fig. 8 Closed-loop setup.

the closed-loop control problem. It has been shown that such an arrangement can be made stable and be capable of controlling the disturbances downstream. There appear to be difficulties in finding the best layout of detectors and actuators because of the near-field footprint of the actuators. In the analytical method the near-field component was separated out so that one could see where the problems occurred. In the closed-loop control problem the near-field of the upstream actuator contaminated the downstream detector signal so that control acted on the sum of the eigensolutions and the near field. If this combined signal is brought to some very low level by the control scheme the eigenmode component may not be adequately controlled and the residue will continue to amplify downstream. It is essential, therefore, to ensure that the control focuses on the eigensolution and rejects the near-field which will decay anyway. Again, increased separation between sensor and actuator is one answer, but this opens up the greater likelihood of instability because of the increased phase shift involved in the feed-back loop. Other solutions involving multiple sensors, or possibly multiple actuators, appear also capable of resolving this difficulty. After some experimentation with different layouts it was found that the pattern of five actuators and five sensors arranged as shown in Figure 9b was successful. The modelling has been done for an isolated impulse excitation in this case. Figure 9a shows the development of the wave packet down the plate without and control while Figure 9b shows the same sequence with control applied. The control used between each detector and the upstream actuator was the standard PID controller. The parameters were adjusted until a satisfactory suppression of waves was obtained and the system shown to be stable.

6 Pseudo-Compliant Surface

It seems to be possible to overcome the difficulty associated with the near-field foot print of the controlling actuators by employing stream-wise arrays of detectors and

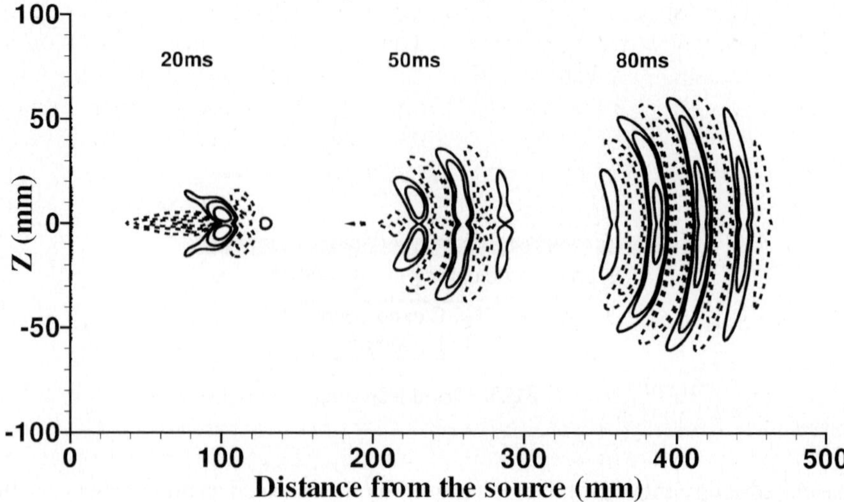

Fig. 9a Skin friction from an impulse without control.

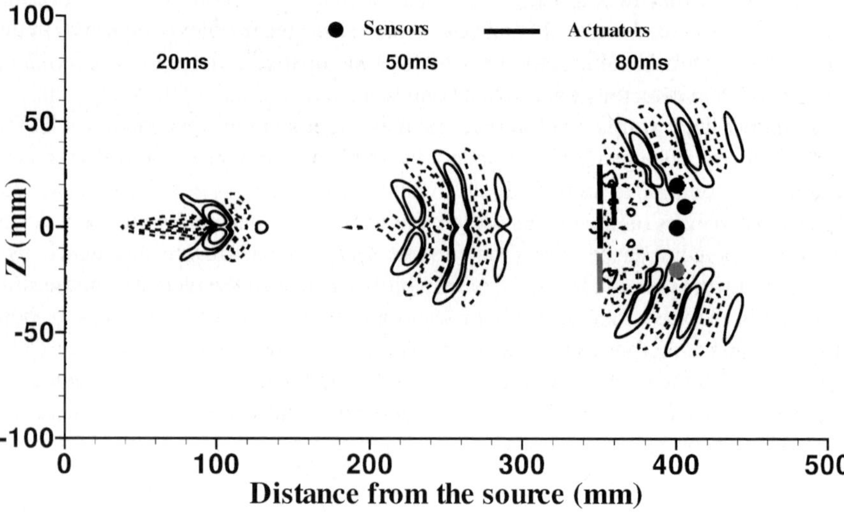

Fig. 9b Skin friction from an impulse with control.

actuators so that the control signal can focus on the fact that the waves that need
to be controlled are travelling downstream at roughly one third of the free-stream
speed. Transducers could be interlaced to produce a zone over which control is active. As the number of elements in the arrays increase the problem approaches that
of the continuum and the behaviour may be likened to that of a compliant surface. It
is clear that the parameters defining the coupling between, say, skin friction and the
normal velocity at the surface can be tailored so that all the eigenmodes are damped.

The behaviour of such a pseudo-compliant surface would then be defined solely by electronics of the feedback control scheme. A controlled zone extending over a couple of Tollmien–Schlichting wavelengths can be modelled by five or six transducers. It would be possible with an electronic control system to act mainly on those amplifying modes that cause transition while ignoring the other fluid/compliant surface interactive modes, like travelling wave flutter or divergence, which can be so troublesome with natural control.

7 Discussion and Conclusions

The control of Tollmien–Schlichting waves by active means has been modelled. Linear modelling of the disturbances created by actuators has enabled various control arrangements to be analysed. In both the open-loop and the closed-loop strategies it appears that the near-field excitation from the control actuators plays a major role in selecting the separation distances between the detector and actuator. The constraints imposed influence the minimum size of an elemental actuator/detector device. Ideally, it would be convenient to make this small so that large numbers of autonomous devices could be deployed where required. By using multiple detectors, or possibly multiple actuators, it may be possible to reduce the effect of the near field and obtain even better results.

In particular, a simple open-loop control arrangement consisting of a row of detectors followed downstream by a span-wise array of actuators has been shown to be capable of reducing the amplitude of the unstable travelling waves to a considerable degree. The control algorithm employed could be simplified from that arising from the full transfer function so that a practical scheme could be considered. Experimental validation was only possible on a single actuator/sensor arrangement. The experiments carried out to validate these calculations for the open-loop case agreed very well with measurements and confirmed the value of the analytical/numerical approach. This showed that an autonomous element could be constructed to provide a local control. A large number of such elemental devices could be used to control large areas of a wing surface.

The closed-loop scheme was explored by means of the analytical models used in the open-loop case. It certainly appears that an arrangement of actuators and sensors can be coupled to a control system that is both stable and can cancel the oncoming disturbances. The schemes so far devised are by no means optimal. It remains to construct the electronic PID controllers so that the models can be validated in a wind tunnel experiment.

Acknowledgements

This paper reports the work that has been done at Queen Mary on the control and suppression of laminar boundary layer disturbance waves by active control using both open-loop and closed-loop control. Much of the work has involved a previous PhD student Dr. Yong Li, and my current PhD student Zhenyu Zhang.

References

1. Bushnell, D.M. and McGinley, C.B., 1989, Turbulence control in wall flows, *Annual Review of Fluid Mechanics* **21**, 1–20.
2. Ho, C.M. and Tai, Y.C., 1998, Micro-electro-mechanical-systems (mems) and fluid flows, *Annual Review of Fluid Mechanics* **30**, 597–612.
3. Gaster, M., 1987, Is the Dolphin a red herring?, in *Proceedings of the IUTAM Symposium*, Bangalore, India, Springer-Verlag.
4. Li, Y. and Gaster, M., 2006, Active control of boundary layer instabilities, *Journal of Fluid Mechanics* **550**, 185–205.

Low-Dimensional Tools for Closed-Loop Flow-Control in High Reynolds Number Turbulent Flows

Joseph W. Hall[1,3], Charles E. Tinney[2], Julie M. Ausseur[1], Jeremy T. Pinier[1], Andre M. Hall[1] and Mark N. Glauser[1]

[1]*Department of Mechanical and Aerospace Engineering, Syracuse University, Syracuse, NY 13244, U.S.A.*

[2]*Labratoire d'Etudes Aerodynamiques, UMR CNRS 6609, Université de Poitiers, France*

[3]*Currently at Department of Mechanical and Engineering, University of New Brunswick, Fredericton, NB, Canada E3B 5A3*

Abstract. A summary of recent experimental research efforts at Syracuse University aimed at active flow control is presented with emphasis placed on the development of low-dimensional tools to facilitate closed-loop control. Results indicate that the near-field pressure in a Mach 0.85 high Reynolds number jet is low dimensional and it is primarily the azimuthal near-field pressure mode 0 that correlates with the acoustic field. The turbulent velocity field in the high-speed jet can also be estimated from the near-field pressure, and is used herein to predict the far-field acoustics. Tools being developed to improve recent successful, high Reynolds number, closed-loop flow control in a NACA 4412 airfoil are also discussed. Together, these results set the framework for active flow control in the high-speed jet with the goal of reducing jet noise.

Key words: Flow control, jets, coherent structures, aeroacoustics, flow separation, low-dimensional techniques.

1 Introduction

Flow control is becoming increasingly feasible in many flows. However much of the previous work has focused on open-loop control strategies and/or low Reynolds number flows. Herein, we present a summary of recent experimental work performed at Syracuse University directed at closed-loop control in two highly turbulent flows: the axisymmetric jet and the separated flow over an airfoil. The first section of this paper will examine the relationship between the near-field pressure surrounding a Mach 0.85 high-speed jet and the far-field acoustic pressure, in the hope of determining a low-dimensional description of the acoustic source. The second portion of the paper then examines the large-scale features of the turbulent flow field using the Proper Orthogonal Decomposition Technique put forth by Lumley [1]. A

J.F. Morrison et al. (eds.), IUTAM Symposium on Flow Control and MEMS, 293–310.
© 2008 *Springer. Printed in the Netherlands.*

low-dimensional description of the turbulent velocity field in the sound producing region of the jet is estimated at each instant using near-field pressure measurements, and ultimately is used to predict the associated turbulent velocity field. The third section presents recent work in progress that directly measures the temporal variation in the velocity field of this jet. New perspectives on closed-loop flow-control techniques using similar active control of flow separation on a NACA 4412 airfoil are presented and discussed.

2 Pressure Correlations in a High-Speed Jet

Lighthill's equation describing the relationship between the turbulent velocity field and far-field acoustic fluctuations has been known for over 50 years [2, 3]:

$$\frac{\partial^2 \rho}{\partial t^2} - c_0^2 \nabla^2 \rho = \frac{\partial^2 T_{ij}}{\partial x_i \partial x_j}, \tag{1}$$

where the Lighthill stress tensor, T_{ij} is defined as,

$$T_{ij} = \rho u_i u_j + (p + c_0^2 \rho)\delta_{ij} - \tau_{ij}. \tag{2}$$

If we confine ourselves to jets where the temperature is comparable to the ambient temperature (where the $p + c_0^2 \rho$ term can be neglected) at high Reynolds numbers (where the viscous stress tensor, τ_{ij}, can also be neglected), the acoustic sources in the jet can be described by

$$T_{ij} = \rho u_i u_j, \tag{3}$$

and the acoustic fluctuations in the far-field for an unbounded flow can be predicted using the well-known solution to Lighthill's equation:

$$p(\mathbf{y}, t) = \int_V \frac{\partial^2 T_{ij}}{\partial x_i \partial x_j} \left(\mathbf{y}, t - \frac{\mathbf{x} - \mathbf{y}}{c_0} \right) \frac{d\mathbf{y}}{4\pi |\mathbf{x} - \mathbf{y}|}. \tag{4}$$

This equation highlights the difficulties encountered by experimentalist in predicting acoustic pressure from turbulence measurements; we do not yet possess the ability to accurately measure the double spatial gradients of the turbulent Reynolds stresses resolved in time simultaneously over the entire sound producing volume of the jet. Alternative aeroacoustic theories in terms of vorticity [4, 5] do exist, but, in general, are just as difficult to experimentally evaluate as the solution to Lighthill's equation.

Owing to the difficulties associated with determining the acoustic pressure fluctuations from the velocity field, considerable research has been devoted to understanding the near-field pressure in high-speed jets in the hopes of understanding how the near-field pressure, which is largely hydrodynamic, relates to the acoustic far field. For flow control purposes, sensing the flow via near field pressure measurements is

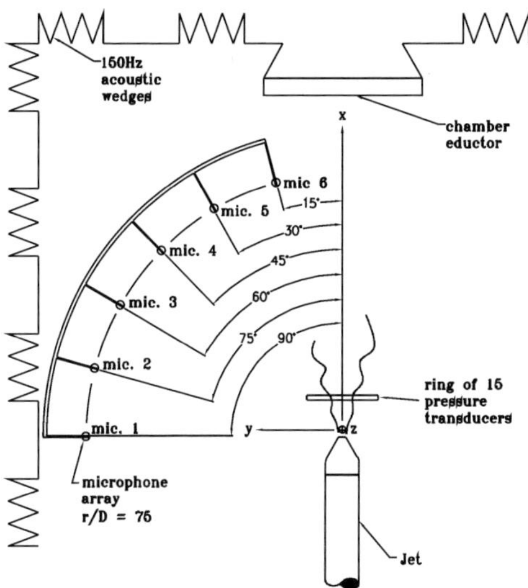

Fig. 1 Schematic of SU anechoic jet facility.

also much more feasible in a practical setting then using unsteady velocity measurements primarily owing to the difficulties associated with obtaining spatially and temporally resolved velocity measurements. The general consensus of previous jet investigations is that the near-field pressure is of lower dimension than the turbulent velocity field [6–9], since the pressure field effectively acts as a spatial filter [10]. Although the spatial coherence of the near-field pressure has been the subject of numerous investigations, previous measurements have never attempted to directly quantify the portion of the pressure field which propagates to the far field. This is done here by measuring the near-field pressure around the periphery of the jet in the noise producing regions using an azimuthal array of pressure transducers with spacing $\Delta\theta$. This allows the near-field pressure to be decomposed into azimuthal modes (m), and then spatially filtered to form a low-dimensional instantaneous reconstruction of the near-field pressure. This spatially filtered pressure signal can then be correlated with the acoustic pressure to examine how the azimuthal coherence of the near-field pressure source affects the acoustic far field. The specifics of this process and the details of the experimental setup can be found in [11, 12].

The measurements were conducted in Syracuse University's fully anechoic chamber, shown in Figure 1. The jet exits a contoured nozzle with centerline velocity of Mach 0.85, corresponding to a Reynolds number of 1×10^6 based on a nozzle diameter of 50.8 mm. The near-field pressure measurements were performed using a ring array of 15 evenly spaced Kulite model transducers positioned 10 mm outside the shear-layer at each downstream position. The jet's far-field acoustic response was measured by 6 G.R.A.S. 1/4 inch condenser microphones mounted $75D$ from the nozzle exit plane along a boom array as shown.

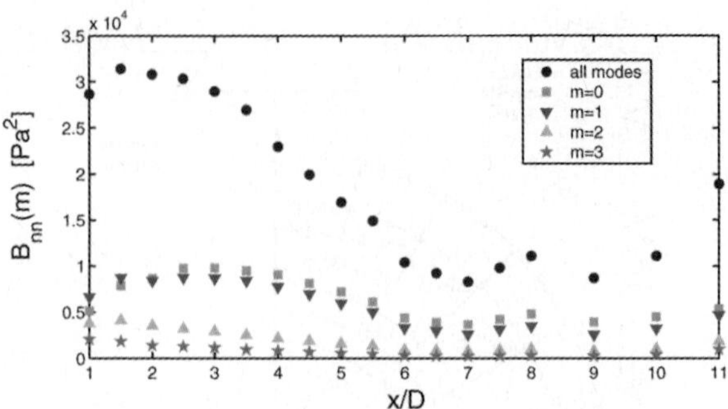

Fig. 2 Downstream growth and decay of the energy contained in the various azimuthal pressure modes.

The azimuthal three-dimensionality character of the near-field pressure at each downstream location is initially examined using two point correlations of the near-field pressure signals,

$$R_{nn}(\Delta\theta,\tau) = \langle p_n(\theta,t)p_n(\theta+\Delta\theta,t+\tau)\rangle. \tag{5}$$

Here, the subscript n is used to denote near-field pressure and f denotes far-field acoustic pressure. These correlations can be Fourier transformed via

$$B_{nn}(m) = \frac{1}{2\pi}\int_0^{2\pi} R_{nn}(\Delta\theta,\tau=0)e^{-i\pi m\Delta\theta}d(\Delta\theta), \tag{6}$$

so that the azimuthal content of the near-field pressure can be examined. The streamwise growth and decay of $B_{nn}(m)$ for azimuthal modes 0 through 3, and for the decay of all modes, the variance, is shown in Figure 2. The contribution from both positive and negative azimuthal modes has been combined to more accurately depict the energy contained in a given structure. Consistent with previous investigations, most of the energy is contained in only the first few modes [6, 13, 14]. In all cases, the contribution from azimuthal mode 0 and 1 are similar and make up the majority of the pressure fluctuations in the flow. Additionally, the streamwise growth and decay of these two modes tends to follow the same downstream trends as the variance, indicating that the downstream behaviour of the full pressure field is related to these two modes. The higher azimuthal modes, such as modes 2 and 3, are much weaker than azimuthal modes 0 and 1 and tend to decay with increasing distance from the jet outlet.

To demonstrate the low dimension of the near-field pressure field from an instantaneous perspective, a reconstruction was performed retaining only selected modes following the methodology outlined by Hall and Ewing [15, 16] i.e.,

$$p^{(rec)}(r,\theta,t) = \hat{p}(r,m,t)e^{im\theta}, \tag{7}$$

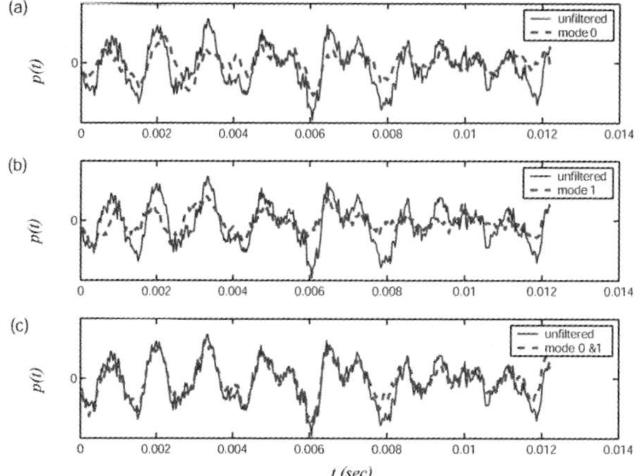

Fig. 3 Comparison of near-field pressure signal at $x/D = 8$ with a reconstruction using (a) azimuthal mode 0, (b) azimuthal mode 1, and (c) both azimuthal modes 0 and 1.

where $\hat{p}(r,m,t)$ is the Fourier coefficient of the pressure in the azimuthal direction. The instantaneous contribution of azimuthal modes 0 and 1 compared to the unfiltered fluctuating pressure signal at $r/D = 8$ at a single pressure transducer are shown in Figure 3. On their own, azimuthal mode 0, or mode 1 accurately recover some of the dynamics of the fluctuating pressure field at a given downstream position. However when combined, these modes accurately recover the dynamics of the near-field pressure at all times. A low-dimensional description of the pressure field using only azimuthal mode 0 and 1 is thus promising for use as a control input, since they capture most of the features of the fluctuating pressure field and because these two modes do not intermittently switch.

The low-dimensional reconstructions of the near-field pressure fluctuations were then correlated with the acoustic pressure in the far field. Typical results for the normalized correlation of azimuthal mode 0, mode 1 and the sum of azimuthal mode 0 and 1, and azimuthal mode 5, with microphone 5 at $x/D = 8$ are shown in Figure 4a. This microphone was at an angle of 30 degrees which is approximately the location where the jet noise radiation peaks. The correlation of all modes (the unfiltered pressure signal) with the far-field acoustic pressure (all modes) is also included. The normalized correlation of azimuthal mode 0 is approximately 60% higher than the correlation of microphone 5 with the full pressure field. This indicates that azimuthal mode 0 is extremely well correlated with the acoustic field and the contribution from the higher azimuthal modes actually reduces the correlation with the far field. For example, the normalized correlation of the higher azimuthal modes with the acoustic far-field pressure is extremely poor. This result demonstrates how large-scale flow events are driving the far field acoustics and moreover suggests that the flow structure that produces a coherent ring pressure mode makes the largest contribution to the far field acoustics.

The time lags associated with the peak in the correlations are consistent with the time required for a pressure wave to propagate from the near-field pressure array for each given downstream location to each far-field microphone. Since the magnitude of the normalized correlation is not dependent on the time lag (providing that the acoustic signal is above the noise floor of the far-field microphones), useful insight into the downstream variation of the acoustic pressure sources in the jet can be gained by examining the streamwise variation of the maxima in the two point correlations, shown in Figure 4b. The peaks in the regular correlation function increase gradually and reach a maximum between $x/D = 6$ and 9, just downstream of the collapse of the potential core in this jet, consistent with the generally accepted location for the dominant source in the jet. Furthermore, the behaviour of azimuthal mode 0 is similar to the regular correlation function, unlike the downstream variation of the correlation of higher azimuthal modes, and indicates that mode 0 is responsible for the downstream behavior in the full correlations.

The above results indicate that the dominant, time-averaged, azimuthal pressure source at a given position is axisymmetric. Michalke and Fuchs [6] also arrived at a similar conclusions, however, this is to the best of our knowledge the first time the correlation of the near field azimuthal pressure modes with the acoustic field has been directly measured. These results also indicate that the higher azimuthal modes only serve to weaken the correlation of azimuthal mode 0 with the far-field acoustic pressure; the cause of this behaviour is presently under investigation. Recently, Hall et al. [12] showed that the poor correlation of the higher near-field azimuthal pressure modes with the acoustic far field was not frequency dependent. Taken together, the present results can be used to guide the development of novel control strategies in the jet based upon spatially filtered measurements of the near-field pressure.

Due care must be taken in interpreting these results, since this is a near-field pressure mode and not a turbulent velocity mode. No direct relationship between pressure and velocity modes is assumed here, particulary given the spatial filtering effect the fluctuating pressure field has upon the velocity field. The direct relationship between the near-field azimuthal pressure modes and the velocity field is presently being examined and will be discussed in the following section of Tinney and Glauser [32]. Furthermore, these measurements do not take into account the streamwise distribution of the potential sources. This is perhaps most easily illustrated by the fact that a lone ring distributed monopole is not capable of emitting a directional source like that detected in the jet. This suggests that the streamwise interaction of the near-field pressure sources is an important feature of the noise source in the jet. The streamwise spatial distribution of the acoustic source in the jet is examined in more detail in the next section.

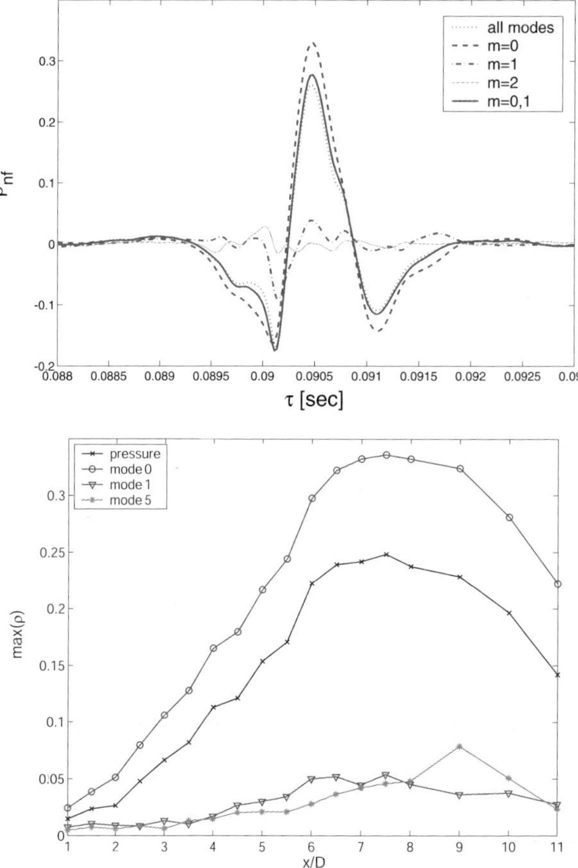

Fig. 4 Normalized two-point, two-time correlations of the near-field pressure with microphone 5 (left) and the downstream variation in the maximum near-field/far-field pressure correlation with microphone 5 (right).

3 Model Estimate of a Mach 0.85 Jet Flow from Near-Field Pressure

The near-field pressure surrounding the jet demonstrates the axisymmetric pressure mode's significant contribution to the far-field acoustics. We now examine the turbulent velocity field estimated from the near-field pressure, and in turn, its contribution to the far-field acoustics.

The complementary techniques of Bonnet et al. [17] are modified to allow for an estimate of the most energetic turbulent features of a subsonic the same Mach 0.85 jet flow via a linear stochastic estimation of the flow's low-order coefficients. The model is low dimensional and is used to predict the far-field acoustics at several ob-

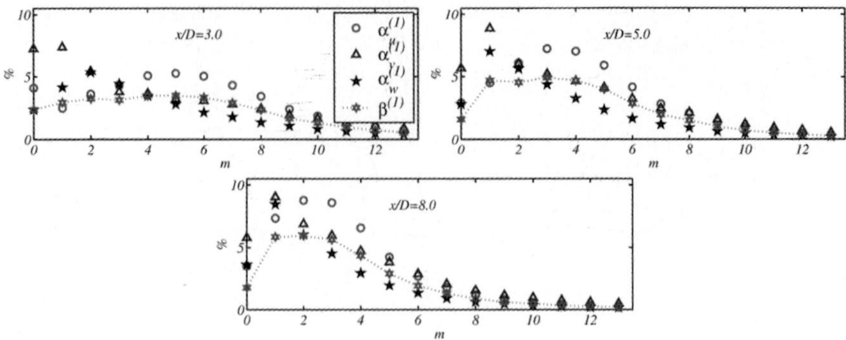

Fig. 5 Fourier-azimuthal mode eigenspectra from the scalar and vector POD of the Mach 0.85 jet.

server positions. Flow field measurements were performed along the (r, θ)-plane of the jet at discrete streamwise locations between $x/D = 3.0$ and 8.0 ($\Delta x/D = 0.25$) using a *Dantec Dynamics* stereo (three-component) Particle Image Velocimetry (PIV) system. Mean and turbulence velocity ratios from this experiment were shown by Ukeiley et al. [18] to agree reasonably well with other jet flow measurements reported in the literature [19–21].

A Proper Orthogonal Decomposition (POD) was performed on the turbulence measurements using a full vector form of the technique as was done recently by Iqbal and Thomas [22]. This included all normal and shear stress terms in the kernel so that terms like $\partial u_i / \partial x_j$ can be computed (for $i, j = 1, 2, 3$) as are necessary for any acoustic analogy. The full (time suppressed) vector form of the integral eigenvalue problem is as follows,

$$\int_R B_{ij}(r, r', x; m) \Phi_j^{(n)}(r', x; m) r' \mathrm{d}r' = \Lambda^{(n)}(x; m) \Phi_i^{(n)}(r, x; m), \tag{8}$$

where the kernel $B_{ij}(r, r', x; m)$ used in the maximization comprised a Fourier transformed ($\vartheta \rightarrow m$), two-point velocity cross-correlation tensor,

$$R_{ij}(r, r', x, \vartheta) = \langle u_i(r, x, \theta, t) u_j(r', x, \theta + \vartheta, t) \rangle. \tag{9}$$

The results are shown in Figure 5 using both scalar ($\alpha_i^{(n)}$) and vector ($\beta^{(n)}$) forms of the technique for the first ($n = 1$) POD mode and the most energetic Fourier-azimuthal modes.

A dynamical estimate of the most energetic events in the flow is performed using an azimuthal array of pressure transducers stationed within the periphery of the hydrodynamic region of the flow (simultaneously with the velocity field) near the nozzle exit and a modified complementary technique. The pressure field in the hydrodynamic regions of the flow are known to comprise a reasonable footprint of the turbulent jet's large scale structure. The procedure comprises a conditional average between the Fourier coefficients of the pressure field (obtained from Fourier-azimuthal decomposition), and the random POD expansion coefficients of the velocity field,

$$\mathbf{S}_{pa}^{(n)}(x;m,-f) = \frac{1}{2\pi} \int_{-\infty}^{\infty} \langle \mathbf{p}^*(m;t-\tau)\mathbf{a}^{(n)}(\beta;x,t_s)\delta_{(m,\beta)} \rangle e^{i2\pi f\tau} d\tau,$$

from which spectral-based estimation coefficients are obtained,

$$b^{(n)}(x;m,f) = \frac{\mathbf{S}_{ap}^{(n)}(x;m,f)}{S_{pp}(m,f)}. \tag{10}$$

Here $S_{pp}(m,f)$ is the ensemble-averaged Fourier-azimuthal pressure cross-spectra. Simplifications to the linear system of Equation (10) can be made by considering the orthogonality of Fourier modes of the pressure and velocity fields and, in turn, avoiding matrix inversion which in some circumstances can be highly ill-conditioned (this is very similar to the Extended-POD technique of Boree [23]).

An estimate of the POD expansion coefficients' temporal frequency is then performed by expanding the estimation coefficients:

$$\tilde{\mathbf{a}}^{(n)}(x;m,f) = b^{(n)}(x;m,f)\mathbf{p}(m,f),$$

from which time-resolved POD-expansion coefficients are easily obtained: $\tilde{\mathbf{a}}^{(n)}(x;m,f) \rightarrow \tilde{\mathbf{a}}^{(n)}(x;m,t)$. The estimate for the time-resolved, fully three-dimensional low-dimensional model is obtained by projecting the time-varying coefficients onto the POD basis followed by a transformation from m to space θ,

$$\tilde{\mathbf{u}}_i(r,x,\theta,t) = \int_{-\infty}^{\infty} \tilde{\mathbf{u}}_i(r,x,t;m)e^{im\theta} dm. \tag{11}$$

The procedure is optimal for estimating conditional events from unconditional sources since the entire procedure is performed by linking the low-order coefficients of both systems, that is, $\mathbf{p}(m;t) \leftrightarrows \mathbf{a}^{(n)}(m;x,t_s)$. Once more, by performing the analysis in the spectral domain, the time-scales of the event estimate are improved.

A sample of the space-time topology of the POD expansion coefficients are shown in Figure 6 for the first two Fourier-azimuthal modes of the POD mode 1. The convective nature of the most energetic flow events are clearly manifest here. It is interesting to point out how the space-time topology of the column mode structure decays around $x/D = 6.0$ where the potential core is known to collapse, whereas the helical mode persists even as far as $x/D = 8.0$.

Estimates of the far-field acoustics are calculated using the LODS model comprising POD modes $n = 1,2$ and Fourier-azimuthal modes $m = 0,1,2$ using equations 2 and 3. Thus, the source field of the LODS low-dimensional model is created using a full Lighthill tensor $i,j = 1,2,3$ and an empirically derived model estimate of a Mach 0.85 jet flow. The source field is computed using a second-order accurate compact finite difference routine. Since the model estimate comprises only the low-order modes of the turbulence field, the higher-order difference scheme (sixth-

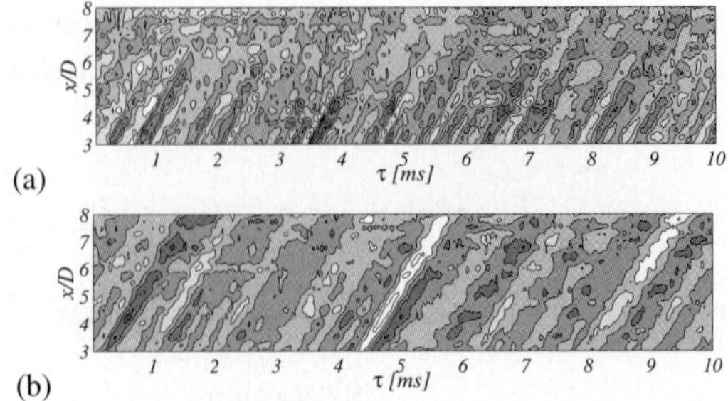

Fig. 6 Space time contour of the estimated time-varying POD expansion coefficients (a)$n = 1, m = 0$, (b)$n = 1, m = 1$.

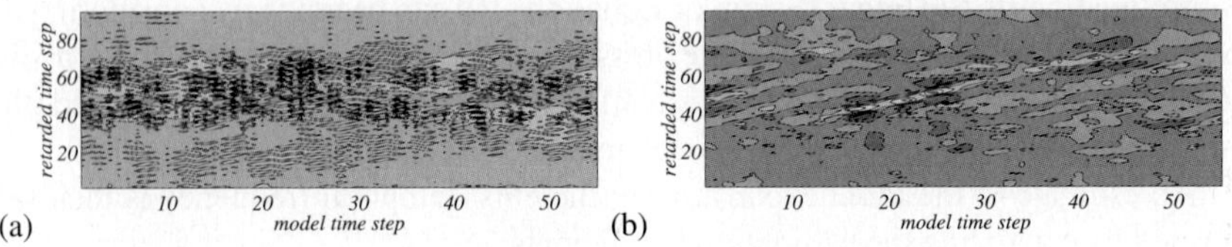

Fig. 7 Retarded time topologies of the far-field pressure ($r/D = 75$) at (a) 90° and (b) 60° relative to the jet axis.

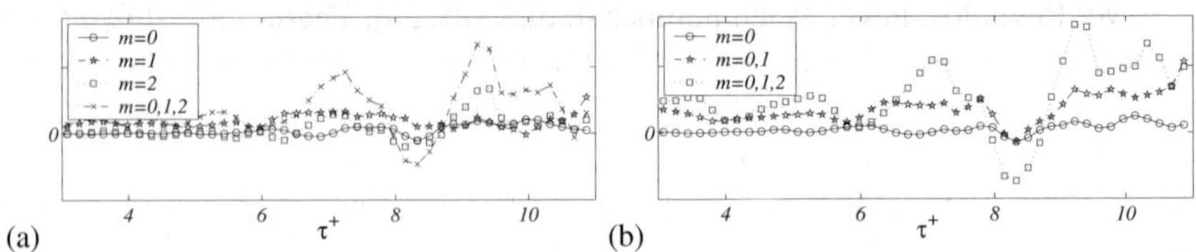

Fig. 8 Estimates of the far-field pressure at 60° using (a) POD mode 1, and (b) POD mode 1+2.

order accurate) was found to have a negligible improvement on the calculation of the source field.

In Figures 7a and b, the far-field acoustics are estimated at 90° and 60°, respectively, for a series of time delays and time steps in the model. At 90°, the time series comprise much higher frequency motions, relative to the far-field signatures at 60°. These results confirm the "large-scale" flow instabilities are responsible for generating high frequency noise at angles normal to the jet axis, as has been the subject of controversy for many years. To correctly model the far-field pressure for every time step of the model, one needs to sum over all instances where the retarded time $(t - \tau^+)$ is constant, suggesting that the streamwise distribution of the acoustic sources is important. After performing this operation, the far-field pressure, using the microphone situated at 60° to the jet axis, is shown for a sample time series using several different POD and Fourier mode combinations. In Figure 8a, only the first velocity POD mode has been used to reconstruct the low-order estimate of the velocity field from which Lighthill's source tensor has been calculated. The results show that the combined contribution of these modes produces a significant acoustic signature in the far field. This can also be seen in Figure 8b using the first two POD modes ($n = 1 + 2$). Since the far-field acoustic pressure has been simultaneously sampled in this investigation, work is presently underway to compare these predictions to the actual pressure field and then to correlate the velocity modes with the acoustic field, similar as done in Section 2 using near-field pressure modes.

4 Dual-Time PIV Investigation of the High-Speed Jet

A complementary investigation was undertaken which enabled temporal derivatives of the velocity field to be measured from dual PIV measurements (see for example [24]). These derivatives are a significant ingredient in the development of low-order dynamical systems models (LODS). With the availability of these models, turbulent large-scale behavior can be predicted, which in turn, provide the necessary tools for creating intelligent feed-back control systems. Moreover, the acceleration field is of great interest from an aeroacoustics perspective since far-field jet noise is believed to be associated with short time and large strain events in the flow which would be accompanied by large outward radial bursts like those originally discovered in the lower Reynolds, lower Mach number studies [25, 26]. These same bursting-like events were identified in the flow estimates of Tinney and Glauser [32] and thus provide greater confidence as to their role in producing noise, and the necessity of these acceleration terms. Finally, dynamical systems can be developed with experimental data to gain a prediction capability in control schemes. In doing so, the time derivative of the velocity is explicitly needed to "train" the low-order dynamical system, as discussed by Ricaud [33].

The experiment is also carried out in the aforementioned facility, with the jet run at a Mach number of 0.6 for reasons of running time capability. The Reynolds number in this case was 690,000. The use of multiple PIV systems was proposed and shown by Kähler and Kompenhans [27] to develop a tool capable of measuring time derivatives of the velocity (i.e. acceleration) as well as spatial derivatives (velocity gradient tensors and vorticity vectors). Two stereoscopic PIV systems are used here to measure all three components of velocity in an identical cross-flow plane perpendicular to the jet axis. Using finite differencing, the acceleration is computed for each pair of velocity fields obtained. The process is reiterated for downstream positions ranging from 3 to 10 jet diameters. Figure 9 shows sample instantaneous velocity fields taken by both systems with a delay of $\Delta t = 25.3$ μs at $x/D = 8$. The resulting out-of-plane component of the acceleration is shown in Figure 10 where local maxima are noticeable in the core region of the jet. These instantaneous events could be the source of strong shear, and consequently, strong acoustic sources. This will be the subject of future work with this database.

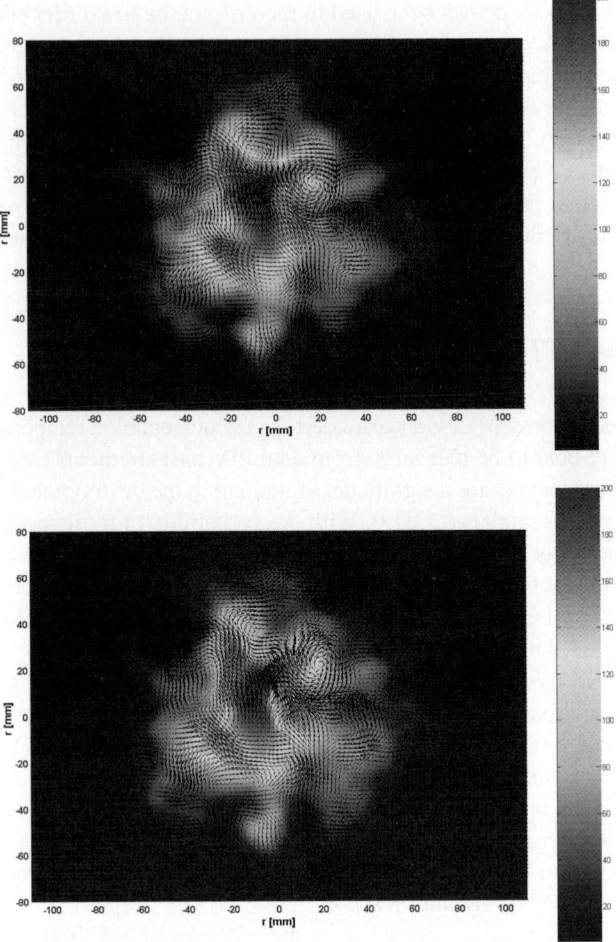

Fig. 9 Instantaneous velocity fields from systems 1 (top) and 2 (bottom), out-of-plane velocity in color contours, $\Delta t = 25$ μs, $M = 0.6$, $x/D = 8$.

5 Control of the Flow Separation over an Airfoil

Similar low-dimensional tools used for the high-speed jet control problem are being developed and implemented for closed-loop control of separation over an airfoil at high angle of attack. This experiment was conducted in the Syracuse University subsonic closed-loop wind tunnel on a NACA 4412 model airfoil. The wing, along with the entire experimental setup described in this section is shown in Figure 11. The airfoil has a 20 cm chord and a 61 cm span. The flow speed was set at $U_\infty = 10$ m/s, and the corresponding Reynolds number based on chord length is Re = 135,000. The experimental velocity measurements were acquired with a Dantec Dynamics

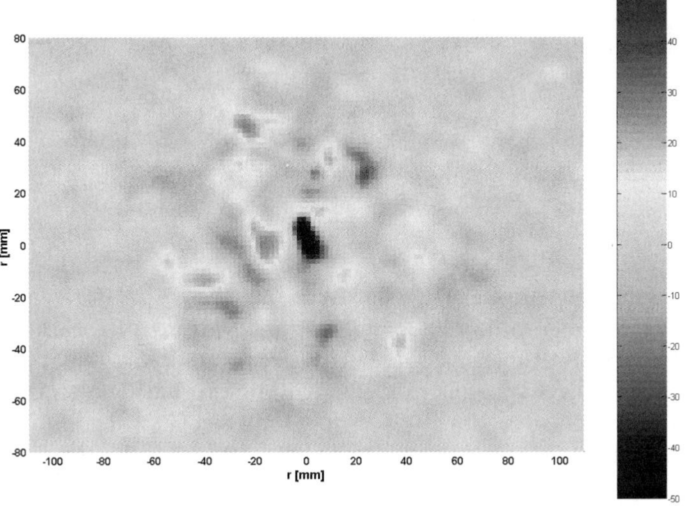

Fig. 10 Instantaneous out-of-plane acceleration field resulting from the finite difference of the fields shown below, $M = 0.6$, $x/D = 8$, $t = 265$.

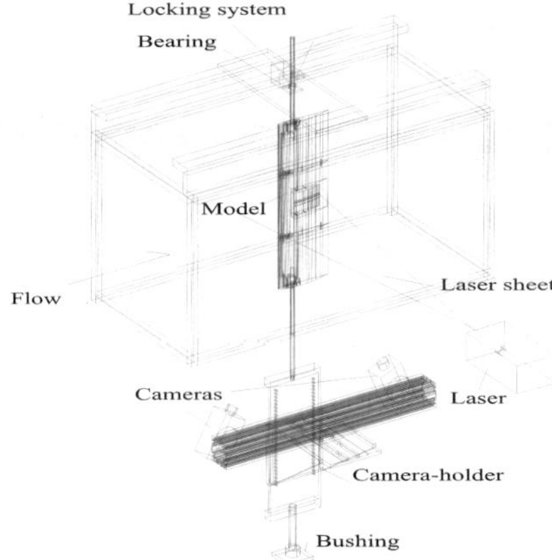

Fig. 11 Overall view of the experimental setup for NACA 4412 airfoil experiment.

stereoscopic PIV system. The design purposely allows the cameras and the wing to move together as the airfoil was pitched providing a fixed measurement window with respect to the airfoil. Online time-resolved flow measurements are available using 11 unsteady pressure sensors embedded along the chord at a mid-span position and evenly spaced between $x/c = 0.29$ and $x/c = 0.78$. All statistics were computed from 1000 statistically independent PIV velocity vector maps and measurements were taken at angles of attack of $\alpha = 10, 12, 14, 16$, and $18°$, with control off and control on.

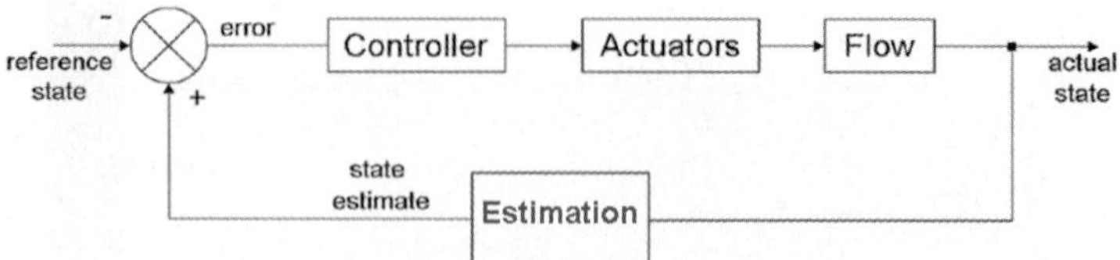

Fig. 12 Schematic of the control diagram.

Our goal here is to estimate the flow state over the NACA 4412 airfoil at each time instant, in order to be able to control the flow using piezoelectric actuators and satisfy our control objective. Our low-cost control objective for this closed-loop approach is to always keep the flow attached to the airfoil. The actuating system consists of 14 oscillatory slot jets near the leading edge of the airfoil, produced by piezoceramic actuators located in individual cavities under the surface of the airfoil. These synthetic jet actuators have been developed in part by the Amitay group [29] at Rensselaer Polytechnic Institute. The major elements of the feedback control are displayed in Figure 12. Using a National Instruments SCXI/PXI signal conditioning/data acquisition platform, we are able to operate the real-time control at 10 kHz.

In order to control the structures of interest in the flow, the estimation and feedback procedures must operate at correspondingly fast time scales. Accessing and processing flow information in time scales less than thousandths of seconds requires the turbulent flow field be described using low-dimensional techniques. Two different mathematical decompositions are considered for the extraction of simple time-dependent coefficients from the complicated turbulent velocity vector fields. The standard POD technique introduced by Lumley [1] which optimally decomposes the velocity field in terms of energy content is used as well as the convection POD [30] that applies the same technique to the convective terms of the Navier–Stokes equation. From the two approaches, we propose to consider the first time-dependent POD coefficient $a^1(t)$ as a potential candidate for the control state variable. The role of such a variable implies that at each instant, $a^1(t)$ should contain information about the flow state, i.e. attached, incipient or stalled. Figure 13 depicts the evolution of the standard deviation of the time-dependent coefficients as the airfoil is pitched. For both POD approaches $a^1(t)$ successfully tracks the changes in the flow, since when the angle of attack is increased, the RMS value increases accordingly; this characteristic is necessary for the control. Indeed, inside the control loop the feedback error is determined by subtracting the actual state from the reference state, which in our case is an attached flow state, therefore requiring that an unwanted drift from the desired state implies an increase in the error and a relevant action from the controller. A simple proportional controller, as implemented by Pinier et al. [28] sends a signal to the actuators that is proportional to the feedback error.

As in the case of the jet control problem, real-time information on the flow state dynamics is not available through direct measurements and therefore is required to be estimated using the instantaneous pressure measurements. The time-resolved

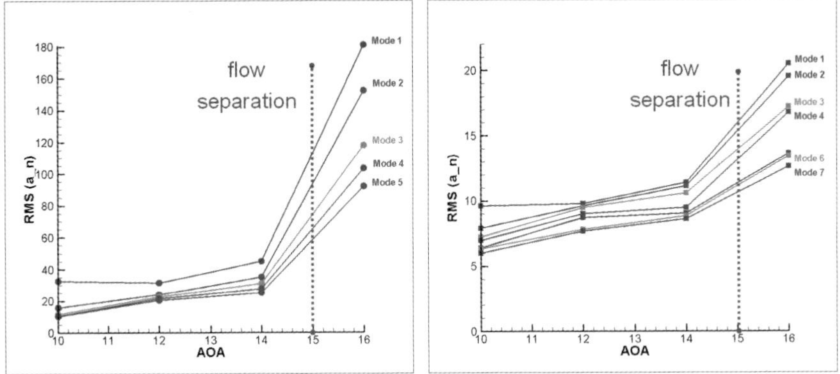

Fig. 13 Evolution of the standard deviation of the POD coefficients with varying angle of attack for different mode numbers; standard POD (left) and convection POD (right).

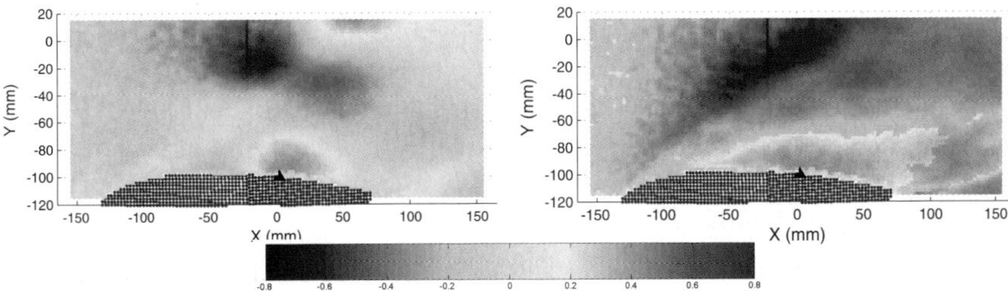

Fig. 14 Single-time (left) and maximum (right) normalized cross-correlations of pressure sensor at $x/c = 0.64$ and U velocity, $\alpha = 16°$, control off.

pressure data and the PIV velocity data are combined offline in a stochastic estimation technique as described in Pinier et al. [28] to then estimate online in real-time the state variable from single-time pressure measurements. Looking at the spatial regions of highest pressure/velocity correlation at the time of the PIV trigger – i.e. when the pressure and the velocity are time-aligned ($\tau = 0$) – one can see the benefits to be gained from including the surrounding time events in the computations. Figure 14 presents the spatial correlations for the single time $\tau = 0$ and compares it to the local maximum correlation.

Strong correlations of up to 60% are found on both plots. The single-time plot displays maximum correlation values in a zone surrounding the corresponding pressure sensor. The spatial extent of the correlations becomes considerably greater when a range of τ containing the PIV trigger is taken into account. This indicates the importance of investigating not only the single time $\tau = 0$ at which velocity and pressure were sampled together, but a larger time period surrounding this τ, as significant additional amount of information about the velocity field will be retrieved from the pressure time series. This is again consistent with the results from the aforementioned jet investigations.

Here, we propose to use the first POD coefficient $a_1(t)$ as our control state-variable. Two distinct decompositions are considered, which both satisfy the control requirements. Both state variables issued from the two methods will be taken to further steps in the experiment in order to determine which one might best qualify. With correlations of the order of up to 60% between the pressure measurements and the first POD coefficient, it is clear that an estimation technique that incorporates multiple time measurements will benefit from the full dynamical information of the flow and will show considerable improvement compared to the single-time method.

6 Concluding Remarks

Owing to the difficulties associated with determining the acoustic sources in the jet with a high Reynolds number flow and a relatively small time scale, a number of low-dimensional tools are presented for the development of active control strategies. In particular, emphasis was placed here on using time-resolved measurements of the unsteady pressure, an integral measure of the turbulent velocity field, to estimate the turbulent velocity field. Unlike measurements of the turbulent velocity field, unsteady pressure measurements are attractive in flow control applications since they are relatively inexpensive, have the ability to be time resolved, and are easily obtainable in many highly turbulent flows.

In the high-speed jet, two low-dimensional control perspectives are presented. In the first, measurements of the near-field pressure are examined and demonstrate the truly low-dimensional nature of the near-field pressure. When correlated with the far-field acoustic pressure, only near-field pressure azimuthal mode 0 displays significant correlation with the far field suggesting, that at any given downstream position, that the dominant time averaged source is axisymmetric. In Section 2, a methodology was developed using instantaneous measurements of the jet lip near-field pressure to estimate the instantaneous turbulent velocity field, and in turn, to predict the far-field acoustics. Here, the instantaneous streamwise evolution of the acoustic sources was examined by applying POD in the streamwise direction. This technique directly yields a transfer function based upon the near-field pressure that could then be implemented in active flow control schemes with the control objective to minimize the jet noise at a point. A novel experiment in the high-speed jet where the fluid acceleration is measured in the high-speed jet using dual PIV systems is discussed in Section 4. In addition to providing heretofore unexplored simultaneous spatial and temporal gradients in the flow, the instantaneous relationship between the near-field pressure and the turbulent velocity field (and the local events that contribute to the jet noise) can be determined. Finally, it is expected that the techniques that build upon previously successful, closed-loop active flow-control techniques for preventing flow separation from an airfoil, will prove useful in the high-speed jet.

Acknowledgments

The authors would like to acknowledge AFOSR funding for the high speed jet noise work and the NACA 4412 work. JWH is grateful for the support of the Natural Sciences and Engineering Research Council of Canada.

References

1. Lumley, J.L.: The structure of inhomogeneous turbulence. In: *Atmospheric Turbulence and Wave Propagation*, A.M. Yaglom and V.I. Tatarski (Eds), Nauka, Moscow (1967).
2. Lighthill, M.: On sound generated aerodynamically: General theory. *Proceedings of the Royal Society* **211** (1952) 564–587.
3. Lighthill, M.: On sound generated aerodynamically: II. Turbulence as a source of sound. *Proceedings of the Royal Society* **222** (1954) 1–32.
4. Powell, A.: Theory of vortex sound. *J. Acoust. Soc. Am.* **36** (1964) 177–195.
5. Howe, M.: Contributions to the theory of aerodynamic sound, with application to excess jet noise and the theory of the flute. *Journal of Fluid Mechanics* **71** (1975) 625.
6. Michalke, A., Fuchs, H.V.: On turbulence and noise of an axisymmetric shear flow. *Journal of Fluid Mechanics* **70** (1975) 179–205.
7. Fuchs, H.: Space correlations of the fuctuating pressure in subsonic turbulent jets. *Journal of Sound and Vibration* **23** (1972) 77.
8. Arndt, R.E.A., Long, D.F., Glauser, M.N.: The proper orthogonal decomposition of pressure fluctuations surrounding a turbulent jet. *Journal of Fluid Mechanics* **340** (1997) 1–33.
9. Ukeiley, L., Ponton, M.: On the near field pressure of a transonic axisymmetric jet. *International Journal of Aeroacoustics* **3**(1) (2004) 43–65.
10. George, W., Beuther, P.D., Arndt, R.: Pressure spectra in turbulent free shear flows. *Journal of Fluid Mechanics* **148** (1984) 155–191.
11. Hall, J., Pinier, J., Hall, A., Glauser, M.: Two-point correlations of the near and far-field pressure in a transonic jet. In: *Proceedings of the Fluids Engineering Division Summer Meeting*, Miami, FL, New York, ASME (July 2006) Paper Number FEDSM2006-98458.
12. Hall, J., Pinier, J., Hall, A., Glauser, M.: A spatio-temporal decomposition of the acoustic source in a Mach 0.85 jet. AIAA Paper 2007-442 (2007).
13. Hall, A., Pinier, J., Hall, J., Glauser, M.: Identifying the most energetic modes of the pressure near-field region of a Mach 0.85 axisymmetric jet. AIAA Paper 2006-314 (2006).
14. Tinney, C., Jordan, P., Guitton, A., Delville, J.: A study in the near pressure field of co-axial subsonic jets. AIAA Paper 2006-2589 (2006).
15. Hall, J.W., Ewing, D.: The development of the large-scale structures in round impinging jets exiting long pipes at two Reynolds numbers. *Experiments in Fluids* **38** (2005) 50–58.
16. Hall, J.W., Ewing, D.: On the dynamics of the large-scale structures in round impinging jets. *Journal of Fluid Mechanics* **555** (2006) 439–458.
17. Bonnet, J.P., Cole, D.R., Delville, J., Glauser, M.N., Ukeiley, L.: Stochastic estimation and proper orthogonal decomposition: Complementary techniques for identifying structure. *Experiments in Fluids* **17**(5) (1994) 307–314
18. Ukeiley, L.S., Tinney, C.E., Mann, R., Glauser, M.N.: Spatial correlations in a transonic jet. *AIAA Journal* **45**(6) (2007) 1357–1369.
19. Bradshaw, P., Ferriss, D.H., Johnson, R.F.: Turbulence in the noise-producing region of a circular jet. *Journal of Fluid Mechanics* **19**(4) (1964) 591–624.
20. Lau, J.C., Morris, P.J., Fisher, M.J.: Measurements in subsonic and supersonic free jets using a laser velocimeter. *Journal of Fluid Mechanics* **93**(1) (1979) 1–27.

21. Jung, D., Gamard, S., George, W.K.: Downstream evolution of the most energetic modes in a turbulent axisymmetric jet at high reynolds number. Part 1. The near-field region. *Journal of Fluid Mechanics* **514** (2004) 173–204.
22. Iqbal, M.O., Thomas, F.O.: Coherent structures in a turbulent jet via a vector implementation of the proper orthogonal decomposition. *Journal of Fluid Mechanics* **571** (2007) 281–326.
23. Boree, J.: Extended proper orthogonal decomposition: A tool to analyse correlated events in turbulent flows. *Experiments in Fluids* **35** (2003) 188–192.
24. Liu, X., Katz, J.: Instantaneous pressure and material acceleration measurements using a four-exposure PIV system. *Experiments in Fluids* **41**(2) (2006) 227–240.
25. Glauser, M.N., Leib, S.J., George, W.K.: Coherent structures in the axisymmetric mixing layer. In: *Turbulent Shear Flows 5*, F. Durst et al. (Eds.), Springer-Verlag, New York (1987) pp. 4.21–4.26.
26. Citriniti, J.H., George, W.: Reconstruction of the global velocity field in the axisymmetric mixing layer utilizing the proper orthogonal decomposition. *Journal of Fluid Mechanics* **418** (2000) 137–166
27. Kaehler, C.J., Kompenhans, J.: Fundamentals of multiple plane stereo particle image velocimetry. *Experiments in Fluids* (Suppl) **29**(7) (2000) S70–S77.
28. Pinier, J., Ausseur, J., Glauser, M., Higuchi, H.: Proportional closed-loop feedback control of flow separation. *AIAA Journal* **45**(1) (2007) 181–190.
29. Glezer, A., Amitay, M.: Synthetic jets. *Annual Review of Fluid Mechanics* **34** (2002) 503–529.
30. Ausseur, J., Pinier, J., Glauser, M.: Flow separation control using a convection based POD approach. In: *3rd AIAA Flow Control Conference*, San Francisco, California. (2006) AIAA-2006-3017.
31. Tinney, C.E., Jordan, P., Delville, J., Hall, A.M., Glauser, M.N.: A time-resolved estimate of the turbulence and sound source mechanisms in a subsonic jet flow. *J. Turbulence* **8**(7) (2007) 1–20.
32. Tinney, C.E. and Glauser, M.N., The modified complementary technique applied to the Mach 0.85 axisymmetric jet for noise prediction. AIAA-2007-3663 (2007).
33. Ricaud, F., Etude de l'identification des sources à partir du couplage de la pression en champ proche et de l'organization instantanée de la zone de mélange de jet (Study of acoustic source identification from the coupling between the near field pressure and the instantaneous organization of the jet mixing layer). Ph.D. Thesis, University of Poitiers, France (2003).

Evolutionary Optimization of Feedback Controllers for Thermoacoustic Instabilities

Nikolaus Hansen[1], André S.P. Niederberger[2], Lino Guzzella[2] and
Petros Koumoutsakos[1]
[1]*Institute of Computational Science,* [2]*Measurement and Control Laboratory,*
ETH Zurich, 8092 Zurich, Switzerland; E-mail: petros@inf.ethz.ch

Abstract. We present the system identification and the online optimization of feedback controllers applied to combustion systems using evolutionary algorithms. The algorithm is applied to gas turbine combustors that are susceptible to thermoacoustic instabilities resulting in imperfect combustion and decreased lifetime. In order to mitigate these pressure oscillations, feedback controllers sense the pressure and command secondary fuel injectors. The controllers are optimized online with an extension of the CMA evolution strategy capable of handling noise associated with the uncertainties in the pressure measurements. The presented method is independent of the specific noise distribution and prevents premature convergence of the evolution strategy. The proposed algorithm needs only two additional function evaluations per generation and is therefore particularly suitable for online optimization. The algorithm is experimentally verified on a gas turbine combustor test rig. The results show that the algorithm can improve the performance of controllers online and is able to cope with a variety of time dependent operating conditions.

Key words: Evolutionary optimization, combustion instabilities, noise.

1 Introduction

Modern gas turbines have to comply with continually more stringent emission regulations (NO_x, CO, etc.). This fact led to the development of lean premixed combustion systems. They operate with excess air to lower the combustion temperature, which in turn decreases NO_x levels. The lean regime however makes the combustor prone to thermoacoustic instabilities, which arise due to a feedback loop involving fluctuations in acoustic pressure, velocity and heat release. Thermoacoustic instabilities may cause mechanical damage, higher heat transfer to walls, noise and pollutant emissions. This phenomenon is observed also in rocket motors, ramjets, afterburners, and domestic burners. One way to substantially reduce such thermo-

J.F. Morrison et al. (eds.), IUTAM Symposium on Flow Control and MEMS, 311–317.
© 2008 *Springer. Printed in the Netherlands.*

acoustic instabilities is *active control* [1, 2]. A feedback controller receives input from pressure sensors and commands a secondary fuel injection. Adjusting the controller parameters into a feasible working regime is an optimization problem with two important properties. First, the stochastic nature of the combustion process leads to a considerable amount of uncertainty in the measurements. Second, changing operating conditions ask for online tuning of the controller parameters.

Evolutionary algorithms are population-based optimization methods, which are considered to be intrinsically robust to uncertainties present in the evaluation of the objective function. The main reason for this robustness is the use of a population [3, 4]. To improve their robustness against noise, either the population size is increased [5, 6] or multiple objective function evaluations of solutions are conducted and an appropriate statistics is taken, usually the mean value. Both methods increase the number of function evaluations per generation, typically by a factor between 3 and 100, which is prohibitive for online applications. Consequently, we suggest an alternative approach to optimize the parameters of a Gain-Delay and an $\mathcal{H}_\infty$ controller online with an evolutionary algorithm. A noise-handling method is introduced that distinguishes between noise measurement and noise treatment. The noise measurement is suited for any ranking-based search algorithm, needs only a few additional function evaluations per generation, and does not rely on an underlying noise distribution. The noise measurement is combined with two noise treatments that aim to ensure that the signal-to-noise ratio remains large enough to keep the evolutionary algorithm in a rational working regime.

The next section introduces the noise-tolerant CMA evolution strategy. Section 3 reports experiments on a test rig with the different controller structures for two operating conditions and Section 4 gives a summary.

2 A Noise-Resistant Evolutionary Algorithm

The evolutionary algorithm serves to minimize a time dependent stochastic objective function L (also loss or cost function)

$$L : \mathcal{S} \times \mathbb{R}^+ \to \mathbb{R}, \quad (\mathbf{x};t) \mapsto f(\mathbf{x};t) + N_f(\mathbf{x};t). \tag{1}$$

The algorithm is based on a $(\mu/\mu; \lambda)$ Covariance Matrix Adaptation (CMA) Evolution Strategy (ES) [7–9] with the default parameters from [9]. The $(\mu/\mu; \lambda)$ CMA-ES is predestined for four reasons. First, it is a *non-elitist continuous domain* evolutionary algorithm. Non-elitism avoids systematic fitness overvaluation [3] and possible subsequent failure. Second, the selection is solely based on the ranking of solutions providing robustness in an uncertain environment. Third, the covariance matrix adaptation conducts an effective and efficient adaptation of the search distribution to the landscape of the objective function. Fourth, the CMA-ES can be used reliably with small population sizes, allowing for a fast adaptation as it is highly desirable in an online application. Here, we introduce a noise-handling (NH) method that can be applied to any ranking based search algorithm and is combined with the CMA-ES into the NH-CMA-ES. The noise handling preserves all invariance properties of the CMA-ES, but biases the population variance when an excessive noise level is detected. The noise measurement and the noise treatment are described in turn.

The noise measurement is based on measured *rank changes induced by reevaluations* of solutions. The algorithm outputs a noise measurement value s and reads

1. Set $L_i^{\text{new}} = L_i^{\text{old}} = L(\mathbf{x}_i)$, $i = 1,\ldots,\lambda$, and let $\mathscr{L} = \{L_k^{\text{old}}, L_k^{\text{new}} | k = 1,\ldots,\lambda\}$, where λ is the number of offspring in the CMA-ES.

2. Compute λ_{reev}, the number of solutions to be reevaluated; $\lambda_{\text{reev}} = f_{\text{pr}}(r_\lambda \times \lambda)$ where the function

$$f_{\text{pr}} : \mathbb{R} \to \mathbb{Z}, \quad x \mapsto \begin{cases} \lfloor x \rfloor + 1 & \text{with probability } x - \lfloor x \rfloor \\ \lfloor x \rfloor & \text{otherwise.} \end{cases}$$

If $r_\lambda \times \lambda < 1$ and $\lambda_{\text{reev}} = 0$ for more than $2/(r_\lambda \times \lambda)$ generations, set $\lambda_{\text{reev}} = 1$ to avoid extremely long sequences without reevaluation.

3. Reevaluate solutions. For each solution $i = 1,\ldots,\lambda_{\text{reev}}$ (assuming the solutions of the population are i.i.d., we can, w.l.o.g., choose *the first λ_{reev} solutions* for reevaluation)

    (a) Apply a small perturbation: $x_i^{\text{new}} = \mathtt{mutate}(\mathbf{x}_i; \varepsilon)$ where $\mathbf{x}_i^{\text{new}} \neq \mathbf{x}_\Leftarrow \Rightarrow \varepsilon \neq 0$. For the CMA-ES, we might apply $\mathtt{mutate}(\mathbf{x}_i, \varepsilon) = \mathbf{x}_i + \varepsilon \sigma \mathscr{N}(0, \mathbf{C})$, where $\mathscr{N}(\cdot)$ denotes a multi-variate normal distribution, and σ and $\mathbf{C}$ are the step-size and the covariance matrix from the CMA-ES, respectively.

    (b) Reevaluate the solution: $L_i^{\text{new}} = L(\mathbf{x}_i^{\text{new}})$.

4. Compute the rank change Δ_i. For each chosen solution $i = 1,\ldots,\lambda_{\text{reev}}$ the rank change value, $\Delta_i \in \{0, 1, \ldots, 2\lambda - 2\}$, counts the number of values from the set $\mathscr{L} \backslash \{L_i^{\text{old}}, L_i^{\text{new}}\}$ that lie between L_i^{old} and L_i^{new}. Formally, we have

$$\Delta_i = \text{rank}(L_i^{\text{new}}) - \text{rank}(L_i^{\text{old}}) - \text{sign}(\text{rank}(L_i^{\text{new}}) - \text{rank}(L_i^{\text{old}})),$$

where $\text{rank}(L_i)$ is the rank of the respective function value in the set $\mathscr{L} = \{L_k^{\text{old}}, L_k^{\text{new}} | k = 1,\ldots,\lambda\}$.

5. Compute the noise measurement, s. Therefore the rank change value, Δ_i, is compared with a limit $\Delta_\theta^{\text{lim}}$. The limit is based on the distribution of the rank changes on a random function L and the parameter θ (see text). Formally, we have

$$s = \frac{1}{\lambda_{\text{reev}}} \sum_{i=1}^{\lambda_{\text{reev}}} (2|\Delta_i| - \Delta_\theta^{\text{lim}}(\text{rank}(L_i^{\text{new}}) - \mathbb{1}_{L_i^{\text{new}} > L_i^{\text{old}}})$$

$$- \Delta_\theta^{\text{lim}}(\text{rank}(L_i^{\text{old}}) - \mathbb{1}_{L_i^{\text{old}} > L_i^{\text{new}}})),$$

where $\Delta_\theta^{\lim}(R)$ equals the $\theta \times 50\%$ of the set $\{|1-R|,|2-R|,\ldots|2\lambda-1-R|\}$, that is, for a given rank R, the set of absolute values of all equally probable rank changes on a random function L (where f and N_f are independent of $\mathbf{x}$).

6. Re-rank the solutions according to their rank sum, i.e. $\mathrm{rank}(L_i^{\mathrm{old}}) + \mathrm{rank}(L_i^{\mathrm{new}})$. Ties are resolved first using the absolute rank change $|\Delta_i|$, where the mean

$$\Delta_i = \frac{1}{\lambda_{\mathrm{reev}}} \sum_{j=1}^{\lambda_{\mathrm{reev}}} |\Delta_j|$$

is used for solutions $i > \lambda_{\mathrm{reev}}$ not being reevaluated, and second, using the (mean) function value.

The parameters are set to $r_\lambda = \max(0.1, 2/\lambda)$, $\varepsilon = 10^{-7}$, and $\theta = 0.2$.

Two noise treatments are used in this paper. First, increase of the evaluation (measuring) time, t_{eval}, for evaluating the controller's performance. Second, increase of the population variance (step-size σ), which can have three beneficial effects: (a) the signal-to-noise ratio is likely to improve, because the population becomes more diverse; (b) the population escapes search-space regions with too low a signal-to-noise ratio, because in these regions the movement of the population is amplified; and (c) premature convergence is prevented. The noise treatment algorithm applied after each generation step uses noise measurement s, and affects step-size σ and evaluation time t_{eval}.

The algorithm reads as follows:

$$\bar{s} \rightarrow (1-c_s)\bar{s} + c_s s$$
$$\text{if } \bar{s} > 0 \qquad \text{\% apply noise treatment}$$
$$\qquad \text{if } t_{\mathrm{eval}} = t_{\max}$$
$$\qquad\qquad \sigma \rightarrow \alpha_\sigma \sigma$$
$$\qquad t_{\mathrm{eval}} \rightarrow \min(\alpha_t t_{\mathrm{eval}}; t_{\max})$$
$$\text{else if } \bar{s} < 0 \qquad \text{\% decrease evaluation time}$$
$$\qquad t_{\mathrm{eval}} \rightarrow \max(t_{\mathrm{eval}}/\alpha_t, t_{\min})$$

Initialization is $t_{\mathrm{eval}} = t_{\min}$ and $\bar{s} = 0$ and the parameters are chosen to be $c_s = 1$, $\alpha_\sigma = 1 + 2/(n+10)$, $\alpha_t = 1.5$, $t_{\min} = 1$ s, and $t_{\max} = 10$ s.

All parameter settings result from the combination of the noise handling with the CMA-ES and simulations on the sphere function. For the combination with different algorithms, a different parameter setting might be useful and necessary.

3 Experimental Results

A lab scale test rig was used for the experiments. Preheated air premixed with natural gas flowed into a downscaled model for the ALSTOM environmental (EV) swirl

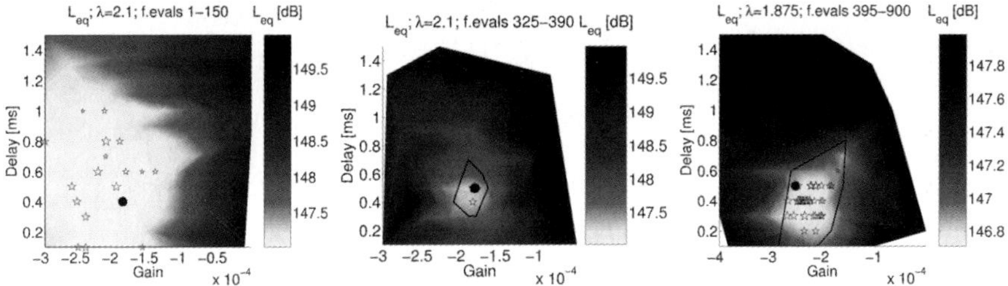

Fig. 1 Cost function landscapes for different time intervals. Pentagrams show the best parameter set obtained from NH-CMA-ES for each generation, the larger they are, the later they have been acquired. The black polygon is the convex hull of all controller parameter values tried in the given time range. Function evaluations, left: 1–150 (0–1300 s); middle: 325–390 (3800–4800 s); right: 395–900 (4900–9800 s). The landscapes are obtained by Delauney triangulation of a second-order polynomial fit to L_{eq} values for individual delay slices.

burner that stabilizes the flame in recirculation regions near the burner outlet plane. The pressure signal was detected by a water-cooled microphone placed 123 mm downstream of the burner. A MOOG magnetostrictive fuel injector installed close to the flame was used as control actuator. The operating conditions were a mass flow of 36 g/s, a preheat temperature of 700 K, and a ratio of actual to stoichiometric air/fuel ratio of $\lambda = 1.875$ and 2.1. Two controller types were investigated: a simple phase-shift or Gain-Delay controller, where gain and delay were optimized by the evolutionary algorithm; and a model-based robust $\mathcal{H}_\infty$ controller where a frequency shift, gain and delay of a previously designed $\mathcal{H}_\infty$ controller [10, 11] are optimized by the evolutionary algorithm.

The cost function to be minimized is the equivalent continuous level of the sound pressure

$$L_{eq} = 10\log_{10}\frac{(p_s^2)_{\mathrm{av}}}{p_{\mathrm{ref}}^2},$$

where $(p_s^2)_{\mathrm{av}}$ is the mean squared pressure and $p_{\mathrm{ref}} = 20\ \mu\mathrm{Pa}$ is the reference pressure. The sound pressure level L_{eq} is acquired from a measurement during t_{eval} seconds with a given controller parameter setting. The total measurement cycle time consists of ramping the controller gain up and down (about 2 s each), pressure data acquisition time $t_{\mathrm{eval}} \in [1,10]$ s (determined by the algorithm), data logging (1 s) and NH-CMA-ES computation time (negligible).

Three cost function landscapes for different time intervals are shown in Figure 1, where the combustor is fired up from ambient temperature (cold start) with an air/fuel ratio of $\lambda = 2.1$ switched to $\lambda = 1.875$ after 4800 seconds, and the Gain-Delay controller is switched on. A trend towards less negative values for the gain with the heating up becomes apparent (left versus middle figure) and the general background noise level rises (indicated by areas getting darker). Also, the parameters evaluated are narrowed down to the small black polygon. The operating con-

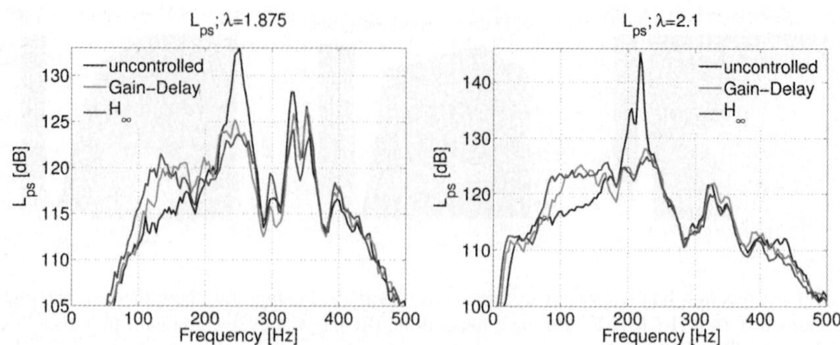

Fig. 2 Comparison of the pressure spectra for the uncontrolled, Gain-Delay controlled and $\mathscr{H}_\infty$ controlled plant. Both controllers are NH-CMA-ES optimized. Left: $\lambda = 1.875$, right: $\lambda = 2.1$.

dition at $\lambda = 1.875$ (right) exhibits less thermal drift. The algorithm finds a new minimum, where the gain is more negative.

Spectra achieved with the optimized Gain-Delay and $\mathscr{H}_\infty$ controllers are compared to the uncontrolled plant in Figure 2. They are shown for the plant which has been running for several hours and is thus heated up. For $\lambda = 1.875$ (left) the Leq of the uncontrolled plant is 148.72 dB, the Gain-Delay controller reduces it to 146.67 dB, while the $\mathscr{H}_\infty$ controller reaches 146.16 dB, which is about 15% less. For $\lambda = 2.1$, the values of L_{eq} are 159.87 dB, 147.48 dB and 147.35 dB, respectively. Here the $\mathscr{H}_\infty$ controller performs only slightly better than the Gain-Delay controller, but the control signal contains about 10% less energy.

4 Summary

This study has investigated feedback controllers for secondary fuel injection used on a test rig designed to study thermoacoustic instabilities. To allow for best controller performance in changing operating conditions, a self-tuning controller is applied. The main difficulty in optimizing the controller parameters is the uncertainty inherent in the pressure measurements. For this reason, a novel noise-handling algorithm is introduced that can be applied to any ranking-based optimization algorithm. The noise-handling algorithm consists of a *noise measurement* and a *noise treatment*, and is applied to the CMA evolution strategy (NH-CMA-ES), where it preserves all invariance properties of the original algorithm. In combination with the CMA-ES, two additional function evaluations per generation are sufficient to establish a functional noise measurement.

Parameters of Gain-Delay and $\mathscr{H}_\infty$ controllers have been optimized online with the introduced NH-CMA-ES while the combustor was running. The experiments show that the algorithm can optimize different controller types and can cope with changing operating conditions and high levels of noise. Model-based $\mathscr{H}_\infty$ controllers perform best, and can be improved further through the use of the NH-CMA-ES. The optimized solutions deviate remarkably from the originally designed solutions and can make up for uncertainties in the model-building and design process, as well as for time-varying plant characteristics.

Acknowledgments

Support by Daniel Fritsche during the experimental phases and assistance by Caroline Metzler for the technical illustrations is gratefully acknowledged. The authors thank David Charypar for the valuable suggestions and Stefan Kern for providing supporting data. Fruitful discussions with Bruno Schuermans and financial support from ALSTOM (Switzerland) Ltd. are gratefully acknowledged.

References

1. Lieuwen, T., Yang, V.: Combustion instabilities in gas turbine engines: Operational experience, fundamental mechanisms, and modeling. *Progress in Astronautics and Aeronautics* **210** (2005).
2. Dowling, A.P., Morgans, A.S.: Feedback control of combustion oscillations. *Annual Review of Fluid Mechanics* **37** (2005) 151–182.
3. Arnold, D.V.: *Noisy Optimization with Evolution Strategies*, Volume 8, Kluwer, Boston (2002).
4. Beyer, H.G., Arnold, D.: Qualms regarding the optimality of cumulative path length control in CSA/CMA-evolution strategies. *Evolutionary Computation* **11**(1) (2003) 19–28.
5. Arnold, D.V., Beyer, H.G.: Local performance of the $(\mu/\mu_I, \lambda)$-ES in a noisy environment. In Martin, W., Spears, W. (Eds.), *Foundations on Genetic Algorithms FOGA*, Morgan Kaufmann (2000) pp. 127–142.
6. Harik, G., Cantu-Paz, E., Goldberg, D.E., Miller, B.L.: The gambler's ruin problem, genetic algorithms, and the sizing of the populations. *Evolutionary Computation* **7** (1999) 231–253.
7. Hansen, N., Ostermeier, A.: Completely derandomized self-adaptation in evolution strategies. *Evolutionary Computation* **9**(2) (2001) 159–195.
8. Hansen, N., Müller, S.D., Koumoutsakos, P.: Reducing the time complexity of the derandomized evolution strategy with covariance matrix adaptation (CMA-ES). *Evolutionary Computation* **11**(1) (2003) 1–18.
9. Hansen, N., Kern, S.: Evaluating the CMA evolution strategy on multimodal test functions. In Yao, X., et al. (Eds.), *Parallel Problem Solving from Nature – PPSN VIII*, Lecture Notes in Computer Science, Vol. 3242, Springer (2004) pp. 282–291.
10. Skogestad, S., Postlethwaite, I.: *Multivariable Feedback Control: Analysis and Design*, John Wiley and Sons Ltd., Chichester, New York (1996).
11. Niederberger, A.S.P., Schuermans, B.B.H., Guzzella, L.: Modeling and active control of thermoacoustic instabilities. In: *Proceedings of 16th IFAC World Congress*, Prague (2005).

Active Cancellation of Tollmien–Schlichting Instabilities in Compressible Flows Using Closed-Loop Control

Marcus Engert[1], Andreas Pätzold[1], Ralf Becker[2] and Wolfgang Nitsche[1]

[1]*Institute of Aero- and Astronautics, Technical University of Berlin, Marchstr. 12–14, 10587 Berlin, Germany; E-mail: marcus.engert@ilr.tu-berlin.de*
[2]*Robert Bosch GmbH, Corporate Sector Research and Advance Engineering (CR/AE33), Robert-Bosch-Str. 19, 71701 Schwieberdingen, Germany*

Abstract. The present paper reports on active cancellation of natural Tollmien–Schlichting (TS) instabilities on an unswept wing in compressible flows. The research concentrates on closed-loop active wave control (AWC) experiments at Mach numbers ranging from 0.2 up to 0.4. These high velocities result in thin boundary layers and therefore in TS frequencies up to 10 kHz. Therefore, the resolution of the applied sensors as well as the amplitude and frequency domain of the actuators are subject to challenging requirements. Additionally, the velocity of the convective TS waves demands a powerful, optimized control algorithm working in real time. The AWC principle applied here delays TS induced transition by stabilizing the linear disturbance waves initiating the laminar-turbulent transition process. This method is based on the wave superposition principle, i.e. the superposition of artificially generated anti-disturbances and the naturally occurring TS disturbances. The energy consumption with this method is considerably lower than the stabilization achieved by manipulating the local mean velocity profile (e.g. boundary layer suction).

Key words: Active control, adaptive filter, Tollmien–Schlichting waves, TS waves, laminar flow control, compressible flows, sensor-actuator systems.

1 Introduction

Many active and passive methods have been investigated in the past to delay laminar-turbulent transition. Most control techniques are aimed at stabilizing the mean velocity profile of the boundary layer (see e.g. Braslow and Visconti [1]). In contrast, the wave-cancellation method assumes that a wavelike disturbance can be linearly cancelled by introducing another wave of the same amplitude and frequency but opposite in phase. Active control of TS waves could be successfully performed with a smart skin consisting of sensors, actuators and controllers. The application of a smart skin may be useful for TS waves in the early linear stage up to the weak

J.F. Morrison et al. (eds.), IUTAM Symposium on Flow Control and MEMS, 319–331.
© 2008 *Springer. Printed in the Netherlands.*

nonlinear stage, especially if they are dominated by two-dimensional TS modes. Such a TS scenario may also be found on a swept wing in the mid-chord region if crossflow instabilities were previously suppressed, for example, by suction in the leading edge region.

Generally, transition in a two-dimensional boundary layer on an unswept wing is dominated by TS instabilities. The natural transition process, which occurs under conditions of very low freestream turbulence levels, can be divided into the following parts: disturbances in the freestream, such as noise and vorticity, enter the boundary layer as steady or unsteady disturbances. This mechanism is usually termed receptivity, following Morkovin [2]. Once the perturbations have found their way into the boundary layer as small amplitude waves, they are either linearly damped, or selectively amplified by frequency while propagating downstream. In their first stage of development, the harmonic instabilities are mainly two-dimensional. All TS modes grow independently, so that the linear superposition with artificially generated counter waves is possible. In the later stage, the development becomes more and more nonlinear and secondary instabilities lead to an increase in the growth of three-dimensional disturbances. These three-dimensional distortions result in lambda vortices and are followed by the turbulent breakdown.

Work over the past two decades has shown the possibility of delaying the laminar-turbulent transition by active control. Reviews of this research are found in [3–5]. Some newer experimental studies in the field of AWC have been performed by Raguse, Evert and Opfer [6–8] at U_∞ up to 15 m/s as well as by Baumann and Sturzebecher [9, 10] at $U_\infty = 17$ m/s. The experiments reported in the present paper are performed in compressible flows ($M = 0.2$ to 0.4) on an unswept generic wing model with a two-dimensional boundary layer in a transonic wind tunnel.

2 Principle of an Active Wave Control System

The principle of an AWC system is displayed in Figure 1. The flow from the left forms a boundary layer on the profile which is perturbed by freestream turbulences, surface roughnesses and noise. Hence, band-limited disturbance waves are generated in the boundary layer which propagate downstream while they are selectively amplified by frequency with respect to the local Reynolds number. These TS waves are detected by the reference sensor array arranged spanwise upstream of the actuator. For detecting these disturbances, it is necessary to use highly sensitive, non-intrusive sensors to detect the tiny TS waves at their early linear stage. The reference signals obtained are fed into a digital controller which models the propagation of the disturbances by means of digital filters. The digital filters are adapted by decorrelating the reference and error sensor signals. The output drives the flush-mounted actuators which introduce appropriate counter waves. This results in a minimization of the amplitudes of the wave field at the downstream array of error sensors.

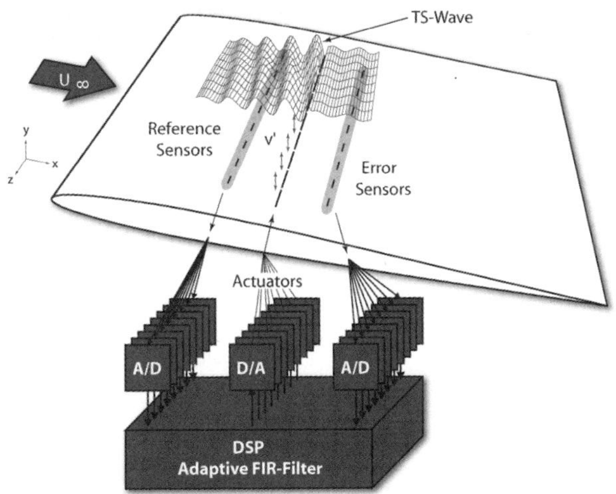

Fig. 1 Principle of an active wave control system [11].

3 Experimental Set-Up

The experiments were conducted in an open loop low-turbulence wind tunnel at $M = 0.2$ to 0.4, as shown in Figure 2. The freestream Mach number is continuously adjustable up to 0.95 resulting in a maximum Reynolds number $Re_l = 1.3 \times 10^5$ (the test section reference length is based on $l = 0.1\sqrt{A_{TS}}$, where A_{TS} is the wind tunnel cross section). Due to the high contraction ratio of 47:1 between the settling chamber and test section, the freestream turbulence intensity is less than 0.15%. The wind tunnel test section has a cross section of 150 mm × 150 mm at the inlet with adjustable top and bottom fibre-glass walls. Appropriate wall adjustments introduce pressure gradients in the flow direction that decelerate or accelerate the natural transition process. This guarantees that transition is always in the range of a surface hot wire (SHW) sensor array, independent of the freestream Mach numbers. Therefore, a single model can be used for a range of Mach numbers.

The unswept test wing (Figure 3) is a 30 mm thick model with a modified NACA 0004 profile at the leading edge and a flap at the trailing edge ($c = 750$ mm). The highly polished aluminium profile is equipped with a rectangular opening located at $0.1 \leq x/c \leq 0.6$ from the model's leading edge for the installation/removal of plate inserts. The pressure distribution is instantaneously measured by 18 pressure orifices equidistantly arranged in the flow direction. The model is horizontally fixed in the center of the two-dimensional adaptive test section.

One plate insert is used as the baseline configuration, and is equipped with a SHW sensor array from $0.12 \leq x/c \leq 0.57$. Generally, SHWs (Figure 4a) are preferred as sensors in the experiments on AWC. This type of sensor was especially designed to detect even the smallest TS waves. A platinum-coated tungsten wire ($\varnothing = 5$ μm) is welded above a narrow cavity (0.075–0.1 mm) flush to the wing's

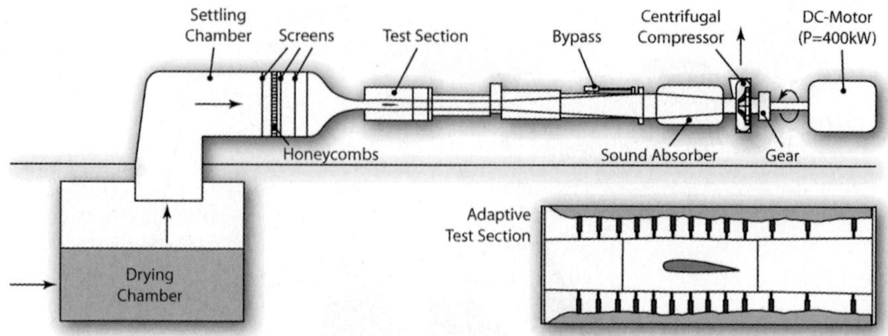

Fig. 2 Transonic wind tunnel test facility.

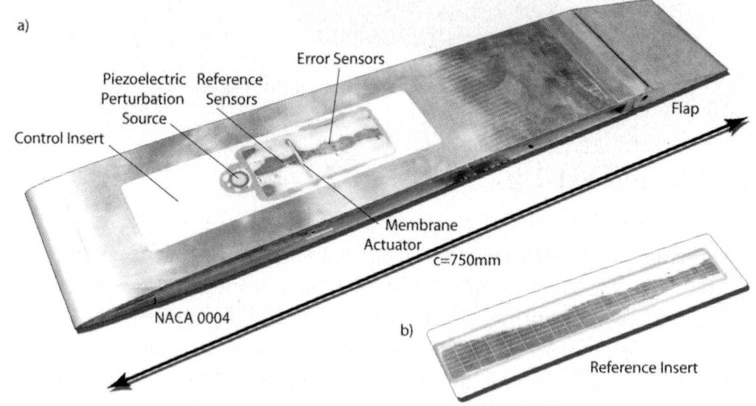

Fig. 3 Generic wing model with control insert.

surface. The cavity can easily be shaped in the copper layer by photo-etching. This arrangement minimizes the heat flux into the wall, therefore resulting in a signal-to-noise ratio which is enhanced by a factor of five compared to a conventional surface hot film. The surface roughness of the sensor/cavity is negligible ($k^+ \ll 5$). The calibration behavior of an SHW shows a typical hot-element characteristic similar to a surface hot film (Figure 4b). Because of its flexibility, it is possible to integrate the sensor into curved surfaces. This potential is used if the sensor and the wing insert are built up together in a negative curved mold as in this case. The SHWs are usually operated by an anemometer in the constant-temperature mode with an overheat ratio of 1.7 [12]. These sensors have been aligned as streamwise and spanwise arrays.

The second plate insert – the control insert (Figure 3) – at $x/c = 0.30$ consists also of an array of SHWs (reference sensors) in spanwise direction. Downstream, the control insert is equipped with flush-mounted actuators ($x/c = 0.33$) that introduce appropriate counter waves. Figure 5a depicts the membrane actuator. A highly flexible membrane is driven by a robust solenoid connected via a T-shaped connect-

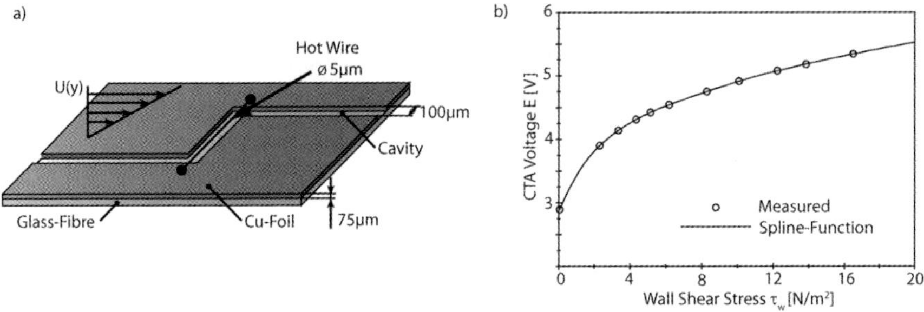

Fig. 4 (a) Principle and (b) typical calibration of a surface hot wire.

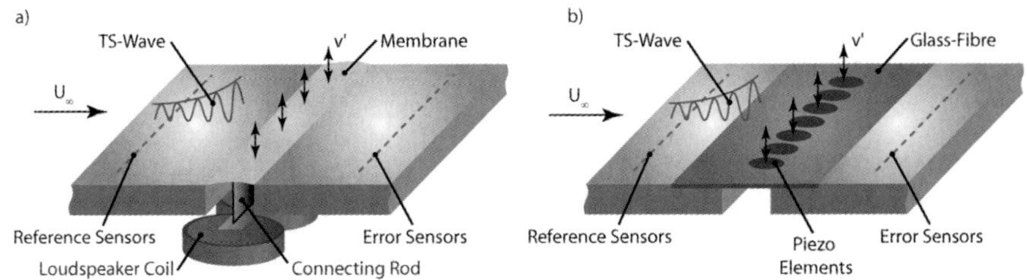

Fig. 5 (a) Membrane actuator and (b) piezoelectric actuator.

ing rod. Further downstream of the actuator ($x/c = 0.36$) another SHW sensor array is used as error sensors.

In addition to the arrangement described above, a perturbation source is installed at $x/c = 0.25$ to introduce controlled artificial perturbations. Considering the fact that the frequency and amplitude introduced perturbations are known. It is possible to test and optimise the controller prior to the experiments with natural TS waves where neither frequency, amplitude nor phase are known. The principle of the perturbation source is shown in Figure 5b. A piezoelectric element is installed underneath a thin glass-fibre reinforced carrier plate. The structure of the piezoelectric element consists of a piezoelectric ceramic plate attached to a metal plate with adhesives. A voltage applied between the electrodes of the piezoelectric element causes mechanical distortion due to the piezoelectric effect. While the piezoelectric element bends in one direction, the metal and glass-fibre plate bonded to the piezoelectric diaphragm do not expand. Thus, when AC voltage is applied across the electrodes, the glass-fibre wall vibrates and introduces oscillations perpendicular to the wall into the flow. The successful application of both described actuator types is dictated by the ability to introduce anti-perturbations in the frequency and wave number domain with respect to the local boundary layer receptivity.

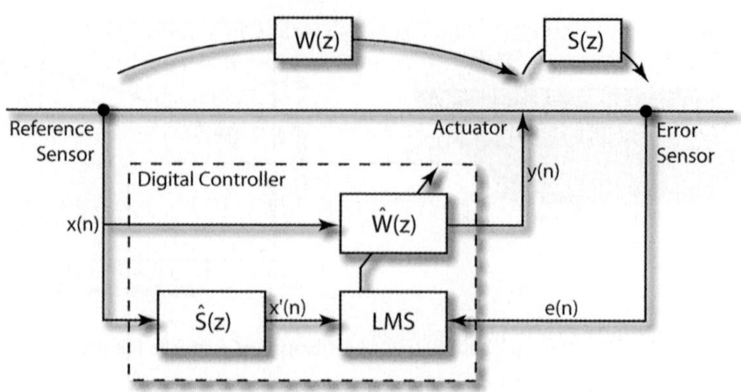

Fig. 6 Block diagram of the feedforward AWC system (physical setup).

4 Control Strategy

The AWC method uses a control algorithm initially applied in the field of active noise control (see e.g. Elliott and Nelson [13]). A common way to approximate the linear transfer behaviour of the TS waves, given by their amplification and propagation in the boundary layer, is to use a Finite Impulse Response (FIR) filter as model. Generally, FIR filters represent the system response to a unit impulse.

The feedforward AWC system is shown in Figure 6. $W(q^{-1})$ denotes the boundary layer transfer behaviour of the TS waves. The purpose of the adaptive controller $\hat{W}(q^{-1})$ is to calculate (based on the reference signal $x(n)$) an appropriate counter wave signal $y(n)$. In order to cancel the primary wave, $\hat{W}(q^{-1})$ has to be adapted such that it is an estimate of the physical path $W(q^{-1})$. Since $W(q^{-1})$ is the unknown system to be identified by the adaption algorithm, see below, it is referred to as adaption path in the following.

The counter wave signal propagated by an actuator combines with the TS wave to create a zone of "silence" in the vicinity of the error sensor. The error sensor measures the residual perturbations, which are used by the Least-Mean-Squares (LMS) algorithm for the adaption of the FIR transfer function $\hat{W}(q^{-1})$ in order to minimize the power of the signal $e(n)$ at the error sensor. Here, the estimate $\hat{S}(q^{-1})$ accounts for the model of the secondary path $S(q^{-1})$ between the output of the controller signal $y(n)$ and the subsequent signal $e(n)$ of the error sensor. The secondary path $\hat{S}(q^{-1})$ comprises the digital to analog converter, the power amplifier, the actuator, the physical behavior of the boundary layer from the actuator to the error sensor, the error sensor, the signal conditioner and the analog to digital converter. Therefore, it even accounts for degeneration processes of single elements, e.g. aging of the actuator. The filtering of the reference signal $x(n)$ through $\hat{S}(q^{-1})$ is required by the fact that the output $y(n)$ of the adaptive path $W(q^{-1})$ is physically filtered through $S(q^{-1})$.

Mathematically, the output signal $y(n)$ is obtained by the discrete convolution $y(n) = \hat{W}(q^{-1})x(n) = \sum_{i=0}^{N-1} \hat{w}_i(n)x(n-i)$ of the reference signal $x(n)$ with the FIR filter $\hat{W}(q^{-1})$. Here, $i = 1 \ldots N-1$ indicates the coefficient and n is the time index. By noting the filter coefficients $\hat{w}_i$ and the input time series of $x(n)$ in vector notation

$$\hat{\mathbf{w}}(n) = \left(\hat{w}_0(n), \hat{w}_1(n), \ldots, \hat{w}_{N-1}(n)\right)^T \tag{1}$$

and

$$\mathbf{x}(n) = (x(n), x(n-1), \ldots, x(n-N+1))^T \, , \tag{2}$$

the convolution reads as the inner product

$$y(n) = \hat{\mathbf{w}}^T(n)\mathbf{x}(n) \, . \tag{3}$$

Here, T denotes the vector transpose. Since the estimated filter coefficients $\hat{w}_i(n)$ in Equation (1) are adapted in every time step using the LMS algorithm described below, they are indicated by n. The convolution in Equation (3) is a weighted moving average of the N past and present input samples. However, since time continuous, physical systems are generally recursive, the FIR filter $\hat{W}(q^{-1})$ is a non-recursive approximation.

Figure 7 illustrates the mathematical formulation of the adaptive system identification scheme for the estimation of the filter coefficients $\hat{w}_i(n)$ as block diagram. The primary path $P(q^{-1}) = W(q^{-1})S(q^{-1})$ comprises the unknown physical system to be identified and the secondary path. The measured residual error $e(n)$ of the estimation can be written as the difference between the desired signal $d(n) = P(q^{-1})x(n)$ and the signal $S(q^{-1})y(n) = \mathbf{s}^T(n)\mathbf{y}(n)$. With Equation (3) the error reads

$$e(n) = d(n) - \mathbf{s}^T(n)\mathbf{y}(n) = d(n) - \sum_{i=0}^{N-1} s_i(n)y(n-i) \tag{4}$$

$$= d(n) - \sum_{i=0}^{N-1} s_i(n)\hat{\mathbf{w}}^T(n)\mathbf{x}(n) \, . \tag{5}$$

The objective of the adaptive control system is to minimize the expected mean squared error $e^2(n)$ (cost function):

$$J(n) = e^2(n) \tag{6}$$

$$= \left(d(n) - \sum_{i=0}^{N-1} \hat{s}_i(n)\hat{\mathbf{w}}^T(n)\mathbf{x}(n)\right)^2 \, . \tag{7}$$

Therefore, the N filter weights $\hat{w}_i(n)$ are adapted in every time step n following the method of steepest descent of the square error signal $J(n)$. The secondary path transfer function $S(q^{-1})$ is replaced by its estimate $\hat{S}(q^{-1})$ in the cost function. The gradient method used for the filter update is called the Filtered-X-Least-Mean-Squares (FXLMS) algorithm (originally introduced by Burgess [14] and Widrow et al. [15]). It is assumed that there is no feedback from the actuator to the reference sensor. The

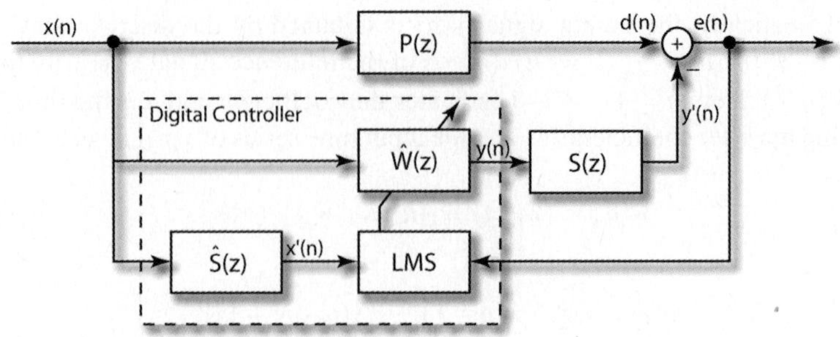

Fig. 7 Mathematical formulation of the adaptive system identification scheme for estimation of the coefficients $\hat{w}_i(n)$ as block diagram.

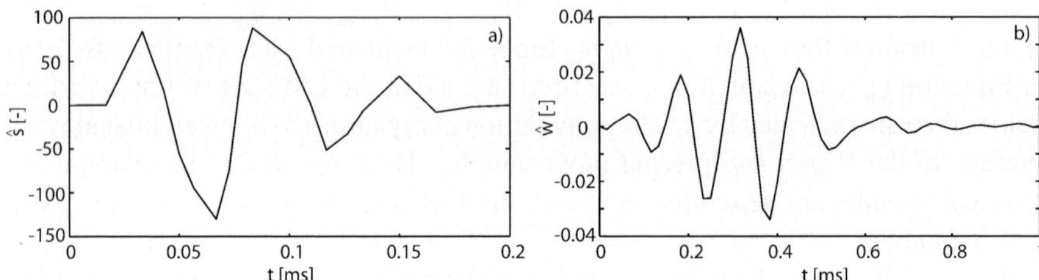

Fig. 8 Estimated filter weights of (a) secondary error path $\hat{S}(q^{-1})$ and (b) unknown system $\hat{W}(q^{-1})$ to be adapted by the FXLMS algorithm.

FXLMS algorithm is widely used in control applications where a transfer function (the secondary path $S(q^{-1})$) exists between the filter output and the error signal.

The FXLMS update equation for the coefficients of $\hat{W}(q^{-1})$ is given as

$$\hat{\mathbf{w}}(n+1) = \hat{\mathbf{w}}(n) + \mu\,\mathbf{x}'(n)\,e(n)\,, \tag{8}$$

where μ is the convergence rate and $x'(n)$ is the reference signal $x(n)$ filtered through the estimated secondary path model $\hat{S}(q^{-1})$. In vector notation this filtering reads

$$x'(n) = \hat{\mathbf{s}}(n)^T\,\mathbf{x}(n). \tag{9}$$

The transfer function $\hat{S}(q^{-1})$ of the secondary path is obtained offline and is either kept fixed during the online operation or is continuously adapted during the operation of AWC.

Typical adapted FIR filters modeling the error path $\hat{S}(q^{-1})$ and $\hat{W}(q^{-1})$ are illustrated in Figures 8a and b. The impulse responses of the FIR filter show the characteristic of a bandpass filter expressing the frequency selective amplification of the TS waves. The delay of the wavy pulse is the modeled propagation time of the TS disturbances.

The digital signal processor system (dSPACE, DS1005 PPC) has direct access to a 32 channel A/D converter (DS2003 A/D Board) and a 32 channel D/A converter

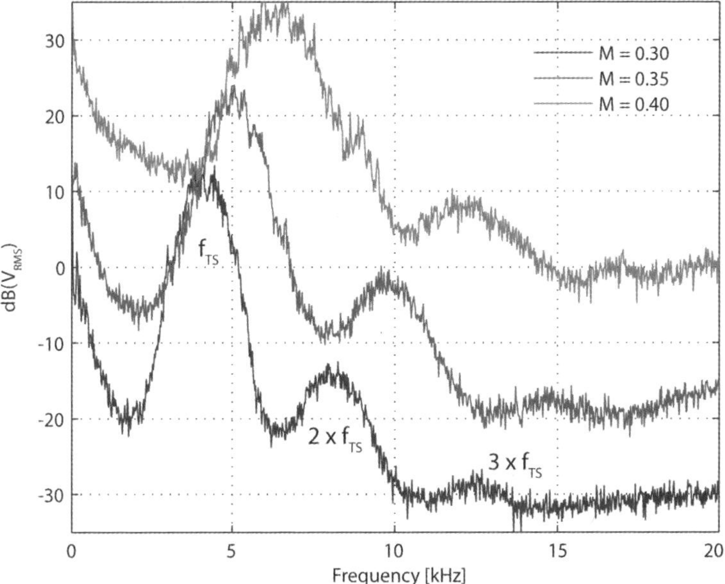

Fig. 9 Typical power-spectra for $M = 0.3$, 0.35 and 0.4.

(DS2103 D/A Board). The DS1005 features an IBM PowerPC 750GX running at 1 MHz. Usually, sampling rates of 60 kHz and filter lengths of 20–30 coefficients for the error path and 60–90 coefficients for the adaptive path are employed for an effective TS wave cancellation. The programming is based on the Matlab Simulink Real-Time Interface acting as the link between the DSP hardware and the development software.

5 Results and Discussion

A first experiment was carried out to identify the relevant TS frequencies and wavelengths dependent on the freestream Mach number. Therefore, the baseline configuration was used to analyse the power-spectra for $M = 0.2$ to 0.4. Figure 9 shows the results for the Mach numbers $M = 0.3$, 0.35 and 0.4. The power-spectra (represented in the decibel scale) clearly exhibit the fundamental TS instability frequency, e.g. for $M = 0.3$ between 2 and 5.5 kHz. Additionally, the first and second higher harmonic frequency ranges are evident as well. In accordance with the theory, the signal power increases with increasing Mach number. The fundamental TS instability frequencies for $M = 0.4$ increase up to 4–8 kHz. Given that the convection velocity of the TS waves is between 30 and 40% of the freestream velocity, some elementary design principles for the actuator can be defined. Aside from the requirement of a smooth actuator surface, the dimension in flow direction is defined

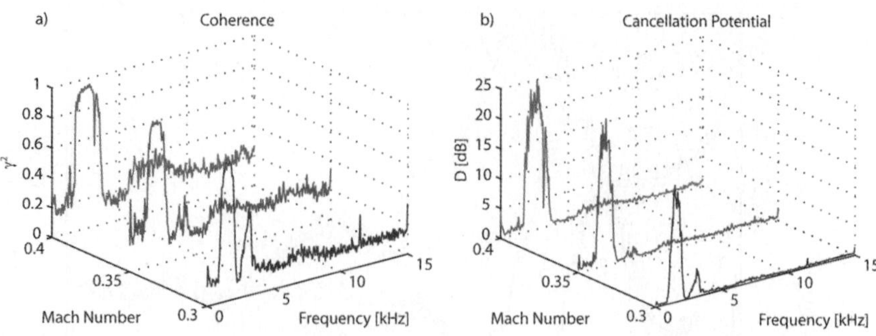

Fig. 10 (a) Coherence-spectra and (b) theoretical cancellation potential for $M = 0.3$, 0.35 and 0.4.

by the TS wavelength. Experience from low speed experiments shows that good results can be achieved with an actuator length in flow direction of approximately $\frac{1}{4}\lambda_{TS}$. That means that for $M = 0.3$ the actuator has to be shorter than 2 mm in flow direction.

Because FIR filters are linear, they can only model linear scenarios. For this reason, every element of the estimation path (amplifier, actuator, boundary layer, SHW, signal conditioner) has to behave linearly as well. Hence, the data were analysed with respect to the coherence-spectra between two signals. To achieve good cancellation results, the coherence – especially in the fundamental TS instability range – has to be as high as possible. A coherence of one means that both signals are fully linearly dependent, while a coherence of zero indicates their independency. For these analyses, two sensors were chosen with a representative spacing comparable to the later control experiments. Figure 10a depicts the coherence-spectra for the Mach numbers $M = 0.3$, 0.35 and 0.4 for sensors at a distance of 30 mm. In the frequency range of the TS instabilities corresponding to Figure 9, the coherence is in the range of 90 to 95% for all Mach numbers. If an actuator with a smooth, non-intrusive surface between the two sensors considered is assumed, the high coherence values promise high cancellation rates. On the assumption that only uncorrelated noise remains once the AWC is operating, an estimation for the optimum annihiliation D can be formulated [13]:

$$D = 1 - \gamma^2 \quad \text{and} \quad D[\text{dB}] = -20\log(1 - \gamma^2),$$

respectively. Figure 10b shows that for $M = 0.3$ an annihiliation of just under 20 dB and for $M = 0.35$ and 0.4 more than 22 dB could theoretically be achieved.

The AWC experiments were conducted between $M = 0.2$ and 0.4. The results for $M = 0.3$ are discussed in the following paragraphs.

If the AWC is operating, the power-spectra of the error signal (Figure 11a) show a remarkably reduced signal power compared to the case without AWC. A mean cancellation of about 7 dB can be observed in the fundamental TS frequency range (2.1–4.2 kHz), which corresponds to a reduction of the TS amplitudes of 55%. In addition, there is a slight reduction in the low-frequency range from 0.5 up to 2.1 kHz.

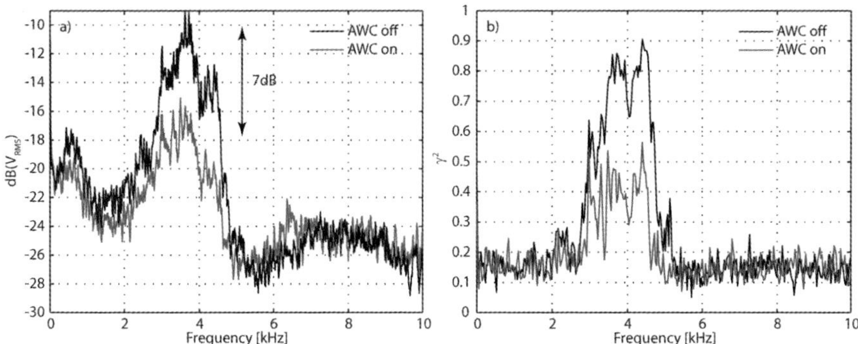

Fig. 11 (a) Power-spectra of the error sensor and (b) coherence-spectra between the reference and error sensor for $M = 0.3$ with AWC on and off.

Although the remaining signal power during the operation of AWC is reduced, the signal power in the fundamental TS frequencies is still enhanced and should allow for further cancellation. Hence, the measurements were analysed with regard to the coherence-spectra between the reference and error sensor with a membrane actuator inbetween for the test cases with and without controller. The comparison of the coherence-spectra in Figure 11b shows the expected decrease of coherent structures from 80% down to 40% with a controller. But already the coherence-spectrum with the AWC system off shows an unsatisfactory distribution. The relatively low and irregular coherence (no distinct plateau) points out that the surface quality of the membrane actuator (even if it is not on) seems to be inadequate. Furthermore, the coherence-spectrum with the AWC system on still shows noticable linear dependencies between the error signal and the reference signal. Due to this weaknesses, two main objectives can be formulated for the future. The actuator surface has to be improved, and the controller has to be optimized.

6 Conclusions and Outlook

For our first AWC experiments under compressible flow conditions, a membrane actuator and a piezoelectric actuator were designed, constructed and integrated in a generic wing model. Their dimension requirements are defined by the preliminary experiments with the simple SHW sensor array to identify the TS frequencies and wavelengths. Both actuator types produce wall-normal velocity fluctuations at a sufficiently large amplitude. For the detection of the TS waves, SHWs are applied.

The developed devices were successfully applied in wind tunnel experiments to cancel TS waves in a two-dimensional compressible laminar boundary layer up to $M = 0.4$ on an unswept wing. A FXLMS algorithm autonomously adapted a transfer function for the path between the reference sensor and the actuator in real-time. The amplitudes of the naturally occurring TS waves at $M = 0.3$ were attenuated down to

45% of their original amplitudes by a single two-dimensional control system. The downstream amplification of the remaining TS waves after control was remarkably smaller.

In further experiments the membrane actuator will be substituted with a piezo-electric actuator as depicted in Figure 5b. Up to now this type of actuator has demonstrated its potential as a perturbation source with the ability to introduce any kind of controlled perturbation. Since the perturbation source is far upstream at a position where the boundary layer is thin and thus sensitive to roughnesses, it is important that this actuator has proven its non-intrusiveness. Therefore, if the piezoelectric surface actuator is installed further downstream at the position of the membrane actuator, the boundary layer will be even less sensitive than it is at the moment.

The optimization process of the controller comprises several parts. The AWC scheme of Figure 6 works well for broadband and narrowband perturbated signals, but it suffers from undesired feedback from the actuator to the reference sensor. If feedback-neutralizing techniques are used, a further transfer function modeling the feedback path is required. Additionally, an implementation of a pre-noise filter for the reference signal is planned. This function utilizes two reference signals instead of the standard method which allows just one reference signal for the adaptive controller.

Acknowledgments

The work on Active Wave Control was funded by AIRBUS within the AIRNET project "Control of Aerodynamic Flows for Environmentally Driven Aircraft" (CAFEDA).

References

1. Braslow, A., Visconti, F.: Further experimental studies of area suction for the control of the laminar boundary layer on a porous bronze NACA 64A010 airfoil. NACA Technical Note 2112, 1950.
2. Morkovin, M.V.: Critical evaluation of transition from laminar to turbulent shear layer with emphasis of hypersonical travelling bodies. AFFDL-TR, 1968, pp. 68–149.
3. Joslin, R.D., Nicolaides, G.E., Erlebacher, G., Hussaini, M.Y., Gunzburger, M.D.: Active control of boundary-layer instabilities: Use of sensors and a spectral controller. *AIAA Journal* **33**(8), 1995, 1521–1523.
4. Kral, L.D., Fasel, H.F.: Numerical investigation of the control of the secondary instability process in boundary layers. AIAA Paper 89-0984, USA, 1989.
5. Thomas, A.S.W.: Active wave control of boundary-layer transition. In: Bushnell, D.M., Hefner, J.N. (Eds.), *Viscous Drag Reduction in Boundary Layers*, Progress in Astronautics and Aeronautics, Vol. 123, AIAA, USA, 1990, pp. 179–199,
6. Raguse, A: Experimentelle Untersuchung zur aktiven dynamischen Beeinflussung des laminar-turbulenten Umschlags in einer Plattengrenzschicht. Ph.D. Thesis, Math.-Nat. Fak., Universität Göttingen, Germany, 1998.

7. Evert, F: Experimentelle Untersuchung zur dynamischen Beeinflussung einer Plattengrenz-schicht. Ph.D. Thesis, Math.-Nat. Fak, Universität Göttingen, Germany, 2000.

8. Opfer, H: Active cancellaction of 3D Tollmien–Schlichting waves in the presence of sound and vibrations. Ph.D. Thesis, Math.-Nat. Fak, Universität Göttingen, Germany, 2002.

9. Baumann, M: Aktive Dämpfung von Tollmien–Schlichting Wellen in einer Flügelgrenz-schicht. Ph.D. Thesis, Institute of Aero- and Astronautics, Technical University of Berlin, Germany, 1999.

10. Sturzebecher, D: Kaskadierte Sensor-Aktuatorsysteme zur aktiven Dämpfung von natürlichen Tollmien–Schlichting Instabilitäten an einem Tragflügel. Ph.D. Thesis, Institute of Aero- and Astronautics, Tech. Univ. Berlin, Germany, 2002.

11. Sturzebecher, D: Active control of boundary-layer instabilities on an unswept wing. In: *Proceedings of the DFG-Concluding Colloquium "Transition"*, Vol. XX of NNFM, *Experiments in Fluids* **31**, Germany, 2001.

12. Sturzebecher, D: The surface hot wire as a means of measuring mean and fluctuating wall shear stress. *Experiments in Fluids* **31**, Germany, 2001.

13. Elliott, S., Nelson, P.: Active noise control. *IEEE Signal Processing Magazine*, 1993, 12–35.

14. Burgess, J.: Active adaptive sound control in a duct: A computer simulation. *Journal of the Acoustical Society of America* **70**, 1981, 715–726.

15. Widrow, B., Schur, D., Shaffer, S.: On adaptive inverse control. In: *Proceedings of the 15th Asilomar Conference on Circuits, Systems and Computers*, Pacific Grove, CA, November 1981, pp. 185–195.

Optimal Boundary Flow Control: Equivalence of Adjoint and Co-State Formulations and Solutions

Ranjan Vepa

Department of Engineering, Queen Mary, University of London, Miles End Road, London E1 4NS, U.K.; E-mail: r.vepa@qmul.ac.uk

Abstract. This paper addresses controversial issues fundamental to the optimal control of aerodynamic flows. Aerodynamic flows being external to wings, the significant region of the flow is in the interface region. In assessing the closedloop performance the relevant performance index must therefore be evaluated exclusively on the wing boundary which is the most significant region for the development of both lift and drag. When this is done the controller may be synthesised relatively easily as it can be shown that the associated optimising co-state equations are identical to the adjoints, which can then be solved by the same methods employed for the Navier–Stokes equations. The control laws may then be deduced by comparing the open and closed loop pressure distributions.

Key words: Optimal flow control, variational methods, reverse flow, reduced order modelling, principal component analysis.

1 The Variational Approach to Flow Control

A variational basis for the derivation of the aerodynamic system equations can be extremely useful when dealing with optimal control design issues. Early efforts to interpret variational formulations of aerodynamic flows were due to Flax [1]. The form of the dynamical equations of fluid motion suggests the possibility that Hamiltons principle for elasto-mechanical systems can be generalised to include fluids as well as solids. Thus one may formulate a generalised Lagrangian density for fluids interacting with the motion of solid boundaries by combining the classical Lagrangians corresponding to the motion and the fluid dynamics. The conditions on the boundary are enforced as constraint equations. The archetypal variational formulation is based on Hamilton's principle but it is by no means the only one. In the context of aerodynamics, it states that the true fluid velocity field causes the time integral of the Lagrangian to have a stationary value for an arbitrary variation of the velocity vector

J.F. Morrison et al. (eds.), IUTAM Symposium on Flow Control and MEMS, 333–337.
© 2008 *Springer. Printed in the Netherlands.*

and pressure fields. Flax enunciated his classic reverse flow theorems that involve solutions of the same planform in forward and reversed main streams and have been particularly useful in estimating and minimising aerofoil drag. The drag minimisation problem could be restated as an optimal control problem. A generic optimal control problem may be dealt with as a variational problem and this naturally leads to the co-state equations which must be solved in conjunction with original state equations to construct the optimal control.

The primary objective of this paper is to lay the theoretical foundations of boundary optimal flow control and establish the close relationship between adjoint and co-state flows. The conditions under which they are identical are identified. The relationship between the classical reverse flows in potential flow theory and the many reverse flow theorems that are employed in potential flow optimisation are interpreted as specialised adjoint and co-state flows. This provides new insights into the nature of the flow optimisation problem and raises the possibility of generalising several important and established concepts based on classical reverse flow theorems. The generalised reverse flow or backward propagating wave type interpretation of the solutions to the adjoint equations facilitates considerable order reduction by the use of simplifying physical models. This also leads to the concept of closed loop order reduction, quite analogous to the methods of order reduction in classical optimal control.

2 Boundary Optimal Flow Control Formulation

The application of optimal control theory to flow control problems is also not new. Although there have been numerous applications in the past, we cite two representative papers here: Joslin et al. [2] and Pralits et al. [3]. Adjoint systems are closely related to the co-states that arise in the development of the optimal control of distributed parameter systems such as those associated with flow, structural and combustion control. Under certain special circumstances, when the performance objective can be evaluated exclusively along the boundary between the wing, aerofoil or lifting surface and the flow, the solution to the optimal control problem can be stated in terms of the solutions to the adjoint system. In these cases the adjoint systems associated with the dynamic model are identical to the co-state equations. Further it is customary to specify the feedback control law in terms of the solution to a matrix Riccati equation or equivalently in terms of the solution to the Hamilton–Jacobi–Issacs equation. However in the case when the adjoint and co-state equations are identical it is possible to find the control law without having to solve either the matrix Riccati equation or the Hamilton–Jacobi–Issacs equation.

The methodology of optimal control may be applied to establish a simple control law for pressure regulation without resorting to the classical process of solving a Riccati-like equation. Our objective is to define the unsteady separation control problem as an optimal control problem with the pressure over a portion of the boundary being the controlling input while the pressure on the remainder of the boundary is the output. The pressure on the control boundary is controlled by an exciter such as a loudspeaker and may be modelled as a finite distribution of momentum sources. As we would like the total pressure to be as close as possible to the base flow pressure (the base flow is assumed to be attached) a suitable performance index or cost function that must be minimised is:

$$J = (r/2) \int_{\delta D} p^2 dS. \tag{1}$$

The domain of integration is the direct sum of the control and output boundary domains. The parameter r is chosen in a special way as is customary in applied optimal control. By applying Gauss's theorem the integral may be written as an integral over the entire domain of the flow. Hence,

$$J = (r/2) \int_{D} \nabla \cdot n p^2 dV, \tag{2}$$

where n is the unit normal to the boundary surface. The optimal control problem can be converted into an unconstrained optimisation problem of higher dimension by the application of Lagrange multipliers. Following methods which are common in optimal control, we seek to minimise J with the additional constraint that small disturbance velocity components and the pressure satisfy the linearised Navier–Stokes equation and the continuity equation. These equations are appended to the performance index in the usual way by means of Lagrange multipliers. An augmented cost function is constructed by appending each of the constraints in this way. However, the form of these equations is that obtained by the variational formulation. These constraint equations are therefore in the form of state equations. The optimality condition results in an equation over the control boundary and another outside it. Outside the control boundary it is precisely the same as the adjoint continuity equation while over the control boundary it may be cast in a slightly different form to give us the control law in terms of the co-states. Over the control boundary,

$$p(x,t) = p_e(x,t) = n \cdot \mathbf{u}^*(-t)/r. \tag{3}$$

Thus the result of the application of the theory of optimal control is an apparently simple control law. However, to implement the control law we must estimate the adjoint velocities, $\mathbf{u}^*(-t)$.

3 Generalisation of Reverse Flow Concept

It is quite clear from the analysis in the previous section that the implementation of the optimal control hinges on the ability to solve the adjoint or equivalently the co-state flow problem. The concept of reverse flow bears a unique relationship to a selfadjoint problem and has served as an extremely useful concept in several potential flow optimis ation and drag minimisation contexts. In these situations although the governing equations are self-adjoint, it is necessary to integrate them backwards both in time and space to solve a typical optimisation problem. The "back propagation" of the flow allows one to determine the control signal "at its source" that is required to achieve the desired closed loop response. This requirement led to the concept of reverse flow and the associated reverse flow theorems followed from the duality between the adjoint and optimal solutions. However the reverse flow concept is more general as it may be applied to the nonlinear inviscid Navier–Stokes equations when the base flow is uniform, quite unlike the adjoint which can be only defined for the linearised Navier–Stokes equation. The reverse flow concept is generalised here and applied to the linearised Navier–Stokes equations. The concept permits the solution of adjoint flow problems by applying standard methods of solution of the Navier–Stokes Equations. The reverse flow problem is obtained by reversing the directions of the spatial and temporal coordinates, in the main flow and base flow. When the adjoint Navier–Stokes problem with the co-ordinates reversed both in the time and spatial domain is considered, where the base flow is assumed to be uniform, reversing the direction of the base flow recovers the original linearised adjoint Navier–Stokes equation. Thus, the adjoint Navier–Stokes equations with the spatial and temporal coordinates reversed may be interpreted as a generalised reverse flow problem, which is governed by a Navier–Stokes like equation thus allowing the solution of the adjoint to be obtained by minimal modification of classical Navier–Stokes solvers.

This results yields the scaled optimal closed loop pressure distribution which could be interpreted as the desirable closed loop pressure distribution in designing a control system and is a central result of this paper. From the closed loop pressure distributions one could in principle compute the closed loop generalised forces such as the lift, the pitching moment and drag, and these quantities may then be employed for synthesising the controls from the computed open loop characteristics. While the comp uted scaled optimal closed loop pressure does not yield the control laws, it serves as a generator of the command inputs to the controller in much the same way as the computed torque method in robotics.

4 Reduced Order Modelling: Principal Component Analysis

Once a desirable optimum pressure distribution is obtained our next objective would be to try and realise such a distribution. One approach is to compute a set of generalised loads corresponding to a set of assumed modal small disturbance or displacement distributions such as a sudden increase in the angle of attack; i.e. extract a set of desirable principal components , based on the method of principal components analysis and obtain the control law by comparison with the corresponding open loop loads.

Unlike discrete-inertia systems, aerodynamic flows are characterised by distributed inertia. Therefore control inputs and measurements may also be distributed anywhere in the flow although it may be possible to only locate them at certain preferable locations on the boundary. Furthermore the location of the control transducers must be selected to ensure that the minimum singular value of the associated controllability or the observability Grammian is sufficiently positive. The former encapsulates the property of flow receptivity while the latter reflects measurability of the measurand. Consequently reduced-order models must be formulated exclusively based on stability considerations and the models can include the unstable and marginally stable modes.

References

1. Flax, A.H., Reverse-flow and variational theorems for lifting surfaces in nonstationary compressible flows, *J. Aero. Sciences* **20**(2), 1953, 120–126.
2. Joslin, R., Gunzburger, M., Nicolaides, R., Erlebacher, F. and Hussaini, M., A methodology for the automated optimal control of flows including transitional flows, *AIAA J.* **35**, 1997, 816–824.
3. Pralits, J., Hanifi, A. and Henningson, D., Adjoint-based suction optimization for 3D boundary layer flows, FFA TN 2000-58, The Aeronautical Research Institute of Sweden, Stockholm, Sweden, 2000.

Optimal Growth of Linear Perturbations in Low Pressure Turbine Flows

Atul S. Sharma[1], Nadir Abdessemed[1], Spencer Sherwin[1] and Vassilis Theofilis[2]

[1]*Imperial College, London, SW7 2BT, U.K.; E-mail: ati.sharma@gmail.com*
[2]*School of Aeronautics, Universidad Politécnica de Madrid, Pza. Cardenal Cisneros 3, E-28040 Madrid, Spain*

Abstract. This paper presents a numerical algorithm for the linearized flow initial value problem involving complex geometries where analytical solution is impossible. The method centres around the calculation of an eigenvalue problem involving the linearised flow and its spatial adjoint, and yields the flow perturbations that grow the most in a prescribed time, the magnitude of that growth and the perturbations after the growth has occurred. Previous work has shown that classical stability analysis of flow past a low-pressure turbine blade gives only stable eigenvalues, which cannot explain transition to turbulence in this flow. The inital value problem for this fan blade is presented and demonstrates significant perturbation growth, indicating that this growth may be the facilitator for transition in this case.

Key words: Transition, transient growth, linear stability analysis, singular value analysis, Arnoldi method, spectral/hp element method.

1 Introduction

The mechanisms of transition from two-dimensional to three-dimensional flow have been extensively investigated for canonical problems such as flat-plate boundary-layers, channel flows and for problems involving more complex geometries, such as the flow past a circular cylinder. For the last case linear stability analysis has been successfully applied to identify the growing eigenmodes responsible for transition and the onset of turbulence [1].

The analysis typically proceeds by considering the growth of small perturbations on a known, laminar solution to the Navier–Stokes equations. A subsequent eigenmode analysis yields a Reynolds number above which an instability occurs. In plane Poiseuille flow, for instance, Tollmien–Schlichting waves grow exponentially

J.F. Morrison et al. (eds.), IUTAM Symposium on Flow Control and MEMS, 339–343.
© 2008 Springer. Printed in the Netherlands.

(in the linear regime) above that critical Reynolds number and transition occurs via a secondary instability resulting from non-linear effects.

In a recent study [2], the flow past a low-pressure turbine blade (LPT) was investigated in order to understand the instability mechanisms in this class of flows. This understanding might be crucial for controlling the laminar boundary layer separation which has experimentally been shown to increase turbine performance. The initial study showed that eigenmodes obtained using classical linear stability analysis are all stable for the entire investigated parameter space comprising the Reynolds number and the spanwise perturbation wavelength. This means that modal analysis with secondary instabilities cannot explain transition in this flow.

This apparent paradox can be understood mathematically in terms of system non-normality. Even when the system linearised about a steady flow is asymptotically stable, it can exhibit large transient growth of the perturbation energy before returning to equilibrium. Physically, this growth is understood to be fed by the transport of energy from the steady flow to the perturbations. Perturbations may then grow in accordance with the linear model at sub-critical Reynolds number, becoming large enough for nonlinearities to become significant. Transition then occurs via secondary instability, 'bypassing' the classical mechanism [3]. In flows such as plane Poiseuille flow, this 'transient growth' has been shown to explain transition [4–6] and previous studies [7] suggest similarities to the flow past the LPT blade. The present study aims to identify modes that are associated with this optimum growth of perturbations as a result of this non-normality.

The initial condition with the most energy growth at time t was found by a calculus of variations method in [4], for Couette flow and for Poiseuille flow. An approach based on the singular value decomposition of the mapping from the initial condition x_0 to the state $x(t)$ at time t is found in [6]. The current work extends that theory by presenting an algorithm allowing calculation of this decomposition for complex geometry flows where an analytical treatment is prohibitively difficult if not impossible.

2 Method

The Schmidt decomposition of $B(t)$ in the linear evolution equation $x(t) = B(t)x_0$ can be expressed in terms of an eigenvalue problem involving $B(t)$ and its adjoint $B^*(t)$. Specifically, the singular values of $B(t)$ are the square roots of the eigenvalues of $B^*(t)B(t)$ and the right Schmidt vectors of $B(t)$ are the eigenvectors of $B^*(t)B(t)$. Where $B(t)$ and its adjoint are evaluated by numerical simulation, this eigenvalue problem can be solved numerically, allowing identification of the Schmidt pairs associated with the largest singular values. The first right Schmidt vector of $B(t)$ gives the initial condition associated with growth of magnitude of the first singular value. The first left Schmidt vector is the final condition that the initial condition evolves into. In this way, the optimal subspace can be found for complex problems

Fig. 1 Initial condition singular vector (Eigenmode of B^*B), $\sigma_1 = 418$. The high value indicates that non-normality is important in this case.

where the operators are not known (or determinable) analytically. These Schmidt pairs are associated with the 'optimals' discussed in [4].

The eigenvalues of the compound operator $B^*(t)B(t)$ can be computed using an Arnoldi algorithm (or simpler non-complex iterative eigenvalue algorithms). The action of B can be expressed by integrating the linearised Navier–Stokes equations via the linear time stepping scheme of [8]

$$x(t + \Delta t) = (\mathbf{I} - \Delta t \mathbf{L})^{-1}(\mathbf{I} + \Delta t \mathbf{N_U})x(t),$$

whose computational implementation is based on a time splitting scheme [9] and is discretised using a spectral/hp element technique to capture the flow solution in the complex geometry. Similarly, the action of $B^*(t)$ can be expressed by integrating over its spatial adjoint system. Time marching those systems allows the computation of the eigenvalues of $B^*(t)B(t)$ to yield the singular values of $B(t)$.

3 Optimal Growth Modes in Low Pressure Turbine Flow

The following results are concerned with the optimal two-dimensional perturbations for the steady base flow past the low pressure turbine blade, whose linear stability has been extensively investigated in previous studies [2, 10]. Considering the complex two-dimensional configuration of the blade shows the mode associated to the largest singular value. A strong energy concentration can be made out in the shear layers around the separation bubble (Figure 1), which is suspected to be the origin of instability and transient growth for this flow. The modes identified in a preceding pseudo-spectrum analysis [10] substantiate the potential for growth, but also raise interesting questions of comparison between the two different types of mode.

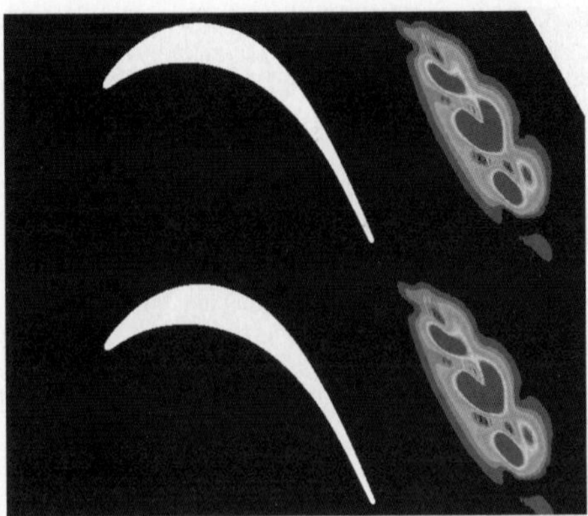

Fig. 2 Final condition SVD-mode (Eigenmode of BB^*).

Considering how the perturbation evolves in a given time, one can seen that the energy of the mode identified here grows significantly as it travels downstream (Figure 2).

Acknowlededgments

A. Sharma is funded by EPSRC, BAE systems through the FLAVIIR project and N. Abdessemed is supported through AFSOR in collaboration with Professor V. Theofilis at the Technical University of Madrid. The authors would also like to acknowledge insightful discussion with Professor Dwight Barkley of Warwick University and Professor David Limebeer at Imperial College London.

References

1. Barkley, D. and Henderson, R., Three-dimensional floquet stability analysis of the wake of a circular cylinder, *J. Fluid Mech.* **322**, 1996, 215–241.
2. Abdessemed, N., Sherwin, S., Theofilis, V., On unstable 2d basic states in low pressure turbine flows at moderate reynolds numbers, AIAA Paper 2004-2541, 2004.
3. Reddy, S., Schmid, P.J., Baggett, J.S. and Henningson, D.S., On stability of streamwise streaks and transition thresholds in plane channel flows, *J. Fluid Mech.* **365**, 1998, 269–303.
4. Butler, K. and Farrell, B., Three-dimensional optimal perturbations in viscous shear flow, *Phys. Fluids* **4**(8), 1992, 1637–1650.
5. Trefethen, L.N., Trefethen, A.E., Reddy, S. and Driscoll, T.A., Hydrodynamic stability without eigenvalues, *Science* **261**(5121), 1993, 578–584.
6. Schmid, P.J. and Henningson, D.S., *Stability and Transistion in Shear Flows*, Springer-Verlag, New York, 2001.
7. Wu, X. and Durbin, P.A., Evidence of longitudinal vortices evolved from distorted wakes in a turbine passage, *J. Fluid Mech.* **446**, 2001, 199–228.

8. Tuckerman, L. and Barkley, D., Bifurcation analysis for timesteppers, in Doedel, E. and Tuckerman, L. (eds.), *Numerical Methods for Bifurcation Problems and Large-Scale Dynamical Systems*, The IMA Volumes in Mathematics and Its Applications, Vol. 119, Springer, New York, 2000, pp. 543–466.
9. Karniadakis, G., Israeli, M. and Orszag, S., High-order splitting methods for the incompressible navier stokes equations, *Journal of Computational Physics* **97**, 1991, 414–443.
10. Abdessemed, N., Sherwin, S. and Theofilis, V., Linear stability of the flow past a low pressure turbine blade, AIAA Paper 2006-3530, 2006.

Simulations of Feedback Control of Early Transition in Poiseuille Flow

John McKernan[1], James F. Whidborne[2] and George Papadakis[3]

[1]*EIP Ltd, 15 Fulwood Place, London WC1V 6HU, U.K.;*
E-mail: john.mckernan@btinternet.com
[2]*Department of Aerospace Sciences, Cranfield University, Cranfield,*
Beds MK43 0AL, U.K.; E-mail: j.f.whidborne@cranfield.ac.uk
[3]*Department of Mechanical Engineering, King's College London, Strand,*
London WC2R 2LS, U.K.; E-mail: george.papadakis@kcl.ac.uk

Abstract. This paper describes simulations of feedback control of small disturbances in laminar Poiseuille flow. A polynomial-form spectral state-space model of the linearised flow with wall transpiration actuation and wall shear-stress measurements is generated, optimal controllers are synthesised, and closed-loop simulations are performed using an independent finite-volume Navier–Stokes solver. In addition, actuation via tangential wall transpiration is investigated, and LMI controllers, which minimise an upper bound on the peak transient energy growth, are synthesised and simulated.

Key words: Navier–Stokes equations, Poiseuille flow, optimal control, flow control.

1 Introduction

It is widely accepted that the amelioration of the high skin friction associated with turbulent flow would lead to valuable reductions in energy expenditure and in carbon emissions from many forms of transportation. Turbulence is characterised by non-linear behaviour on many interacting length and time scales, but the early stages of transition to turbulence occur on large length and slow time scales, and are amenable to mathematical modelling via linearisation and the synthesis of modern controllers with guaranteed closed-loop properties.

Such an approach is also independent of phenomenological descriptions of transition, as noted by Hogberg and Bewley [1], and the importance of the linear terms in maintaining turbulence is emphasized by Kim [2]. Furthermore, any complete relaminarisation of turbulence would need to control the early transition regime, as would full control of the turbulence cascade. Indeed, control of the early transition regime may obviate the need for control of turbulence itself, although distributed control via MEMs devices would still be involved. This paper develops the sem-

J.F. Morrison et al. (eds.), IUTAM Symposium on Flow Control and MEMS, 345–348.
© 2008 *Springer. Printed in the Netherlands.*

inal work of Bewley and Liu [3] on systems theoretic control of early transition in plane Poiseuille flow. A polynomial-form state-space model has been developed, optimal controllers synthesised, and closed-loop simulations performed using an independent full Navier–Stokes solver. Furthermore, actuation via tangential wall transpiration has been investigated, and controllers that minimise an upper bound on the peak transient energy growth have been synthesised.

2 Method

A polynomial-form spectral state-space model of periodic linearised plane Poiseuille flow in velocity-vorticity formulation has been developed from Reddy's hydrodynamic stability code [4, p. 489], by introducing actuation by rate of change of wall transpiration velocity, and wall shear-stress measurements [5]. The state-space model takes the form

$$\dot{\mathbf{x}} = \mathbf{Ax} + \mathbf{Bu}$$

$$\mathbf{y} = \mathbf{Cx} \tag{1}$$

where $\mathbf{x}$ are the state variables, $\mathbf{u}$ are the transpiration inputs, $\mathbf{y}$ are the shear-stress measurements, and $\mathbf{A}, \mathbf{B}, \mathbf{C}$ are constant system, input and output matrices respectively. $\mathbf{A}$ is free from spurious eigenvalues in this approach. Figure 1 shows the two configurations considered: streamwise and spanwise wavenumbers $\alpha = 1, \beta = 0$ respectively (which yields Tollmien–Schlichting waves), and $\alpha = 0, \beta = 2.044$ (which yields streamwise vortices with the largest energy growth).

The model has been shown to be consistent with the interpolating-form model of Bewley and Liu [3] with control by wall transpiration velocity. Several recombined Chebyshev series methods of applying the Dirichlet and Neumann boundary conditions have been investigated, and a novel extension for the Neumann boundary condition, in conjunction with derived preconditioning, has been shown to produce the best numerical conditioning of the Laplacian term [6].

Optimal state-feedback (Linear Quadratic Regulator, or LQR) controllers $\mathbf{u} = -\mathbf{Kx}$ have been synthesised for the state-space model (1), which minimise the time integral of the perturbation energy ($\mathbf{x}'\mathbf{Qx}$) plus weighted control effort ($\mathbf{u}'\mathbf{Ru}$), for actuation by both wall-normal and tangential transpiration (both of zero net mass-flow). Tangential transpiration allows the possibility of alternative MEMS technologies to fluid jets, such as rollers. State-feedback controllers that minimise upper bounds on the peak transient energy and control effort have also been synthesised, via linear matrix inequality (LMI) methods as described by McKernan [6] and Whidborne et al. [7]. Since the state variables are unavailable, optimal observers have also been synthesized, which provide estimates $\hat{\mathbf{x}}$ of the state variables with minimised estimation error expectation. Output-feedback controllers $\mathbf{u} = -\mathbf{K}\hat{\mathbf{x}}$ have been formed by coupling the observers with the state-feedback controllers.

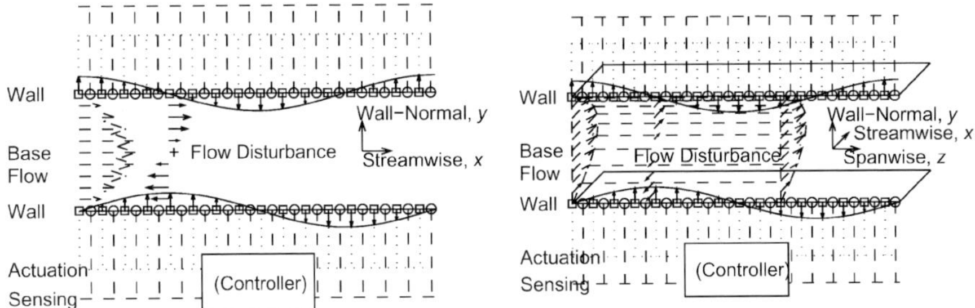

Fig. 1 Control of Poiseuille flow via transpiration, (left) $\alpha = 1$, $\beta = 0$, (right) $\alpha = 0$, $\beta = 2.044$.

Simulations were performed using the linear solvers of Matlab, and an independent finite-volume non-linear Navier–Stokes solver, modified to work in terms of perturbations about the base flow for improved accuracy on small perturbations [6]. Initial perturbations which produce the worst subsequent transient energy growth, as derived by Butler and Farrell [8], were employed, since high transient energy growth is believed to lead to non-linearity and ultimately transition to turbulence.

3 Results and Conclusions

For small initial perturbations, as compared to the transition thresholds of Reddy et al. [9], the spectral linear and finite volume non-linear simulations agreed for both open- and closed-loop configurations. The state-feedback controllers were able to stabilise Tollmien–Schlichting waves (at Reynolds number $R = 10^4$, based on centreline velocity and channel half-height), and reduce the transient energy growth of stable streamwise vortices (at $R = 5 \times 10^3$), although comparatively large amounts of transpiration fluid were locally required. For larger initial perturbations, the flow saturated and the state-feedback controllers continued to stabilise the flow. Despite numerical difficulties with the simulation worst initial conditions, there were indications that tangential wall transpiration may result in lower worst transient energy growth than wall-normal transpiration, on streamwise vortices. The performance of the LMI and the LQR controllers were seen to be similar, indicating that for the configurations investigated, controllers which minimise the time integral of transient energy do as well at minimising the peak transient energy as controllers which minimise an upper bound on the peak transient energy. Some open- and closed-loop non-linear simulation results are shown in Figure 2.

Output feedback controllers were also able to stabilise small flow perturbations, and reduce the transient energy growth, but not as well as the state-feedback controllers. On larger perturbations the flow generally saturated and was successfully controlled, although on some occasions the observers over-estimated the states and the flow was destabilised.

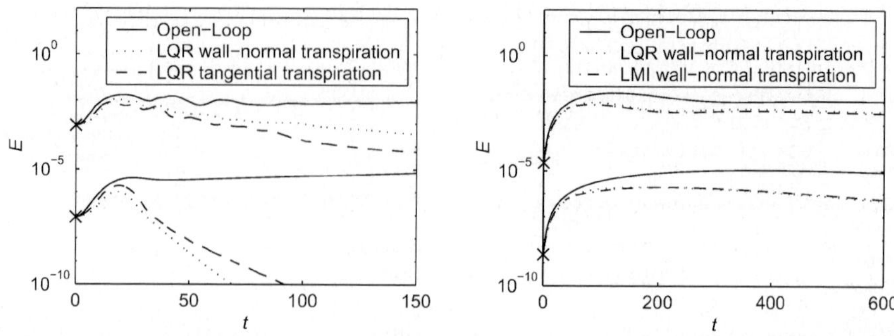

Fig. 2 Open- and state-feedback closed-loop non-dimensionalised transient energy E vs time t, for Tollmien–Schlichting waves (left) and streamwise vortices (right), from both small and large 'worst' initial perturbations ($\times$). The transition threshold for streamwise vortices is approximately 2.6×10^{-7} [9, p. 292].

References

1. Hogberg, M., Bewley, T.R.: Spatially localized convolution kernels for feedback control of transitional flows. In: *Proc. 39th IEEE Conference on Decision and Control*, Sydney, Australia, December 12–15 (2000) 3278–3283.
2. Kim, J.: Control of turbulent boundary layers, *Physics of Fluids* **15**(5) (2003) 1093–1105.
3. Bewley, T.R., Liu, S.: Optimal and robust control and estimation of linear paths to transition. *Journal of Fluid Mechanics* **365** (1998) 305–349.
4. Schmid, P.J., Henningson, D.S.: *Stability and Transition in Shear Flows*, Applied Mathematical Sciences, Vol. 142. Springer-Verlag, New York (2001).
5. McKernan, J., Papadakis, G., Whidborne, J.F.: A linear state-space representation of plane Poiseuille flow for control design: A tutorial. *The International Journal of Modelling, Identification and Control* **1**(4) (2006) 272–280.
6. McKernan, J.: Control of plane Poiseuille flow: A theoretical and computational investigation. PhD Thesis, Department of Aerospace Sciences, School of Engineering, Cranfield University (2006).
7. Whidborne, J.F., McKernan, J., Papadakis, G.: Minimal transient energy growth for plane Poiseuille flow. In: *Proc. UKACC International Conference Control 2006 (ICC2006)*, August 30–September 1, Glasgow, UK (2006).
8. Butler, K.M., Farrell, B.F.: Three-dimensional optimal perturbations in viscous shear flow. *Physics of Fluids* **4**(8) (1992) 1637–1650.
9. Reddy, S.C., Schmid, P.J., Baggett, J.S., Henningson, D.S.: On stability of streamwise streaks and transition thresholds in plane channel flows. *Journal of Fluid Mechanics* **365** (1998) 269–303.

A Switched Reduced-Order Dynamical System for Fluid Flows under Time-Varying Flow Conditions

Howard H. Hamilton[1,*], Andrew J. Kurdila[2] and Anand K. Jammulamadaka[2]

[1]*Munitions Directorate, U.S. Air Force Research Laboratory, Eglin AFB, FL 32542, U.S.A.; E-mail: howard.hamilton@eglin.af.mil*
[2]*Department of Mechanical Engineering, Virginia Polytechnic Institute and State University, Blacksburg, VA 24061, U.S.A.*

Abstract. We develop a collection of reduced-order models for fluid systems that operate under time-varying flow and actuation parameters. A switched dynamical system is formulated that is a combination of reduced-order models valid for specific parametric subspaces and discrete switching logic. An open-loop simulation of the switched dynamical system on a two-sided driven cavity demonstrates the system's ability to capture the evolution of the flow and input parameters in the full-order model.

Key words: Reduced-order models, dynamical systems, cavity flow.

1 Introduction

This publication presents results on the development of a switched dynamical system, composed of a family of reduced-order models, to approximate the dynamics of a fluid flow under varying flow and input parameters. This work is motivated by the desire of researchers in the fluid mechanics and control systems communities to create suitable dynamical models for real-time control. Proper orthogonal decomposition (POD) has been applied to the full-order Navier–Stokes equations to create low-dimensional systems that represent the dominant flow physics. POD-based models have been shown to effectively capture the essential physics of the flow, but only for a single flow regime. Moreover, several researchers (e.g. Burns et al. [1]) have shown that a strict reduction of the flow physics to their most energetic modes is not sufficient for closed-loop control. Ravindran [2], Fahl [3] and Jørgensen et al. [4] have proposed techniques such as adaptive schemes and sequential POD to extend the capability of existing reduced-order models.

---

* This research was conducted while the author held a National Research Council (NRC) Research Associateship at the U. S. Air Force Research Laboratory.

J.F. Morrison et al. (eds.), IUTAM Symposium on Flow Control and MEMS, 349–352.
© 2008 *Springer. Printed in the Netherlands.*

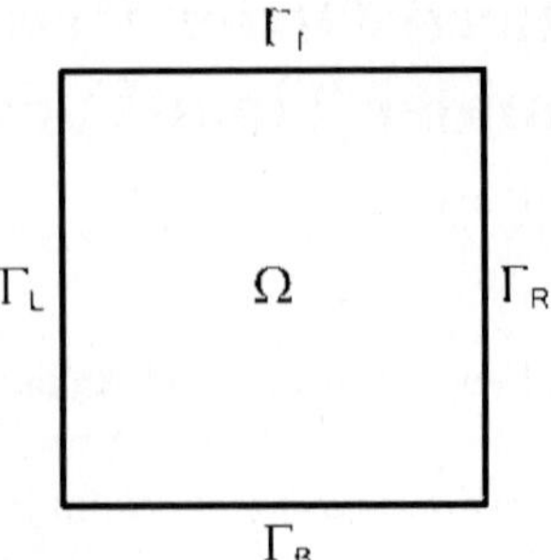

Fig. 1 Schematic of a two-dimensional flow domain.

2 Modelling Framework

Consider flow in a domain Ω, illustrated in Figure 1, that is governed by the two-dimensional incompressible Navier–Stokes equations:

$$\left.\begin{array}{c} \dfrac{\partial \mathbf{V}}{\partial t} + \dfrac{1}{St}\displaystyle\sum_{i=1}^{2} V_j\dfrac{\partial \mathbf{V}}{\partial x_j} + \nabla p = \Theta\Delta\mathbf{V} \\[2mm] \nabla\cdot\mathbf{V} = 0 \end{array}\right\} \qquad \forall (x,u)\in\Omega\times[0,T]. \qquad (1)$$

Its velocity field $\mathbf{V}$ is parameterized by a finite set of flow and/or input variables that span a parametric space Λ. We want to develop a model of the flow physics $\dot{\mathbf{x}} = A_j(t)\mathbf{x} + B_j(t)\mathbf{u}$; $j(t)\in\mathscr{J}$, where $\mathbf{x}$, $\mathbf{u}$, and $j(t)$ represent the state, control, and an index function, respectively. Each dynamical system corresponds to a single contiguous region $\mathscr{R}_j$ and holds on constant intervals of $j(\cdot)$. To create the switched model we define the regions $\mathscr{R}_j$, which represent the subspaces of Λ, their corresponding reduced-order dynamics $A_{j(\cdot)}$ and $B_{j(\cdot)}$, and the switching rules between regions.

We first assume that there exists a set of velocity fields $\mathbf{V}_i$ for an unordered list λ_i (i.e. elements) in the parametric space. Assuming that each field can be written as

$$\mathbf{V}_i = \mathbf{V}_m + \mathbf{V}_0 h(t) + \mathbf{V}_c\beta(t) + [\psi_i]\alpha_i(t),$$

where $h(t)$ is an exogenous function, $\beta(t)$ a control function, ψ_i the subspace basis and α_i its modal coefficient vector, we apply a metric that measures the angle between two subspaces that span their respective velocity fields:

$$\mathbf{V}_i\cdot\mathbf{V}_j = \|\mathbf{V}_i\|\,\|\mathbf{V}_j\|\cos\theta_{ij}. \qquad (2)$$

A point-to-point comparison between the elements of the parametric space produces a matrix of subspace angles relative to a single set of parameters. Such a matrix can be visualized as an "image" whose "voxels" correspond to the elements of the parametric space. Temporal cross-correlations are taken between pairs of "im-

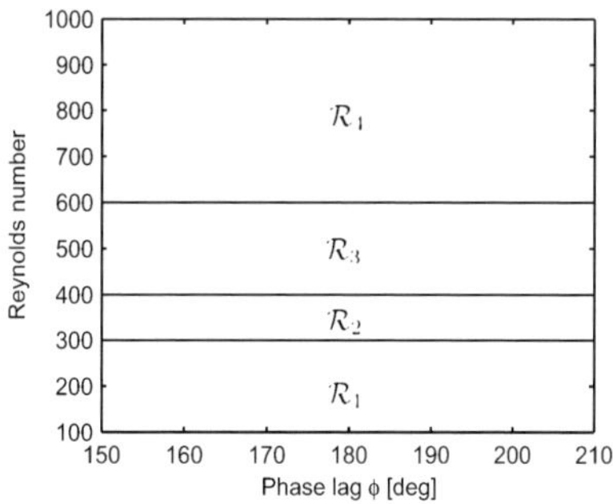

Fig. 2 Partitioning of parametric space for switched model of driven cavity.

age" data, and elements of graph theory are applied to develop clusters of related elements based on a median value of the cross-correlation and a minimum threshold value. These clusters are defined as the regions $\mathcal{R}_j$. Within each region, a subspace basis is selected and the reduced-order dynamical system formed.

It is now left to develop a hypothesis testing algorithm that governs switching between dynamical systems. Over a time interval $t \in [0;T]$, we assume that there exist r switching intervals defined by $\tau_j \in [t_j \; t'_j], t'_j = t_{j+1}, t'_N = T; j = 1 \ldots r$, and that there exist continuous exogenous disturbances and control inputs across switching instants. At the end of each switching instant, the velocity field is compared to a library of steady-state flow solutions over $\mathcal{R}_j$ (resulting in a 2-norm error measure), and the subspace is selected which corresponds to the region where the error measure is smallest. Continuity of the velocity vector across switching intervals is maintained by adding an extra term to the mean velocity expression and driving it to zero during the switching instant.

3 Simulation Results

We apply these algorithms to the modelling of the flow in a controlled two-sided driven cavity parameterized by Reynolds number Re and the phase differential ϕ between the periodic translational velocities of the upper and lower walls. In this simulation, the translational velocities of the walls are identical ($V = 1$ m/s), $Re \in [100; 1000]$ and $\phi \in [150; 210]$ deg. The modeling procedure successfully partitions the parameter space (Figure 2) and implements the switching strategy between reduced-order models. An open-loop simulation of the resulting switched dynamical system (illustrated in Figure 3) demonstrates its ability to capture the evolution of the flow and input parameters in the full-order model.

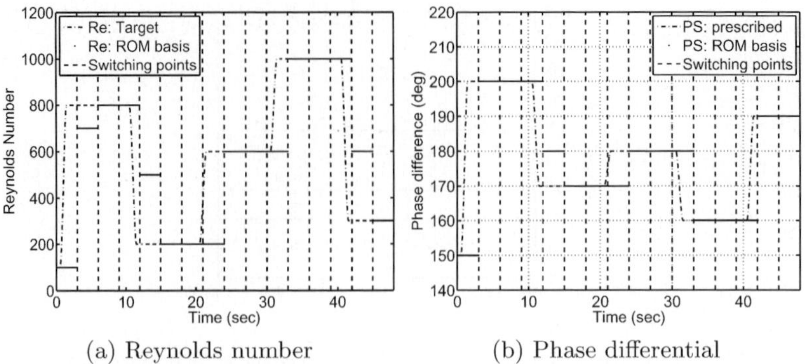

(a) Reynolds number   (b) Phase differential

Fig. 3 Open-loop results for switched model of driven cavity.

4 Conclusions

We have developed a new procedure that formulates a switched dynamical system for the modelling for fluid systems under varying flow and actuation parameters. Open-loop simulations of the switched model of a two-dimensional geometry demonstrated its ability to adequately approximate the parametric trajectories of the full-order model. Further work will involve a more systematic selection of reduced-order bases, the application of hybrid estimation algorithms, and the synthesis of controllers for the switched system.

References

1. Burns, J.A., King, B.B., Rubio, D.: On the design of feedback controllers for a convecting fluid flow via reduced order modeling. In: *Proceedings of IEEE International Conference on Control Applications*, Kohala Coast, Hawai'i (1999).
2. Ravindran, S.S.: Reduced-order adaptive controllers for fluid flows using POD. *Journal of Scientific Computing* **15**(4) (2000) 457–478.
3. Fahl, M.: Trust-region methods for flow control based on reduced-order modeling. PhD Thesis, Universität Trier, Germany (December 2000).
4. Jørgensen, B.H., Sørensen, J.N., Brøns, M.: Low-dimensional modeling of a driven cavity flow with two free parameters. *Theoretical and Computational Fluid Dynamics* **16** (2003) 299–317.

Strategies for Optimal Control of Global Modes

Olivier Marquet, Denis Sipp and Laurent Jacquin
ONERA, Departement of Fundamental and Experimental Aerodynamics, 8 rue des Vertugadins, 92190 Meudon, France; E-mail: olivierket@yahoo.fr

Abstract. The aim of this paper is to expose two different strategies for the optimal control of a three-dimensional global mode in a two-dimensional recirculation bubble. The formulation of the optimal control problem, that consists to reduce the energy growth of the global mode, depends on the characteristics of the actuation – unsteady and three-dimensional or steady and two-dimensional. A gradient-based optimization procedure is used and the gradient is evaluated using the adjoint of the stability equations in the former case and the adjoint of the stability equations as well as the adjoint of the base flow equations in the latter case.

Key words: Optimal control, recirculation bubble.

1 Introduction

A recirculation bubble is a prototype flow to investigate the transition process in highly non-parallel open shear flows. The increase of computional capacities has only recently allowed to investigate this transition process in the framework of a global stability analysis. A recent study on the stability property of a two-dimensional steady recirculation bubble shows that the first unstable global mode in terms of the Reynolds parameter is a three-dimensional stationary mode [2]. The goal of this paper is to control the growth of such a global mode in the case of a recirculation bubble forming in a S-duct configuration. In the last decade, flow control investigations have been focused on applying modern control theory to determine the optimal control of fluid flows [1]. In the context of disturbance growth control, this approach has been sucessfully used to determine optimal controllers in parallel flows [5] or slightly non-parallel flows as for the boundary layer [3,6]. The question of the optimal control of global mode growth has been investigated using reduced order model of the flow dynamics [4].

Here, we present the optimal control strategy of the global mode growth through a velocity control on the wall and without any reduction of the flow dynamics. In the present configuration, the flow bifurcates at a critical Reynolds number from a

J.F. Morrison et al. (eds.), IUTAM Symposium on Flow Control and MEMS, 353–357.
© 2008 Springer. Printed in the Netherlands.

stationary two-dimensional state, known as the base flow, towards an exponentially growing three-dimensional state as long as the linear approximation is valid. The bifurcation may thus be characterized by a break of the invariance of the flow in the spanwise direction and in time, as long as only the linear regime is investigated. Such a distinction between the flow symmetries before and after the bifurcation implicitly suggests the design of the actuator. If a three-dimensional actuator is considered, it should act on the disturbance flow, since it respects its symmetry, whereas a two-dimensional actuator should act on the base flow and be steady to respect its symmetry. In this paper, we present the formalism of the optimal control theory developed to reduce the growth of the three-dimensional unstable global mode of a recirculation bubble, by focusing on the distinction that arises from the choice of a steady two-dimensional or an unsteady three-dimensional actuation.

The motion of a viscous fluid is governed by the non-dimensional incompressible Navier–Stokes equations. In the framework of a global stability analysis, the velocity and pressure fields are decomposed into a two-dimensional base flow $(\mathbf{U},P)(x,y) = (U,V,0,P)$ and a three-dimensional disturbance flow $\varepsilon(\mathbf{u}',p')$, where ε is the amplitude of the perturbation. The base flow is assumed to be a solution of the stationary two-dimensional incompressible Navier–Stokes equation

$$(\nabla \mathbf{U}).\mathbf{U} = -\nabla P + \mathrm{Re}^{-1}\nabla^2 \mathbf{U}, \ \nabla.\mathbf{U} = 0, \tag{1}$$

where $\nabla = (\partial_x, \partial_y, \partial_z)$ is the gradient operator. At leading order in ε, the three-dimensional perturbation $(\mathbf{u}',p')$ is a solution of the unsteady Navier–Stokes equation linearized around the base flow. Since the base flow is stationary and homogeneous in the spanwise direction z, the three-dimensional perturbations may be looked for as a Fourier mode

$$(\mathbf{u}',p') = \frac{1}{2}((\tilde{\mathbf{u}},\tilde{p})(x,y,t)\exp[ikz] + \text{c.c.}),$$

where k is the real spanwise wave number and $(\tilde{\mathbf{u}},\tilde{p}) = (\tilde{u},\tilde{v},i\tilde{w},\tilde{p})$ the associated real Fourier mode. By introducing this Fourier decomposition into the linearized Navier–Stokes equation shows that $(\tilde{\mathbf{u}},\tilde{p})$ is a solution of the initial value problem

$$\frac{\partial \tilde{\mathbf{u}}}{\partial t} + (\nabla_k \tilde{\mathbf{u}}).\mathbf{U} + (\nabla \mathbf{U}).\tilde{\mathbf{u}} = -\nabla_k \tilde{p} + \frac{1}{\mathrm{Re}}\left(\nabla^2 - k^2\right)\tilde{\mathbf{u}}, \ \nabla_{-k}.\tilde{\mathbf{u}} = 0, \tag{2}$$

$$\tilde{\mathbf{u}}(t=0) = \tilde{\mathbf{u}}_0, \tag{3}$$

where the three-dimensional gradient operator is $\nabla_k = (\partial_x, \partial_y, -k)$. The control denoted $\mathbf{u}_c = (u_c,v_c,w_c)$ is a velocity imposed at the control wall, denoted Γ_c and is expressed as

$$\mathbf{u}_c = \mathbf{U}_c(x,y) = (U_c,V_c,0) \text{ on } \Gamma_c, \tag{4}$$

$$\mathbf{u}_c = \frac{\varepsilon}{2}(\tilde{\mathbf{u}}_c(x,y,t)\exp[ikz] + \text{c.c.}) \text{ on } \Gamma_c, \tag{5}$$

if a stationary two-dimensional control (4) acting on the base flow equation is considered, or if an unsteady three-dimensional control (5) is applied on the evolution equation of the perturbations (2). The objective of the control is to reduce or suppress the growth of the energy of the unstable global mode, when the instantaneous energy of the transverse perturbations is defined as

$$E(t) = \frac{1}{2} \int_\Omega \left(\tilde{u}^2 + \tilde{v}^2 + \tilde{w}^2 \right) dx dy. \tag{6}$$

2 Optimal Control through Disturbance Flow Modification

In this section, the control is three-dimensional, unsteady and defined by (5) with $\tilde{\mathbf{u}}_c = (\tilde{u}_c, \tilde{v}_c, \tilde{w}_c)$. The objective is to minimize the energy growth of the perturbation at a final time T balanced by the energetic cost of the control over the time interval $[0, T]$, which is written using a functional expressed as

$$J_1(\tilde{\mathbf{u}}_c) = \frac{E(\tilde{\mathbf{u}}(T))}{E(\tilde{\mathbf{u}}_0)} + \frac{l_1^2 \varepsilon^2}{2T} \int_0^T \int_{\Gamma_c} \left(\tilde{u}_c^2 + \tilde{v}_c^2 \right) dt d\Gamma, \tag{7}$$

where the initial disturbance $\tilde{\mathbf{u}}_0$ is equal to the leading global mode, denoted $\hat{\mathbf{u}}_1$ in the following, and l_1 is a weight parameter. The minimum of this functional J_1 is looked for using a gradient-based algorithm and the gradient is expressed as

$$\frac{dJ_1}{d\tilde{\mathbf{u}}_c} = \frac{l_1^2 \varepsilon^2}{T} \tilde{\mathbf{u}}_c - \tilde{p}^\dagger \mathbf{n} - \frac{1}{\mathrm{Re}} \left(\nabla \tilde{\mathbf{u}}^\dagger \right) . \mathbf{n}, \text{ on } \Gamma_c, \tag{8}$$

where adjoint variables $(\tilde{\mathbf{u}}^\dagger, \tilde{p}^+)$, called thereafter the adjoint perturbations, have been introduced. A variational approach, not detailed here, enables to determine that the adjoint perturbations are solutions of the backward in time evolution equations

$$-\frac{\partial \tilde{\mathbf{u}}^\dagger}{\partial t} - (\nabla_k \tilde{\mathbf{u}}^\dagger).\mathbf{U} + (\nabla \mathbf{U})^t . \tilde{\mathbf{u}}^\dagger = \nabla_k \tilde{p}^\dagger + \frac{1}{\mathrm{Re}} \left(\nabla^2 - k^2 \right) \tilde{\mathbf{u}}^\dagger, \nabla_{-k} . \tilde{\mathbf{u}}^\dagger = 0 \tag{9}$$

starting from the final condition $\tilde{\mathbf{u}}^\dagger(t = T) = \tilde{\mathbf{u}}(t = T)/E(\tilde{\mathbf{u}}_0)$.

The optimal unsteady three-dimensional control is then computed through an iterative method consisting in solving successively (2) and (9) to estimate the gradient (8) and the control at the next iteration of the iterative procedure.

3 Optimal Control through Base Flow Modification

In this section, the control is two-dimensional, steady and defined by (4). A time-modal decomposition of the perturbations is introduced to describe the disturbance flow dynamics

$$(\tilde{\mathbf{u}}, \tilde{p})(x,y,t) = \frac{1}{2}\left((\hat{u},\hat{v},\hat{w},\hat{p})(x,y)\exp[\sigma t] + \text{c.c.}\right), \tag{10}$$

where $(\hat{\mathbf{u}}, \hat{p}) = (\hat{u},\hat{v},\hat{w},\hat{p})$ is a global complex eigenmode, whose complex growth rate is σ. The real part of σ represents the temporal growth rate and its imaginary part is the frequency of the global mode. Once the decomposition (10) is introduced into the time evolution equation (2), the equations governing the perturbation has the form of a generalized eigenvalue problem

$$\sigma\hat{\mathbf{u}} + (\nabla_k\hat{\mathbf{u}}).\mathbf{U} + (\nabla\mathbf{U}).\hat{\mathbf{u}} = -\nabla_k\hat{p} + \frac{1}{\text{Re}}\left(\nabla^2 - k^2\right)\hat{\mathbf{u}}, \nabla_{-k}.\hat{\mathbf{u}} = 0. \tag{11}$$

The long-time behavior of the disturbance flow is described by the leading global mode $\hat{\mathbf{u}}_1$ defined as the eigenmode associated to the eigenvalue of the largest real part $\sigma_1 = \max_\sigma(\text{Re}(\sigma))$. To minimize the energy growth at the horizon time weighted by the energy of the control, a functional J_2 is defined as

$$J_2(\mathbf{U}_c) = \exp[(\sigma_d + \sigma_d^*)\,T] + \frac{l_2^2}{2}\int_{\Gamma_c}\left(U_c^2 + V_c^2\right)d\Gamma \tag{12}$$

and the gradient of this functional with respect to the control is expressed as

$$\frac{dJ_2}{d\mathbf{U}_c} = l_2^2\mathbf{U}_c - P^\dagger\mathbf{n} - \frac{1}{\text{Re}}(\nabla\mathbf{U}^\dagger).\mathbf{n} \tag{13}$$

where the adjoint perturbations $(\hat{\mathbf{u}}^\dagger, \hat{p}^\dagger)$, as well as the adjoint base flow $(\mathbf{U}^\dagger, P^\dagger)$ have been introduced. Comparing this approach to the previous one for which the control does not modify the base flow which is determined once for all, the steady two-dimensional control modifies the base flow solution of (1) in order to reduce the disturbances solution of the generalized eigenvalue problem (11). As a consequence, the expression of the gradient depends on both the adjoint perturbations, solution of equations similar to (11)

$$\sigma^*\hat{\mathbf{u}}^\dagger - (\nabla_k\hat{\mathbf{u}}^\dagger).\mathbf{U} + (\nabla\mathbf{U})^t.\hat{\mathbf{u}}^\dagger = \nabla_k\hat{p}^\dagger + \frac{1}{\text{Re}}\left(\nabla^2 - k^2\right)\hat{\mathbf{u}}^\dagger, \nabla_{-k}.\hat{\mathbf{u}}^\dagger = 0, \tag{14}$$

and the adjoint base flow solution of the linear equations

$$-(\nabla\mathbf{U})^t.\mathbf{U}^\dagger + (\nabla\mathbf{U}^\dagger).\mathbf{U} = -\nabla P^\dagger - \text{Re}^{-1}\nabla^2\mathbf{U}^\dagger + \mathbf{S}(\hat{\mathbf{u}}, \hat{\mathbf{u}}^\dagger), \nabla.\mathbf{U}^\dagger = 0, \tag{15}$$

with a normalization condition of the adjoint global mode

$$\int_{\Omega} \hat{\mathbf{u}}^* . \hat{\mathbf{u}}^\dagger d\Omega = -2 \exp[(\sigma_d + \sigma_d^*)T]T. \tag{16}$$

The source term on the right-hand side of (15) depends on the global and adjoint global modes $\mathbf{S}(\hat{\mathbf{u}}, \hat{\mathbf{u}}^\dagger) = -(\nabla\hat{\mathbf{u}})^t . \hat{\mathbf{u}}^\dagger + (\nabla_k \hat{\mathbf{u}}^\dagger) . \hat{\mathbf{u}}$ and couples the adjoint base flow equations (15) with the perturbation (11) and adjoint perturbation equation (14).

References

1. Bewley, T.R.: Flow control: New challenges for a new Renaissance. *Progess in Aerospace Science* **37** (2001) 21–58.
2. Barkley D., Gomes M.G.M. and Henderson R.D.: Three-dimensional instability in flow over a backward-facing step. *J. Fluid Mech.* **473** (2002) 167–190.
3. Corbett P. and Bottaro A.: Optimal control of nonmodal disturbances in boundary layers. *Theoret. Comput. Fluid Dynam.* **15** (2001) 65–81.
4. Hœpffner J., Åkervik E., Ehrenstein U. and Henningson D.S.: Control of cavity-driven separated boundary layer. In *Proceedings of the Conference on Active Flow Control*, Berlin (2006).
5. Högberg M., Bewley T.R. and Henningson D.S.: Linear feedback control and estimation of transition in plane channel flow. *J. Fluid Mech.* **481** (2003) 149–175.
6. Högberg M. and Henningson D.S.: Linear optimal control applied to instabilities in spatially developing boundary layers. *J. Fluid Mech.* **470** (2002) 151–179.

APPLICATIONS

Modeling and Development of Synthetic Jet Actuators in Flow Separation Control Application

Quentin Gallas

Renault SAS, Research Department, 78288 Guyancourt, France;
E-mail: quentin.gallas@renault.com

Abstract. This paper presents the application of lumped element modeling for the modeling and design of a synthetic jet actuator. The reduced-order model is first reviewed and the basic dynamic behavior discussed. Quantitative design goals for a specific flow control application are then translated into desirable actuator characteristics, and used to solve the optimal design synthesis problem. The actuator built from the specifications given by the model is finally characterized via hot-wire anemometer (HWA) and compared with lumped element modeling (LEM) prediction. Ultimately, the goal of this work is to achieve drag reduction flow control using the synthetic jet actuator embedded in the afterbody of a car vehicle.

Key words: Synthetic jet, lumped element modeling, design tool, flow control.

1 Introduction

Active flow control techniques are currently being investigated at the advanced-project stage of automobile development in order to provide engineers with alternative solutions that will reduce the aerodynamic drag of vehicles, their fuel consumption and greenhouse gas emissions. In this context, active control by means of a synthetic jet is one of the possibilities currently being evaluated at Renault's research department. Several experimental and numerical works have already proven the efficiency of the synthetic jet control technique in various flow separation applications. In the field of automobile aerodynamics, active control techniques have been typically developed to reduce the aerodynamic drag associated with the wake flow topology of the geometry and the dynamics of the longitudinal and transversal vortices in the wake. The potential of using synthetic jets to reduce the aerodynamic drag on a simplified 2D vehicle has been recently numerically demonstrated by Leclerc et al. [1]. Yet, for the deployment of such practical flow control applications, appropriate tools are needed for the modeling and design of suitable actuators.

J.F. Morrison et al. (eds.), IUTAM Symposium on Flow Control and MEMS, 361–364.
© 2008 *Springer. Printed in the Netherlands.*

Recently, lumped element modeling (LEM) has been combined with equivalent circuit representations to estimate the nonlinear dynamic response of a synthetic jet as a function of device dimensions and material properties [2, 3]. These models have provided good agreement between predicted and measured frequency response functions and thus are suitable for use as design tools.

This paper presents the application of LEM technique for the modeling and design of a synthetic jet actuator. First, the reduced-order model is reviewed and the basic dynamic behavior discussed. Quantitative design goals for a specific flow control application are then translated into desirable actuator characteristics, and LEM is used to solve the optimal design synthesis problem. The actuator built from the specifications given by the model is finally characterized via hot-wire anemometer (HWA) and compared with LEM prediction. Ultimately, the goal of this work is to achieve drag reduction flow control using the synthetic jet actuator embedded in the afterbody of a car vehicle (see Renault concept car Altica, 2006).

2 Design Problem and Analysis

This section presents the lumped element model along with key features of the actuator dynamics. Once the design goals are defined, the actuator is characterized and compared with the model prediction.

2.1 Lumped Element Model & Equivalent Circuit Model

At low frequencies, where the characteristic length scales of the governing physical phenomena are much larger than the largest geometric dimension, the governing partial differential equations of the dynamic system can be "lumped" into a set of coupled ordinary differential equations. The resulting lumped-parameter system can then be represented as an equivalent electrical circuit possessing idealized discrete circuit elements and conjugate power variables for the equivalent voltage and current. This approach provides a simple method to estimate the non-linear dynamic response of a synthetic jet actuator for design and control-system implementation purposes.

Figure 1 shows an equivalent circuit representation of a piezoelectric-driven synthetic jet actuator, where the lumped parameters represent generalized energy storage elements (i.e., capacitors and inductors) and dissipative elements (i.e., resistors). Model parameter estimation techniques, assumptions, and limitations are discussed in [2]. The frequency response function of the circuit is then derived to obtain an expression for Q_j/V_{ac}, the jet volume flow rate during the expulsion part of the cycle per applied voltage.

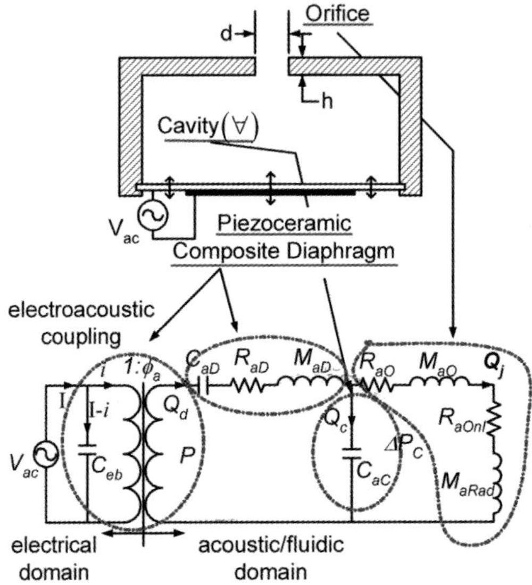

Fig. 1 Equivalent circuit model of a piezoelectric-driven synthetic jet actuator [2].

2.2 Actuator Characterization & Model Verification

In order to reduce the aerodynamic drag coefficient of a car vehicle, several control authority schemes can be investigated. It is likely that separated flow behind a car geometry acts as a nonlinear-multifrequency closed-flow system. In such a system, and in addition to the 3D effects, the vortex shedding frequency of the shear layer instability formed at the aft of the geometry (of Kelvin–Helmholtz type) and that of the wake instability (of Von Karman type) may interact with each other in a non-linear fashion. Here, the actuator quantities of interest chosen are the momentum coefficient and reduced frequency, while the design objective is to achieve broad-band control authority in the low frequency range (DC to 1.5 kHz).

Based on these specifications, LEM is applied to design a synthetic jet array [2, 3]. It consists of a sealed rectangular cavity to which circular piezoelectric disks are fitted on each side, where a total of 10 disks are used. The hardware is similar to that of used by Gallas et al. [2]. A thin slot (0.5 mm wide by 200 mm long) at the top of the cavity permits oscillatory fluid flow.

The frequency response to sinusoidal excitation, i.e. peak centerline velocity during maximum expulsion of the actuator is plotted in Figure 2. The LEM prediction is compared to HWA data taken slightly above the orifice slot centerline.

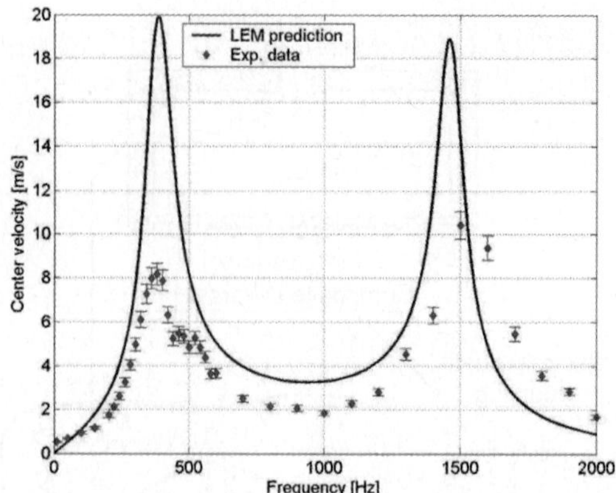

Fig. 2 Centerline velocity at the time of maximum expulsion during a cycle as a function of frequency. Experimental data are compared with the model prediction.

References

1. Leclerc, C., Levallois, E., Gallas, Q., Gilliéron, P., and Kourta, A., Phase Locked Analysis of a Simplified Car Geometry Wake Flow Control Using Synthetic Jet, FEDSM2006-98469, 2006.
2. Gallas, Q., Holman, R., Nishida, T., Carroll, B., Sheplak, M., and Cattafesta, L., Lumped Element Modeling of Piezoelectric-Driven Synthetic Jet Actuators, *AIAA Journal* **41**(2), 2003, 240–247.
3. Gallas, Q., Wang, G., Papila, M., Sheplak, M., and Cattafesta, L., Optimization of Synthetic Jet Actuators, AIAA Paper 2003-0635, 2003.

Feedback Control Using Extremum Seeking Method for Drag Reduction of a 3D Bluff Body

Jean-François Beaudoin[1], Olivier Cadot[2], José Eduardo Wesfreid[3] and Jean-Luc Aider[1]

[1] *PSA Peugeot-Citroën, Department of Research and Innovation, 2 route de Gisy, 78943 Vélizy-Villacoublay, France; E-mail: jeanfrancois.beaudoin@mpsa.com*
[2] *Unité de Mécanique, ENSTA, Chemin de la Hunière, Palaiseau, France; E-mail: cadot@ensta.fr*
[3] *PMMH UMR 7636-CNRS-ESPCI, 10 rue Vauquelin, 75231 Paris Cedex 5, France*

Abstract. The flow around a modified Ahmed body with a curved rear section is studied at $Re > 2.10^6$ and a line of vortex generators (VG) is used as actuators. By varying the angle α of the VG we observe an optimal value of α defining a minimum of aerodynamic drag and correspondingly a maximum base pressure coefficient. As this optimum value is shown to be Reynolds dependent we use an extremum control strategy for the system to find this optimal condition autonomously. It consists of the synchronous detection of the response measured in either the base pressure signal or in the drag and a slow sinusoidal modulation of the angle of the VG. It is finally demonstrated that the closed-loop system is robust and reacts successfully to unpredictable changes in the external flow conditions.

Key words: Feedback loop control, drag reduction.

1 Introduction

From a general point of view, closed-loop control is supposed to drive in real time a dynamical system on a predetermined trajectory in the phase space. In many practical situations, one can use feedback control to allow a system to keep autonomously its optimal set point. In the case of the control of turbulent flows whose external conditions are stable (constant free-stream velocity for instance), one often tries to control the dynamics of the coherent structures: vortex distribution in a bluff-body wake [1], location of a separation point [2], or longitudinal structures in a turbulent boundary layer [3]. In these cases, the detection of the structures and the activation of the actuator must be realized as fast as the characteristic times of the structures. It implies a control system including the sensor(s), the actuator(s) and the algorithm efficient enough to deal with high frequency inputs and outputs. In turbulent flows, which is our case, the characteristic frequencies are large (a few hundred Hertz) and would require very fast actuators such as MEMS (Micro-Electro-

J.F. Morrison et al. (eds.), IUTAM Symposium on Flow Control and MEMS, 365–372.
© 2008 *Springer. Printed in the Netherlands.*

Mechanical Systems, [4, 5]) for experimental control. Nevertheless, there are many realistic and important situations where *external conditions* are likely to change: the incidence of a wing, modification of the free-stream velocity, or the yaw angle of a bluff-body. These phenomena occur on much longer time scales than those of the flow fluctuations and have a major influence on the global dynamics of the flow. Considering this point of view, one can see that it becomes realistic to use more "conventional" mechanical actuators for flow control in the case of *time dependent external conditions*. We will use this point of view in the following to demonstrate the efficiency of feedback control on our system.

The search for the appropriate algorithm is different depending on wether we know the governing equations (or reduced-order model) of the dynamical system or only a few of its properties. In the first case, it is possible to anticipate the effect of the control on the dynamics of the system and then to foresee its "longer time" evolution: we then speak about *predictive control* [1, 6, 7]. This method is the most efficient one since it may provide an optimal feedback law. However, it requires the knowledge of the future of the system which seems to be unrealistic for turbulent flows (especially for experiments). We then choose to turn to another approach, called *adaptive control* [8], in which we only need to know the state of the system at each time step. We then try to modify it on-line using the extremum seeking control scheme [9–11] , in order to take into account the modification of the external conditions. The scheme has already been successfully applied for an academic configuration (backward facing step) in [10, 11]. The aim of the present work is to extend this *adaptive control* to a 3D bluff body with eventual applications for automotive aerodynamics.

2 Experimental Set-up

2.1 Description of the Model

The geometry of the bluff-body is inspired by the Ahmed body [12]. We keep the front part of the Ahmed model but choose a curved rear section: its longitudinal cross-section is a circular arc of constant radius (see Figure 1). Because of this geometry, we expect the creation of an unsteady and Reynolds dependant separation line moving over the rear section. The model is 290 mm high and 340 mm wide, corresponding to a quarter-scale vehicle. Its total length L is 900 mm. The curvature radius of its rear part is 450 mm. The model is raised on four 40 mm high streamlined struts in order to avoid separation on the wheels.

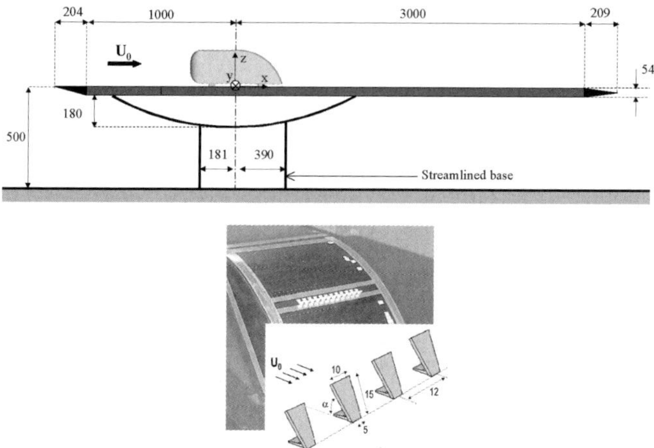

Fig. 1 Top: Bluff body geometry and experimental facility. The raised floor provides a determined boundary-layer. The illustration is not to scale. Bottom: Close-up view of the vortex generator line located at the rear part of the model. All dimensions are given in millimeters.

2.2 Wind Tunnel and Experimental Techniques

All the measurements are performed in our in-house wind tunnel. It is an open wind tunnel with a closed test section. The test section is $2.1 \times 5.2 \times 6$ m long. The operating flow conditions are the following:

- free-stream velocity from $U_0 = 20$ m/s to 40 m/s
- zero yaw angle
- Reynolds number from $Re = U_0 L / \nu = 1.2 \times 10^6$ to 2.4
- turbulence intensity $= 1.3\%$

The blockage coefficient, defined as the ratio of the projected frontal area of the model over the test-section surface, has to be as low as possible (typically not higher than 10%). In our case it is less than 1.2% so that blockage effects are negligible.

A line of 12 aligned vortex generators is placed at $s = 220$mm (see [13] for full details) from the beginning of the curved rear part (Figure 1). The angle α of the vortex generators (defined as the angle between the VGs and the tangent of the body) is controlled by a DC servomotor and can be varied from $\alpha = 0°$ (i.e. no actuator) to $90°$.

To evaluate the efficiency of the VG on the aerodynamic forces we use a six-component aerodynamic balance. It is an underfloor-type balance located under the raised floor. The model is mounted on top of the raised floor, with each individual wheel lying on a separate platform. We will only discuss the measurements of the drag, i.e. D. Since global measurements may not be possible in real conditions, we also measured an indicator of the drag given by the local base pressure coefficient.

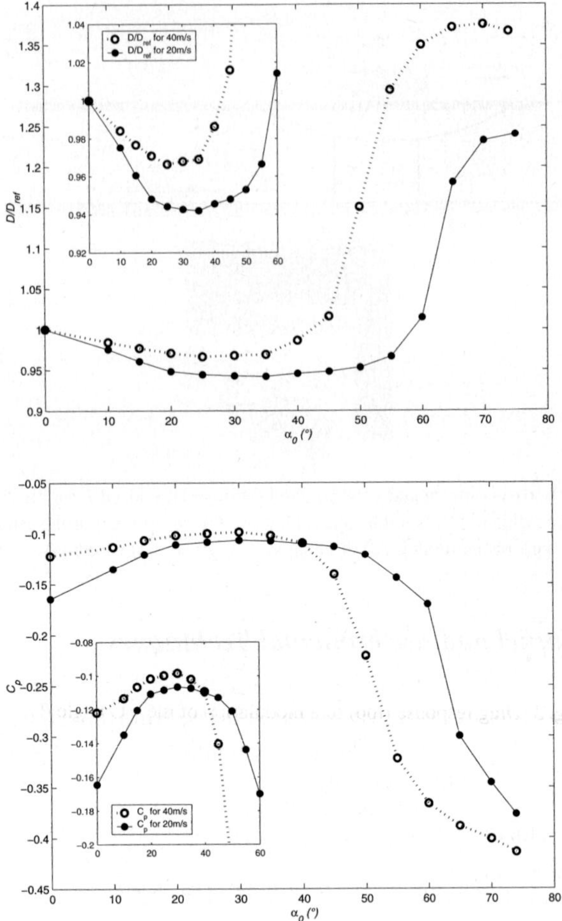

Fig. 2 Drag (top) and base pressure coefficient (bottom) vs. the VG line angle for $U_0 = 40$ m/s (open circles) and $U_0 = 20$ m/s (filled circles).

A pressure tap is located behind the VG line at $s = 240$ mm halfway between the side edges of the model. The base pressure coefficient C_p is defined as:

$$C_p = \frac{p - p_0}{\frac{1}{2}\rho U_0^2}. \tag{1}$$

Labview software is used to drive the angle of the VG line and to acquire either the drag or the base pressure. For closed loop control experiment, the acquisition and the driving of the VG line angle are performed in single point mode at a rate of 20 Hz.

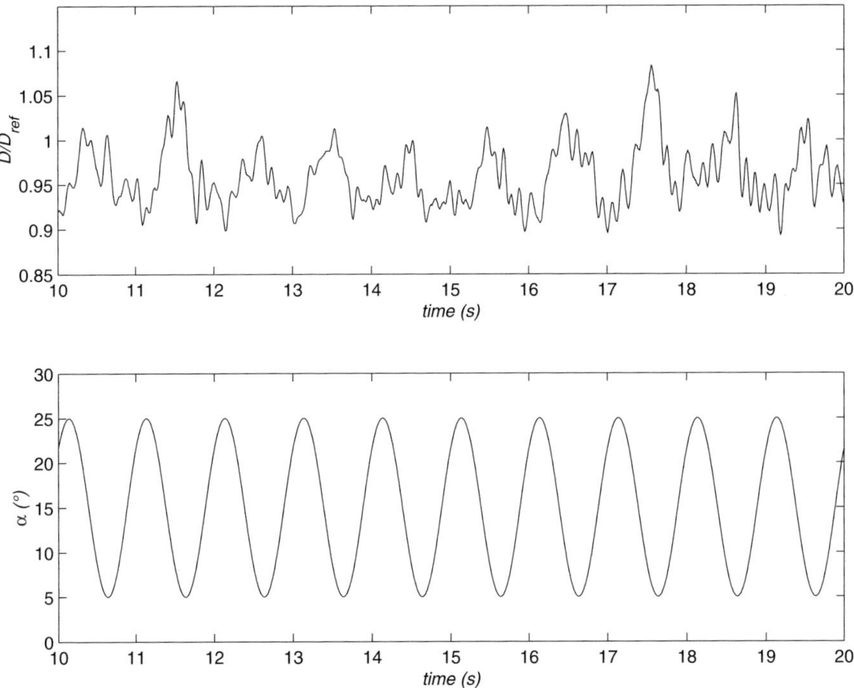

Fig. 3 Drag response (top) to a modulation of the VG angle (bottom).

3 Measurements and Discussion

3.1 Open Loop Control and Gradient Estimator

Drag and base pressure coefficients were measured for different angles α of the VG lines and two external free-stream velocities, 20 m/s and 40 m/s. In Figure 3, we can see that there is an α, depending on the free-stream velocity, for which the drag is minimum or equivalently the base pressure maximum. These angles that define optimal configurations for lower drag are $\alpha = 30°$ at 40 m/s and $35°$ at 20 m/s.

Figure 2 shows the drag response to a slow modulation ($f_m = 1$ Hz, $a = 10°$) of the VG angle around a fixed value $\alpha_0 = 15°$:

$$\alpha(t) = \alpha_0(t) + a\cos(2\pi f_m t). \tag{2}$$

The modulation that is obtained in the responses will be used as an estimator [11] of the gradient $\partial J/\partial\alpha$ for $J = D$ or C_p:

$$\frac{\partial J}{\partial\alpha} \approx \frac{A_J(f_m)}{a} \, \mathrm{sign}(\cos(\phi_\alpha(f_m) - \phi_J(f_m))). \tag{3}$$

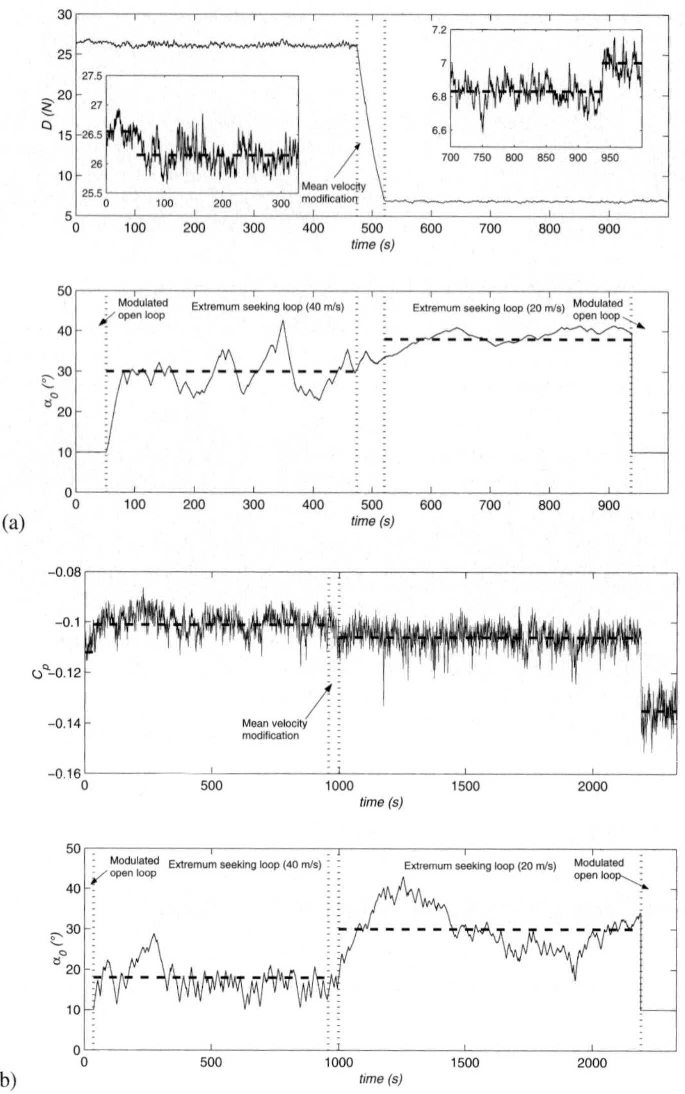

Fig. 4 Drag (a) and base pressure coefficient (b) vs. the VG line's angle for $U_0 = 40$ m/s (open circles) and $U_0 = 20$ m/s (filled circles).

The phase relationship $\phi_\alpha(f_m) - \phi_J(f_m)$ between the excitation and the response gives the sign of the gradient. In Figure 2, the phase shift is about π meaning that the gradient is negative as it is observed in Figure 3 around $\alpha_0 = 15°$ for the drag. We generally observe the estimated gradient sign to be very robust, however the amplitude at the modulation frequency $A_J(f_m)$ in the response that should be proportional to the gradient modulus is not accurate and strongly depends on the turbulent fluctuations in the measurements.

3.2 Closed Loop: Extremum Seeking Control

We use an extremum-seeking control strategy which is efficient in the case where a non-linear plant has an extremum. It consists of using a real-time gradient optimization method with the following feedback law:

$$\frac{d\alpha}{dt} = -K\frac{\partial J}{\partial \alpha} \tag{4}$$

where K is a positive or negative gain depending whether the extremum is a minimum (case for the drag) or a maximum (case for the base pressure), and its magnitude is chosen by the experimentalist. The gradient is estimated as explained above and using FFT computation [11].

In Figure 4, we show the result of the closed loop control under changes in the free-stream velocity from 40 m/s to 20 m/s. The system reacts successfully in the case of the drag measurements. Both extrema found by the system ($30°$ and $35°$, displayed in the figures by horizontal dashed lines) correspond to the one measured in the open loop control experiment displayed in Figure 3. However, in the case of the base pressure measurements, the extremum is systematically underestimated. The discrepancy is due to the stronger fluctuations that are associated with local measurements like base pressure, compared global measurements like drag, which are intrinsically spatially filtered.

4 Conclusion

We believe that the present method may be suitable for other applications and we demonstrate that it overcomes the main difficulties associated with the closed-loop control of turbulent flows. We are confident that such adaptive schemes could be a powerful tool for some industrial applications.

References

1. B. Protas and A. Styczek. Optimal rotary control of the cylinder wake in the laminar regime, *Phys. Fluids* **14**(7), 2002, 2073–2087.
2. Y. Wang, G. Haller, A. Banaszuk, and G. Tadmor. Closed-loop Lagrangian separation control in a bluff body shear flow model, *Phys. Fluids* **15**(8), 2003, 2251–2266.
3. J. Kim. Control of turbulent boundary layers, *Phys. Fluids* **15**(5), 2003, 1093–1105.
4. C. M. Ho and Y. C. Tai. Review: MEMS and its applications for flow control, *J. Fluids Eng.* **118**, 1996, 437–447.
5. C. M. Ho and Y. C. Tai. Micro-electro-mechanical-systems (MEMS) and fluid flows, *Annu. Rev. Fluid Mech.* **30**, 1998, 579–612.
6. T. R. Bewley. Flow control: New challenges for a new Renaissance, *Prog. Aerospace Sci.* **37**, 2001, 21–58.

7. T. R. Bewley, P. Moin, and R. Temam. DNS-based predictive control of turbulence: An optimal benchmark for feedback algorithms, *J. Fluid Mech.* **447**, 2001, 179–225.

8. M. Krstić, I. Kanellakopoulos, and P. V. Kokotovic. *Nonlinear and Adaptive Control Design*, J. Wiley, New York, 1995.

9. M. Krstić and H. H. Wang. Stability of extremum seeking feedback for general nonlinear dynamic systems, *Automatica* **36**, 2000, 595–601.

10. J. F. Beaudoin, O. Cadot, J. L. Aider, and J. E. Wesfreid. Bluff-body drag reduction by extremum seeking control, *J. Fluids Struc.* **22**, 2006, 973–978.

11. J. F. Beaudoin, O. Cadot, J. L. Aider, and J. E. Wesfreid. Drag reduction of a bluff-body using adaptive control methods, *Phys. Fluids* **18**, 2006, 085107.

12. Ahmed, S. R. Influence of base slant on the wake structure and drag of road vehicles, *J. Fluids Eng.* **105**, 1983, 429–434.

13. J. F. Beaudoin. Contrôle actif d'écoulement en aérodynamique automobile, PhD Thesis, Ecole des Mines de Paris, 2004.

Flow Control in Turbomachinery Using Microjets

Sven-J. Hiller[1], Tobias Ries[2] and Matthias Kürner[2]
[1]*MTU Aeroengines, Dachauer Straße 665, Munich, Germany;*
E-mail: sven-juergen.hiller@muc.mtu.de
[2]*Institut of Propulsion, University of Stuttgart, Pfaffenwaldring 6, Stuttgart,*
Germany; E-mail: {tobias.ries, matthias.kuerner}@ila.uni-stuttgart.de

Abstract. The paper gives a brief overview about the challenges and the concepts for future aeroengine applications based on microjets. Technical requirement and difficulties of the potential applications of Active Flow Control (AFC) in turbomachines are discussed. The importance of an active stabilized turbocompressor for an efficiency increase is described. Future concepts of advanced turbocompressors with the application of active and passive flow control, partially based on MEMS devices are briefly sketched. Finally, the numerical methods necessary to unveil the physical background of AFC are presented.

Key words: Active flow control, turbomachinery, compressor, microjet, detached eddy simulation, scale-adaptive simulation.

1 Introduction

The rising oil price and the jeopardy of global warming in junction with the annual increase in civil aircraft transportation requires the introduction of new technologies into aero engines for further fuel saving. The aircraft is one of the few transportation systems depending completely on oil as the primary energy source. Although the search for alternative fuels (synthetic kerosene) is underway, a stepwise introduction of alternative concepts like hydrogen, hybrid or natural gas driven power units as done currently in the automotive industry is not possible for aircrafts. The only feasible way is a significant reduction of fuel consumption per passenger and per flight. Two groups of experts [1, 2] estimated the targets for future aircraft and aeroengine applications. Both groups identified a 50% reduction of CO_2 per passenger kilometer over the next 15–20 years.

However, today's aeroengine is a very high optimized turbomachine. During the last years some significant improvements and a dramatic efficiency increase was achieved due to the better understanding of the flow physics inside a turbomachine

J.F. Morrison et al. (eds.), IUTAM Symposium on Flow Control and MEMS, 373–380.
© 2008 Springer. Printed in the Netherlands.

and the intensive application of state-of-the-art numerical methods. Advanced airfoil design methods provide a better control over the shock position, laminar-turbulent transition, secondary passage flow, etc. It is common practice today to get component efficiencies in excess of about 90%.

On the other hand, an aeroengine is optimized for few specified guarantee points (e.g. Aero Design Points ADP). The ADP describes the technical parameters an aeroengine has to achieve, such as thrust, specific fuel consumption (SFC), etc. It does not take into account the different application profiles of the final customer – the airliner. There is no difference in engine design for an engine dedicated for long range or for short range applications. As a consequence, an engine design is a technical compromise for very different application profiles.

A new way to achieve further fuel saving is the design of adaptable engines with a large and efficient operation range which provides fuel saving over the complete flight mission. The most feasible way to realize this is the introduction of active controlled engine components.

2 Active Controlled Aeroengine

In general, active control methods can be applied at very different engine parts and range from the engine support structure (e.g. active bearings) to the engine gas path (e.g. active core). An overview about possible applications of active flow control methods is given in [3]. The paper describes the wide spectra of control concepts and concludes that almost all engine components (intake, fan, compressor, turbine, combustion chamber, exit nozzle, etc.) can be equipped with active control devices. While some of the concepts operate pure mechanically (e.g. mechanical driven active tip clearance control, vortex generators), other concepts apply fluidic effects (e.g. actuation jets, air injection).

The benefit of an active controlled compressor can be described best on the basis of a compressor map (Figure 1). A compressor runs steady-state at the working line (dashed line). A certain distance from the surge line at constant mass flow rates (so-called surge margin) is necessary to enable a stable operation of the compressor.

Due to the nature of the compressor physics, the surge margin at part speed is significant smaller then that at high speed. A shift of the surge line (raised surge line marked red) by any stability improvement method provides sufficient surge margin increase at part speed to cover transient engine runs (green acceleration curve). The raised new working line enables the compressor to operate at higher efficiency (dotted lines).

To summarise, a stability improvement method is beneficial in two ways: (i) a better operability of the compressor to tolerate distortions and, (ii) an efficiency benefit due to the raised working line.

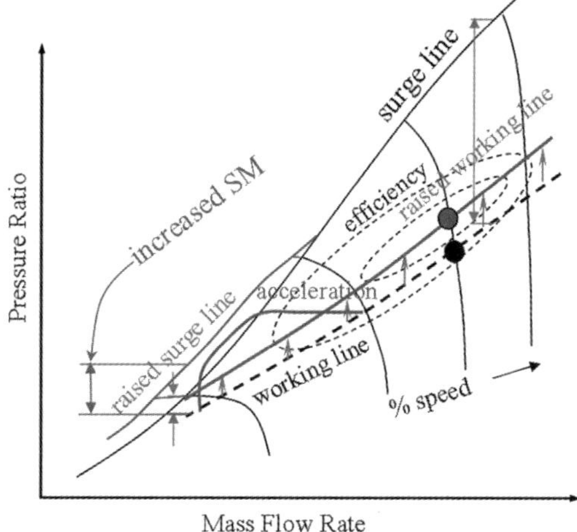

Fig. 1 Compressor map.

3 Compressor Stability

The compressor stability plays a vital role in the design of high efficient compressors. The higher blade or stage loading increases the jeopardy of earlier airfoil stall and, consequently, compressor stall and surge in the case of throttling (Figure 1). Three different methods are known to increase the surge margin:

- casing treatment,
- tip injection, and
- tip clearance control.

While casing treatment [4] is a passive method and consists of mechanical inserts into the compressor casing, the latter two methods can be integrated into an active control system. Tip clearance control is successfully applied for high and low pressure turbines. The installation of a similar technique for compressors is under investigation. The tip injection method consists of an array of nozzles (10 to 20) placed circumferentially at the inner casing in front of the rotor. Each nozzle generates a jet towards the rotor blade tip clearance and influences the clearance flow. It is shown [5–7] that steady air injection is capable of stabilizing the compressor. The method extends the stable operation range of a compressor significantly towards lower mass flow and increases the surge margin. However, the injected high pressure air must be taken from a rear stage. This method may decrease the overall efficiency of the compressor because at least a fraction of the core mass flow (about 1–5%) must be compressed twice. But the benefit in operability compensate the deficit. The technique, although it is working, is not implemented in engines in service. A further reduction of the injected mass flow is necessary to increase the

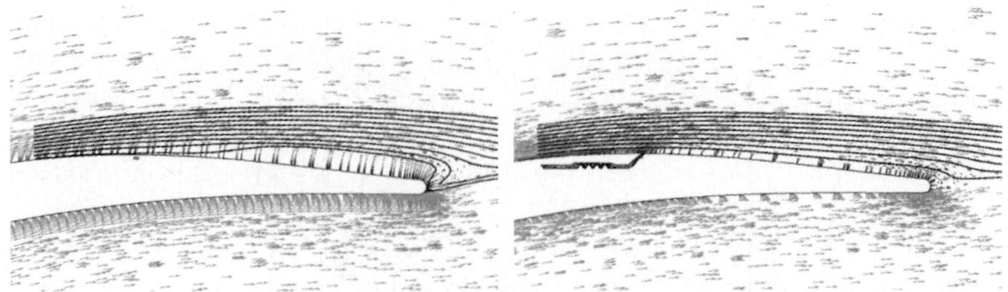

Fig. 2 High pressure compressor blade without (left) and with (right) flow actuation by a microjet.

attractiveness of the method for next generation engines. Modulated and controlled air injection promises the required further mass flow reduction [7].

4 Micro Valves

A more advanced injection technique uses micro-sized valves built by MEMS technologies [8, 9]. Microvalves based on magnetostriction are capable to exceed the mechanical characteristics of piezoelectric or comparable devices [10].

A large number of low-cost micro valves (about 400) embedded in the compressor casing influences the rotor tip clearance flow directly. A control system manages and drives the micro valve array according the signals received from a compressor surge detection sensor.

The described micro valves can also be applied to prevent or at least delay the flow separation on the suction side of a compressor blade. A pulsing or steady micro jet driven by a micro valve energizes the boundary layer flow locally. The pulsing micro jet supplied by external air provides more kinetic energy to the boundary layer than comparable zero-net flow devices or synthetic jets. Although the design of the micro valve is more complicated compared to synthetic jets devices, the technology is more appropriate for high Mach number applications. Due to the micro-sized design and the magnetostrictive effect, the valves can operate at a frequency range attractive for turbomachinery applications (see Section 5).

Figure 2 shows the interaction between a steady blowing micro jet and a boundary layer separation. The turbulent separation on the suction side of a compressor blade is suppressed almost completely. The jet hole in this configuration is smaller than the typical boundary layer scale.

5 Flow Actuation Devices for Aeroengines

The technical requirements for flow actuation devices applicable for aeroengines have to take into account the harsh environment in which the engine operates. Not only are the temperature and pressure levels occurring inside the engine challenging, but also the reliability and robustness. Aeroengine parts have to withstand significant additional mechanical, thermal and chemical loads (e.g. erosion by dust, sand and saline aerosol).

The benefit of micro-sized flow actuation devices is the small dimension which adds only little additional weight to the engine. Reliability and robustness have to be considered in the context of the likely length of engine service which is in the order of 20 000 hours. The environmental temperature for the actuation device is between $-50°C$ and $600°C$, approximately. The first application of fluidic actuation devices are realistic for the low temperature engine components (e.g. intake, low and intermediate pressure compressor). It is not necessary that a specific device design has to cover the complete temperature range, but the actuator technology should be applicable over a certain temperature range. According to the flow phenomena to be influenced, the actuation devices need to cover a range of frequencies in the spectrum: (i) a flow phenomenon which scales with the shaft speed (e.g. stalled blade passages) requires a device operating at about 200–300 Hz, (ii) a flow phenomenon linked to the blade passing frequency requires a device operating at about 5–20 kHz, (iii) a flow phenomenon scaled with the vortex shedding frequency (of a thin compressor blade) requires about 50–100 kHz.

6 Requirement for Future Flow Actuation Devices

Future flow actuation concepts consist of a sensor, an actuator and a processing unit with implemented feed-back control laws. An appropriate philosophy for integrating a network of distributed actuation devices and techniques with the engine control system can be found in [11, 12].

In almost all cases, external power is required for the operation of the actuation device. However, the additional tubes and wires introduce more complexity and weight to the engine and are additional sources of possible malfunctions. Therefore, an autonomous power supply system which drives the actuation device is desirable. The aeroengine itself delivers numerous sources for small local power generation. The rotor blade motion, the high local pressure, velocity and temperature gradients inside an turbomachine, the high vibration forces and the high pressure fluctuation can be used as a power source for the device. The appropriate energy conversation techniques need to be developed, probably based on MEMS technologies.

7 Numerical Techniques

The study of flow control mechanisms is still subject of intensive experimental and numerical investigations [13, 14]. The applied numerical methods range from the fast and well-known Reynolds Averaged Navier–Stokes method (RANS) to the ex-

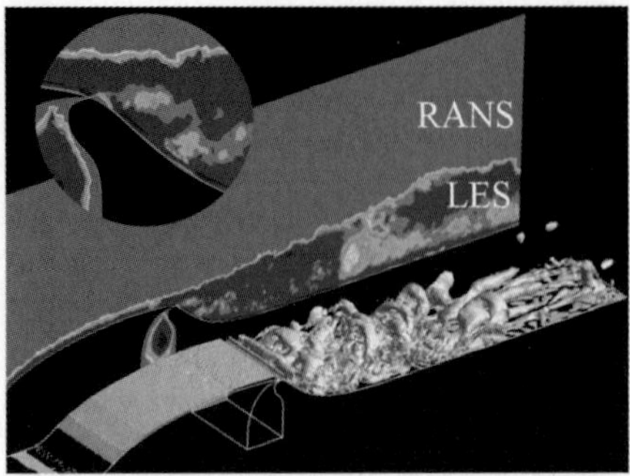

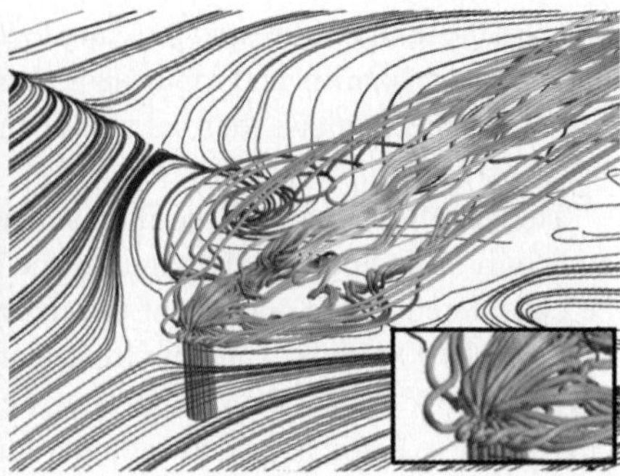

Fig. 3 Channel flow with flow actuation through a slot.

Fig. 4 Pulsing microjet installed in a compressor blade.

pansive Large Eddy Simulation (LES) and Direct Numerical Simulation (DNS). The numerical results indicate that pure eddy-viscosity models (RANS) fail to predict the phenomena sufficiently accurate. The micro-sized flow actuation devices change the turbulence structure dramatically in a way not covered by the mathematical derivation of eddy-viscosity models (isotropic turbulence). LES and DNS methods can be considered as an alternative. However, the computational effort for turbulent flow prediction is not acceptable within today's industrial environment.

An encouraging compromise between enhanced turbulence simulation techniques and computational effort is the so-called zonal approach which combines a LES-type model with a RANS-type model. The method is also known as Detached Eddy Simulation (DES) [15]. Some criteria implemented in the numerical model determine which method is applied in the different zones (Figure 3). Zones in which the turbulent vortex motion dominates are predicted by a LES-type model. The remaining domain is simulated by an appropriate RANS model. An advanced zonal approach model was developed by Menter [16] and is known as Scale-Adaptive Simulation (SAS). The benefit of the zonal approach is the moderate computational effort compared to a full LES simulation and a significant enhancement of the numerical resolution of the turbulent vortex structures.

The numerical simulation of a pulsing jet embedded in a high-speed compressor blade with the SAS model is shown in Figure 4. Local mesh refinement in junction with the SAS model enables the prediction of the fluctuating horseshoe vortex around the pulsing jet and the dominant vortex structures downstream.

Although, the achievements of today's simulation techniques are impressive, the inaccuracy and incompleteness of the numerical models must be assessed individually by experimental data. However, the advances in simulation technique are a valuable source to provide a better understanding of the flow phenomena involved in active flow control.

8 Conclusions

Active flow control is capable to stabilize an aeroengine compressor. The active compressor stabilization enables a shift of the compressor working line to a region of higher efficiency which provides a contribution to the fuel saving requested for future aeroengines.

The application of micro jets generated by micro valves plays a dominate role for future flow control applications. Two basic configurations are promising: (i) an array of microjets embedded in the suction side of blade, and (ii) an array of microjets installed in the compressor casing. Both methods are capable of increasing the operability and the surge margin of a compressor.

An integrated system for future flow control applications consists of a sensor, an actuator, a control unit and an autonomous power supply system.

Advanced turbulence simulation techniques like DES and SAS provide the best compromise between computational effort and resolved flow details in an industrial environment.

References

1. European Aeronautics: A Vision for 2020, Report of the group of personalities, January 2001, Published by the European Commission
2. *Review of NASA's Aerospace Technology Enterprise*, National Research Council, The National Academies Press, Washington, D.C., 2004.
3. Hiller, S.-J., Hirst, M., Webster, J., Ducloux, O., Pernod, P., Touyeras, A., Garnier, E., Pruvost, M., Wakelam, Ch. and Evans, S., ADVACT – A European Program for Actuation Technology in Future Aeroengine Control Systems, in *3rd AIAA Flow Control Conference*, 5–8 June 2006, San Francisco, CA, AIAA 2006-35110, 2006.
4. Wilke, I., Kau, H.-P. and Brignole, G., Numerical aided design of a high-efficient casing treatment for a transsonic compressor, in *ASME Turbo Expo 2005: Power for Land, Sea and Air*, June 6–9, 2005, Reno, NV, USA, GT2005-68993, 2005.
5. Suder, K.L., Hathaway, M.D., Throp, S.A., Strazisar, A.J. and Bright, M.M., Compressor stability enhancement using discrete tip injection, *ASME Journal of Turbomachinery* **123**, 2001, 14–23.
6. Kefalakis, M. and Papailiou, K.D., Active flow control for increasing the surge margin of an axial flow compressor, in *ASME Turbo Expo 2006: Power for Land, Sea and Air*, May 8–11, 2006, Barcelona, Spain, GT2006-90113, 2006.
7. Scheidler, S.G., Mundt, Ch., Mettenleiter, M., Hermann, J. and Hiller, S.-J., Active stability control of the compression system in a twin-spool turbofan engine by air injection, in *The 10th International Symposium on Transport Phenomenon and Dynamics of Rotating Machinery*, March 7–11, 2004, Honolulu, Hawaii, USA, ISROMAC10-2004-073, 2004.
8. Pernod, P., Preobrazhensky, V., Merlen, A., Ducloux, O., Deblock, Y., Talbi, A. and Tiercelin, N., MEMS for flow control: Technological facilities and MMMS alternatives, in *Flow Control and MEMS*, Proceedings of the IUTAM Symposium held at the Royal Geographical Society, 19–22 September 2006, J. Morrison et al. (Eds.), Springer, Dordrecht, 2008.
9. Ducloux, O., Deblock, Y., Talbi, A., Pernod, P., Preobrazhensky, V. and Merlen, A., Magnetically actuated microvalves for active flow control, in *Flow Control and MEMS*, Proceedings of the IUTAM Symposium held at the Royal Geographical Society, 19–22 September 2006, J. Morrison et al. (Eds.), Springer, Dordrecht, 2008.

10. Huber, J.E., Fleck, N.A. and Ashby, M.F., The selection of mechanical actuators based on performance indices, *Proc. R. Soc. Lond. A* **453**, 1997, 2185–2205.
11. Garg, S., Controls and health management technologies for intelligent aerospace propulsion systems, National Aeronautics and Space Administration, Glenn Research Center, NASA/TM-2004-212915, 2004.
12. NATO Research and Technology Organisation, AVT-128 More Intelligent Gas Turbine Engines, Report under preparation (expected 2007/2008).
13. Rumsey, C.L., Gatski, T.B., Seller III, W.L., Vatsa, V.N. and Viken, S.A., Summary of the 2004 CFD validation workshop on synthetic jets and turbulent separation control, in *2nd AIAA Flow Control Conference*, 28 June–1 July 2004, Portland, Oregon, USA, AIAA 2004-2217, 2004.
14. Hiller, S.-J. and Seitz, P.A., The interaction between a fluidic actuator and main flow using SAS turbulence modeling, in *3rd AIAA Flow Control Conference*, 5–8 June 2006, San Francisco, CA, AIAA 2006-3678, 2006.
15. Spalart, P.R., A young person's guide to detached-eddy simulation grids, NASA/CR-2001-211032, 2001.
16. Menter, F.R. and Egorov, Y., A scale-adaptive simulation model using two-equation models, in *43rd AIAA Aerospace Science Meeting and Exhibit*, January 10–13, 2005, Reno, NV, USA, AIAA 2005-1095, 2005.

ONERA/IEMN Contribution within the ADVACT Program: Actuators Evaluation

E. Garnier[1], M. Pruvost[1], O. Ducloux[2], A. Talbi[2], L. Gimeno[2], P. Pernod[2], A. Merlen[2] and V. Preobrazhensky[2]

[1]*ONERA, Applied Aerodynamics Department, 29 avenue de la Division Leclerc, 92322, Châtillon Cedex, France; E-mail: eric.garnier@onera.fr*
[2]*Joint European Laboratory LEMAC: IEMN-DOAE-UMR CNRS 8520, LML-UMR CNRS 8146, Ecole Centrale de Lille, BP 48, 59652 Villeneuve d'Ascq Cédex, France*

Abstract. This paper summarizes the ONERA/LEMAC contribution within the work-packages 2 and 3 of the EU project ADVACT. This activity is dedicated to the evaluation of pulsed jets based on magnetic actuation principle on a generic separated flow. It is supported by some advanced numerical work based on RANS/LES coupling.

Key words: Separation control, MEMS, pulsed jets.

1 Introduction

This paper summarizes the ONERA/LEMAC contribution within the work-packages 2 and 3 of the EU project ADVACT which is dedicated to the "Development of Advanced Actuation Concepts to Provide a Step Change in Technology Used in Future Aero-Engine Control Systems". The project is lead by Rolls Royce and groups of the main European engine manufacturers and universities or research centres. One of the purposes of the aforementioned workpackages is to evaluate the potential of pulsed jet actuators developed in particular by LEMAC on configurations of increasing complexity. The latter one being a high speed cascade flow experiment planed at VKI.

The ONERA contribution focuses on the evaluation of actuators performances on a very basic experiment of flow separation only due to pressure gradient. This choice is motivated by the fact that, in a compressor, the pressure increase can cause separation. The conception of an experiment in which the flow separates on a flat plate is not trivial and many RANS numerical simulations were necessary to define the experimental set-up. The general idea is to generate a pressure gradient by a modification of the wall curvature of one side of an existing low velocity Eiffel type wind tunnel (see Figure 1). The air is sucked up on the curved side through perforated holes to prevent separation. Consequently, the flow separates on the opposite

J.F. Morrison et al. (eds.), IUTAM Symposium on Flow Control and MEMS, 381–386.
© 2008 *Springer. Printed in the Netherlands.*

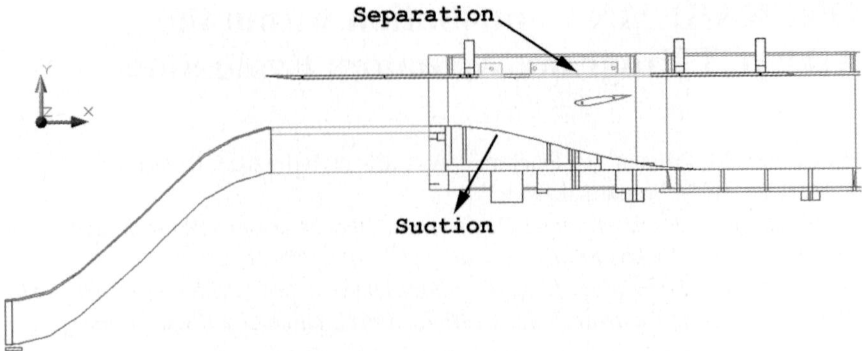

Fig. 1 Experimental setup (the wind tunnel is only partially shown).

side of the flat plate. A NACA airfoil is located at about 8 cm from the flat plate side to cause the reattachment.

The inflow velocity is 30 m/s and the wind tunnel height before the diffuser is 0.15 m. The wind tunnel span is 0.3 m. The Reynolds number based on the boundary layer displacement thickness is about 4000 at the actuator location. The suction velocity deduced from the measured flow rate is about 0.6 m/s and applies on the first third of the diffuser. The experimental characterisation means are mainly on one and two components hot-wire and stereo PIV. Pressure taps and friction sensors have not been implemented yet. The first attempts to suppress separation have been performed using mechanical counter-rotating vortex generators. Their height is about $0.37\delta_0$ with $\delta_0 = 1.5$ cm.

Their geometric characteristics are as follows:

Height (h)	Length	Spacing between the two VGs trailing edges	Spacing between each couple of VGs	Sideslip angle
$0.37\delta_0$	$2h$	$2.5h$	$6h$	$18°$

Five pairs of these VGs are located on an interchangeable 80 mm wide plate. A picture of the plate receiving counter-rotating vortex generators is presented in Figure 2.

Actuators are located at about $14\delta_0$ before the separation point. Former attempts with actuators located 80 mm downstream ($9\delta_0$ before the separation point) have only shown a slight actuator effect on the flow. With the present location, the longitudinal velocity field acquired with two components PIV presented in Figure 3 demonstrates a strong influence of the VGs which are able to fully reattach the flow.

Figure 4 presents the longitudinal velocity in a normal plane located in the separated zone of the baseline configuration. The results were acquired using one-dimensional hot-wire measurements. Consequently, the sign of the velocity is not accessible. One can notice that, on the extent of the measurement domain (one-third of the wind tunnel span), the three-dimensional effects in the flow remains reason-

Fig. 2 Vortex generators in the wind tunnel.

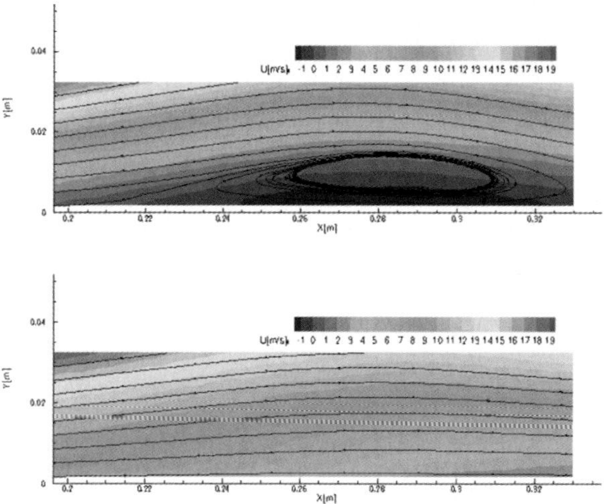

Fig. 3 Longitudinal velocity field (PIV). Top: baseline flow. Bottom: VGs controlled flow.

ably weak even in the baseline configuration which is strongly separated. On the controlled case, evidences of the VGs wakes are not observable.

Additionally, pulsed jets designed and manufactured by LEMAC/(IEMN-LML) within this project have been tested in place of the VGs. These micro-valves, described in more details in separate papers [1, 2] are based on a magnetostatic actuation principles. A magnet attached to a polymer membrane is actuated to close/open a channel connected to the input and output holes located on each side of the micro-valve. The hole output diameter is 1 mm. With an inflow pressure of 1.5 bars, an output velocity of about 90 m/s can be reached at a frequency of a few hundreds Hz. In the ONERA experiment, 8 microjets actuators with a spanwise spacing of 1.5 cm have been used. They are oriented with pitch and skew angles with

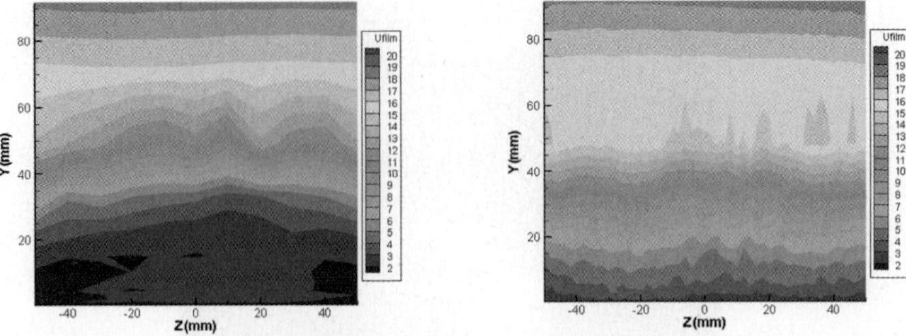

Fig. 4 Longitudinal velocity in a normal plane located in the separation zone. Left: uncontrolled flow. Right: flow controlled using mechanical VGs.

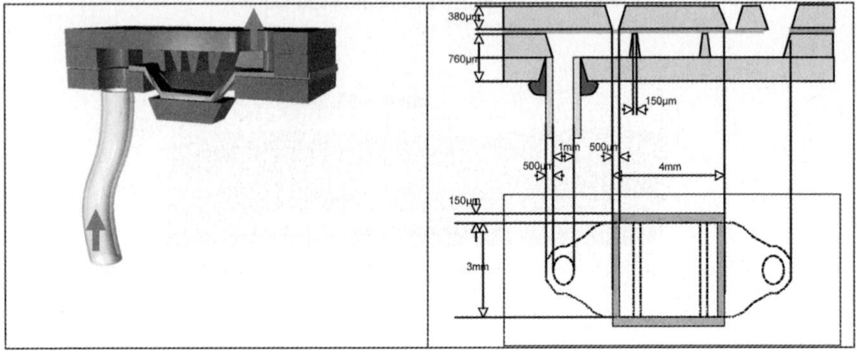

Fig. 5 Principles and dimensions of the LEMAC pulsed jet actuators.

respect to the longitudinal axis equal to 45 degrees. A sketch giving the principles and dimensions of the actuators are given in Figure 5.

These actuators have been successfully implanted in the wind tunnel. Figure 6 shows the backside of the array of actuators. The coils and the present pressure distribution system are visible on this picture. This system which feed the 8 different valves from a unique pressure source ensures that the blowing is homogeneous on each valve.

In terms of results, the longitudinal velocity field observed in the symmetry plane (Figure 7) is similar to the one obtained with vortex generators and the flow fully reattaches. These results have been obtained with a frequency of 70 Hz (the reduced frequency based on the separation length is about 1.2) and a jet inflow pressure of 1.35 bars which leads to a velocity ratio between the jets and the inflow velocity of about 2. For a velocity ratio of 1, the flow remains separated but the separation heigth and the backflow intensity is divided by three.

These actuators are still in development and, for the next generation, the airtightness will be improved by a new arrangement of the coils. Additionally, one has to mention that the micro-jets have functioned during hours (typically 10 hours are ne-

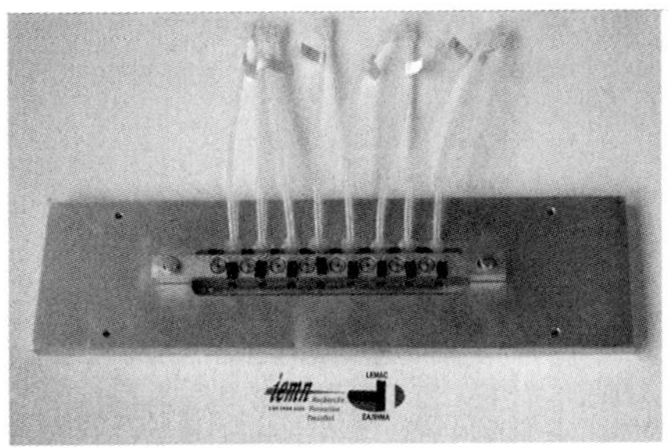 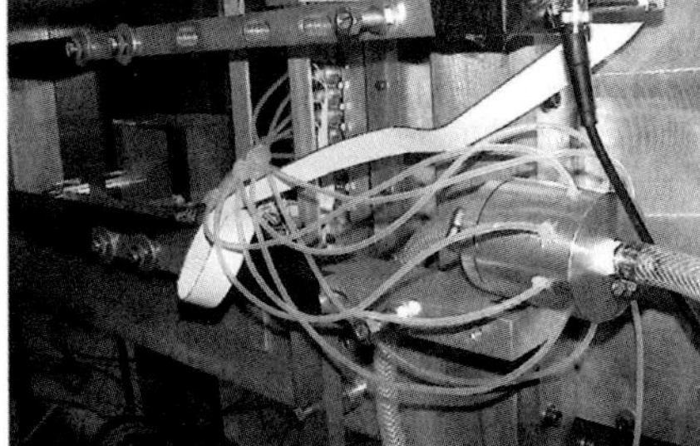

Fig. 6 Picture of the backside of the pulsed jet showing the coils (left) and the pressure distribution system (right).

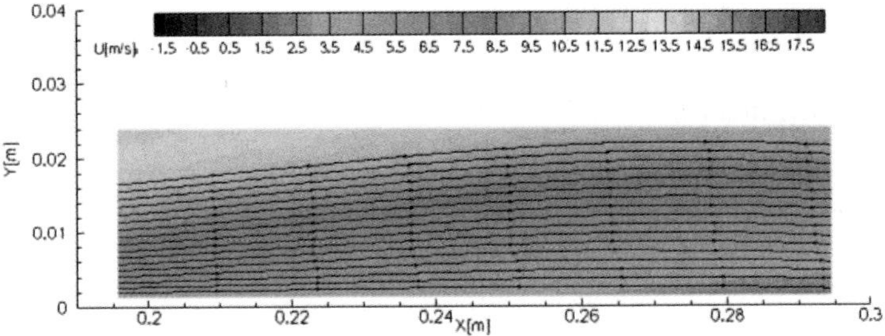

Fig. 7 Longitudinal velocity field (PIV) for the pulsed jets controlled flow.

cessary when sensing a large plane with hot wire technique) without experiencing reliability issues.

The experimental approach is supplemented with a numerical work. For this separated flow, the idea is to use advanced numerical methods, like Large Eddy Simulations that have demonstrated their accuracy in such situation but at a high cost in terms of computational resources. Additionally, the chosen setup requires the computation of the whole wind tunnel in order to obtain the proper pressure gradient on the plate. A RANS/LES coupling has then been used to bridge these contradictory constraints. A multi-block solver was used and each block can be either of RANS or LES type. The distribution of RANS and LES domains is presented in Figure 8 together with an instantaneous view of the flow. Realistic turbulent fluctuations coming from an additional boundary layer LES are injected at the entrance of the LES domain. Before spending considerable computational resources on this computation, it has been recently checked that the vertical velocity on the line $y = 0.1$ m obtained in the RANS computation is in agreement with the one measured experimentally. This means that the suction boundary condition chosen in the computation works reasonably well. The full computation without control is presently running and the effect of the pulsed jets will also be investigated numerically.

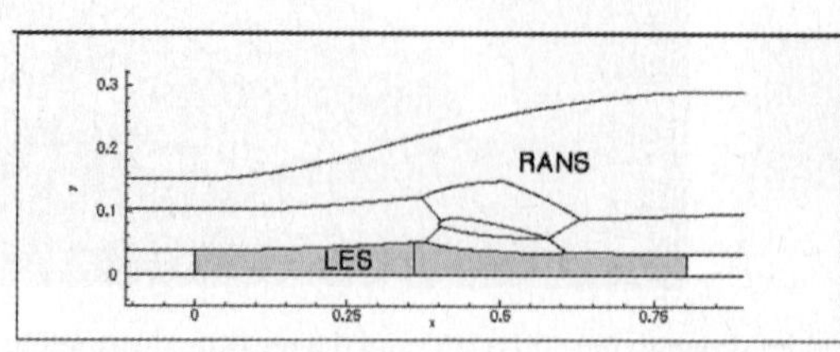

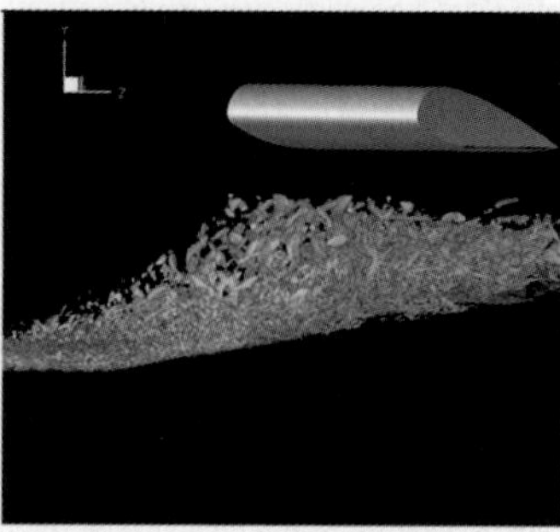

Fig. 8 RANS/LES computation (domain repartition and flow visualisation in the LES zone using the Q criteria colored by the longitudinal velocity).

Acknowledgements

This work has been supported by the European Union within the STREP ADVACT. J.C. Monnier and C. Fatien from ONERA are warmly acknowledged for providing the PIV results.

References

1. Ducloux, O., Deblock, Y., Talbi, A., Pernod, P., Preobrazhensky, V., Merlen, A., Magnetically actuated microvalves for active flow control, in *Flow Control and MEMS*, Proceedings of the IUTAM Symposium held at the Royal Geographical Society, 19–22 September 2006, J. Morrison et al. (Eds.), Springer, Dordrecht, 2008.
2. Pernod, P., Preobrazhensky, V., Merlen, A., Ducloux, O., Talbi, A., Gimeno, L., Tiercelin, N., MEMS for flow control: technological facilities and MMMS alternatives, in *Flow Control and MEMS*, Proceedings of the IUTAM Symposium held at the Royal Geographical Society, 19–22 September 2006, J. Morrison et al. (Eds.), Springer, Dordrecht, 2008.

Control of Flow-Induced Vibration of Two Side-by-Side Cylinders Using Micro Actuators

Baoqing Li[1,2], Yang Liu[2], K. Lam[2], Wen J. Li[3] and Jiaru Chu[1]
[1]*Department of Precision Machinery and Instrumentation, University of Science and Technology of China, Hefei, Anhui, China*
[2]*Department of Mechanical Engineering, The Hong Kong Polytechnic University, Hong Kong; E-mail: mmyliu@polyu.edu.hk*
[3]*Department of Automation and Computer Aided Engineering, Chinese University of Hong Kong, Hong Kong*

Abstract. The control of the vibration of two side-by-side cylinders in a cross flow has been experimentally studied using micro actuators. Three spacing ratios, $T/d = 1.2$, 1.8 and 3.0, are studied, where T is the center-to-center distance, and d is the diameter of the cylinder. The experiments show that the micro excitation can effectively reduce the flow-induced vibration for T/d of 1.2 and 3.0, when the excitation frequency and actuator location are optimized.

Key words: Flow-induced vibration, control, PZT actuator.

1 Introduction

One approach to tackle the control of flow-induced vibration of high-rise buildings in close proximity is to develop an understanding of the interactions of near-wake flow and the motions of the buildings. The basic physics of this problem could be gleaned from developing an effective control method for simple geometric bluff bodies. Cylinders in a cross flow are one of the most basic and revealing cases in the general subject of fluid-structure interaction, and the cylinder vibration as a result of fluid forcing and the resulting effects on the surrounding flow add to the complexity of the problem. The effective control of flow-induced vibration of cylinders would provide insight and understanding toward the control of flow-induced vibrations of large structures.

2 Experimental Details and Methods

The experimental setup is shown in Figure 1a. Experiments were conducted in a closed- circuit wind tunnel which has a 2.4 m-long work area with a 0.6 m $\times$ 0.6 m

J.F. Morrison et al. (eds.), IUTAM Symposium on Flow Control and MEMS, 387–391.
© 2008 *Springer. Printed in the Netherlands.*

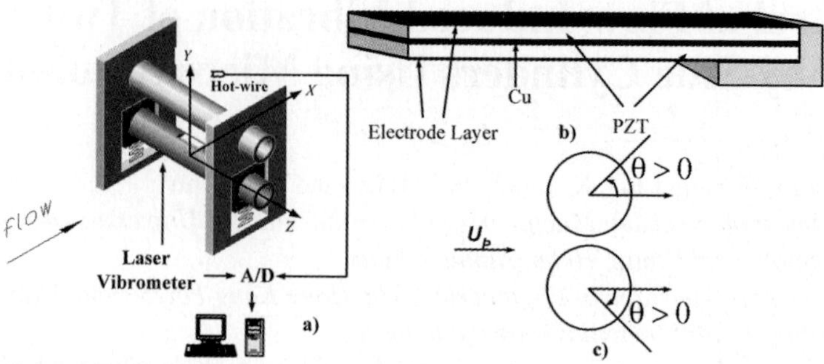

Fig. 1 Schematic illustration of (a) setup, (b) the actuator, (c) definition of angle.

square cross section. Two identical cylinders each with an embedded actuator are placed side-by-side in a cross flow. One of the cylinders is rigidly supported on both ends, while the other is spring-supported through a block on each end. The ratio of the working length of actuator to cylinder is 3% and the cross section of the actuator is 1 mm × 0.5 mm. When driven at voltage of 200 V, the vibrating amplitude of the end is about 40 μm over a wide range of 0–600 Hz. The location of the actuator is defined as in Figure 1c. A Polytec Series 3000 dual beam laser vibrometer was used to measure the vibrating velocity of the elastically supported cylinder and a single hot-wire was used to measure the flow velocities.

3 Results and Discussion

Figure 2 shows the representative control results at different excitation frequencies for three spacing ratios. The experimental Reynolds numbers Re ($= U_\infty d/\nu$) are 9272, 6700 and 3935, and the angles presented are $\theta = 95°$, $97.5°$ and $95°$, respectively, where d is the diameter of cylinder, U_∞ is the free-steam velocity, ν is the kinematic viscosity, and θ is the angular position of the actuator. The reduced natural frequency f_n^*, the reduced shedding frequency f_s^* and the reduced excitation frequency f_e^* hereinafter are normalized by $f^* = fd/U_\infty$. Measurements show that the cylinder's oscillation induced by the actuator's reaction is of the order of 10^{-1} μm in static air. This oscillation could be ignored, compared to the amplitude of the FIV.

For $T/d = 1.2$, the shedding frequency from the two cylinders (which together act as a single bluff body) is $f_s^* = f_{n1}^* = 0.092$. The cylinder's vibration is decreased when f_e^* is within 1.282–2.929, which is equal to $14f_s$–$32f_s$. The cylinder's vibration is reduced particularly when the excitation frequency is around $28f_s$.

For $T/d = 1.8$, due to the biased flow, the vibration is not so significant even at a higher Reynolds number. The shedding frequency (f_{s1}^*) near the upper spring-supported cylinder is 0.123, and is the same as f_{n1}^*. The other shedding frequency

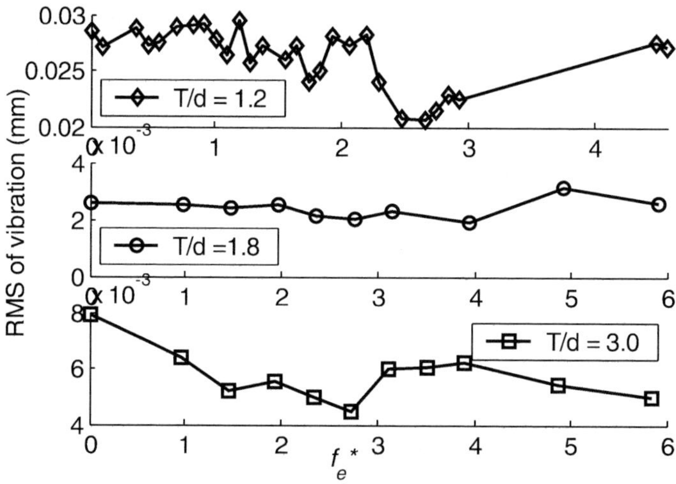

Fig. 2 Root mean square of vibrating displacement at different excitation frequencies.

(f_{s2}^*) is 0.320. However, the biased flow pattern switches intermittently from being directed towards one cylinder to the other, leading to a much weaker control effect.

For $T/d = 3.0$, there exists a symmetric flow with synchronized vortex shedding. The two vortex shedding frequencies are identical (0.210). The vibration decreases significantly at $f_e^* = 2.727$ and 5.843, respectively, which is equivalent to $13f_s^*$–$28f_s^*$.

Figure 3 shows the control is most effective near the separation region for the spacing ratios of 1.2 and 3.0. The power spectral density (PSD) of vibration at the natural frequency (E_{yn}) can represent the vibrating energy. It shows that the effective region ($\theta = 95°$– $100°$) is near the separation point, which agrees roughly with the results by Hsiao et al. [1] and Fujisawa et al. [2]. The small difference may be due to the influence of two side-by-side cylinders. At $\theta = 100°$ for $T/d = 1.2$, the reduction reaches its maximum at 7 dB, equal to an 80% reduction of energy.

The excitation also influences the interaction between the structure and flow field. As shown in Figure 4, when the excitation is at a frequency of $f_e^* = 2.655$ from $t = 6$ s, the two signals are shifted slightly out of phase. During the excitation period (6–38 s), the two signals switch intermittently between being in-phase and out-of-phase, and the vibrating amplitude is suppressed significantly. When the excitation stops at $t = 38$ s, the two signals change back to being in-phase again, and the vibrating amplitude increases again. This indicates that the flow-induced vibration may be controlled by influencing the fluid-structural interaction.

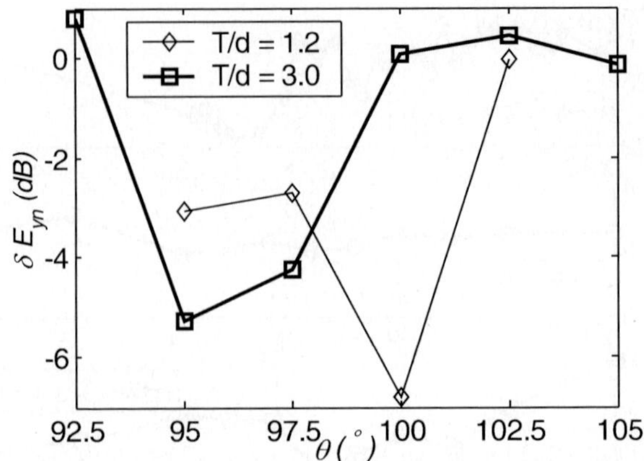

Fig. 3 The reduction of E_{yn} at different excitation locations ($f_e^* = 2.7$).

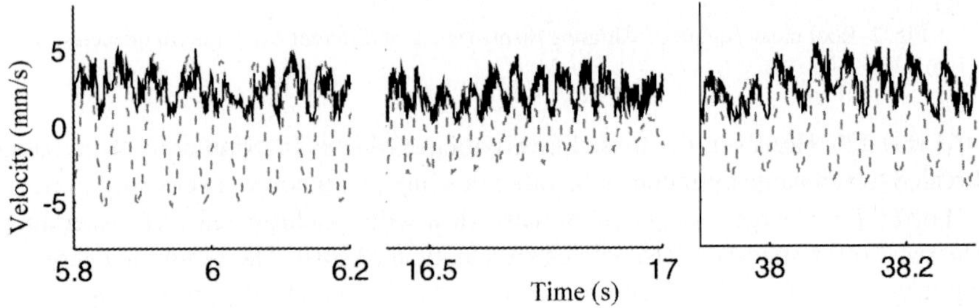

Fig. 4 Comparison of velocity oscillations between vibrating velocity of cylinder and flow velocity (dotted line: vibrating velocity; real line: flow velocity. $T/d = 1.2$, $\theta = 95°$).

4 Conclusions

A type of micro actuator is used in the flow-induced control of two side-by-side cylinders. The experimental results lead to the following conclusions:

1. The flow-induced vibration is significantly for $T/d = 1.2$ and 3.0 decreased when the excitation frequency and the actuator position are optimized. Particularly for $T/d = 1.2$, vibration energy can be decreased up to 80% at $\theta = 100°$ and $f_e^* = 2.7$. There exists a relationship between the effective excitation frequency and the shedding frequency, and the excitation is most effective near the separation region.
2. The fluid-structure interaction can be suppressed greatly by the micro excitation.

Acknowledgements

Support given by the Research Grants Council of the Government of the HKSAR under Grant No. PolyU 5305/03E, and by The Hong Kong Polytechnic University under Central Research Grant No. A-PE53 is gratefully acknowledged.

References

1. Hsiao F.B., Shyu J.Y.: Influence of internal acoustic excitation upon flow passing a circular cylinder. *Journal of Fluids and Structures* **5** (1991) 427–442.
2. Fujisawa N., Takeda G.: Flow control around a circular cylinder by internal acoustic excitation. *Journal of Fluids and Structures* **17** (2003) 903–913.

Improvement of the Jet-Vectoring through the Suppression of a Global Instability

Vincent G. Chapin[1], Nicolas Boulanger[1] and Patrick Chassaing[2]

[1]*Fluid Mechanics Department, ENSICA, 1 place E. Blouin, 31056 Toulouse, France; E-mail: vincent.chapin@ensica.fr*

[2]*Institut National Polytechnique de Toulouse (INPT), 31029 Toulouse Cedex 4, France*

Abstract. The behaviour of the near-field region of a vertical rectangular jet of aspect ratio 4:1 controlled by a rotating cylinder placed on the jet major-axis is investigated experimentally using a new design facility. The objective is to investigate flow control strategies of a rectangular jet based on instability manipulation. It is found experimentally that the controlled jet exhibits a similar behaviour to the one described theoretically and numerically by Hammond and Redekopp [1, 2] on bluff-body wakes with higher control efficiency when the global instability mode is suppressed.

Key words: Flow-vectoring, thrust-vectoring, flow control, global instability, rectangular jet, rotating cylinder.

1 Introduction

We present experimental results about jet-vectoring using a rotating cylinder placed on the jet major-axis near the exit plane ($z/h = 1.3$, $y/h = 0$), as shown in Figure 1. An interpretation of these results is proposed within the theoretical framework of Hammond and Redekopp [1, 2] concerning the relationship between the control efficiency and the presence of global instability modes. Finally, it is concluded that a highly efficient rectangular-jet vectoring can be obtained when the global instability mode present in the controlled flow is suppressed by using a supercritical rotating speed.

J.F. Morrison et al. (eds.), IUTAM Symposium on Flow Control and MEMS, 393–396.
© *2008 Springer. Printed in the Netherlands.*

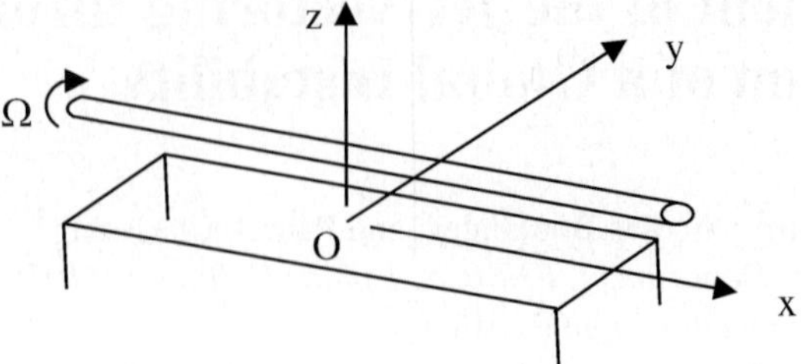

Fig. 1 Jet flow configuration.

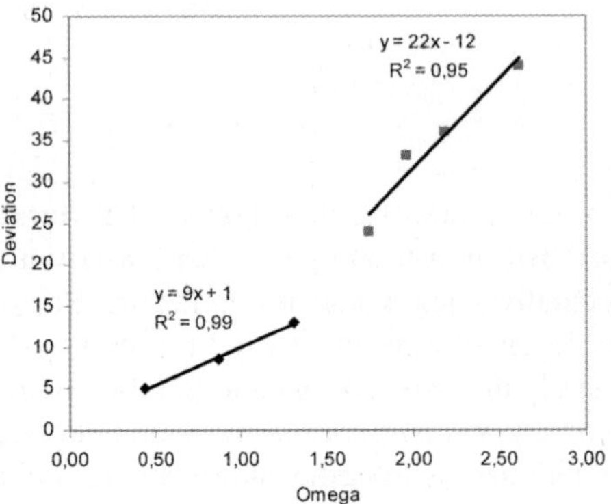

Fig. 2 Mean deviation angle of the jet versus rotating speed of the cylinder Ω^*. Linear regression equations and correlation coefficient are given below and above the critical speed Ω_c^*.

2 Results

The control methodology proposed by Hammond and Redekopp [2] will be tested here. Briefly speaking, they show that when a global instability mode is present in the flow, it is necessary to suppress it to be able to develop an efficient control.

In our experiment, the rectangular jet is known to be convectively unstable, according to the theoretical concepts reviewed by Huerre and Monkewitz [3]. By introducing a cylinder on the jet centreline, an absolute instability region of critical size develops behind the cylinder, which results in the von Karman vortex shedding (a well known global mode). Also, following the theoretical scenario, it should be necessary to suppress this global mode to enhance the flow control effectiveness.

Numerical simulations of a rotating cylinder in a uniform flow, by Mittal and Kumar [5], have shown that the von Karman street disappears for supercritical rotating speed ($\Omega^* = \Omega d / 2U_\infty > 2$). Since, in the present situation, the cylinder is located in a jet of centreline velocity U_j, various rotating speed have been applied to investigate the existence of a threshold Ω_c^* specific to our flow.

The response of the controlled jet to the rotating speed $\Omega^* = \Omega d / 2U_j$ is given in Figure 2, in terms of the mean deviation angle. A critical rotating speed may be

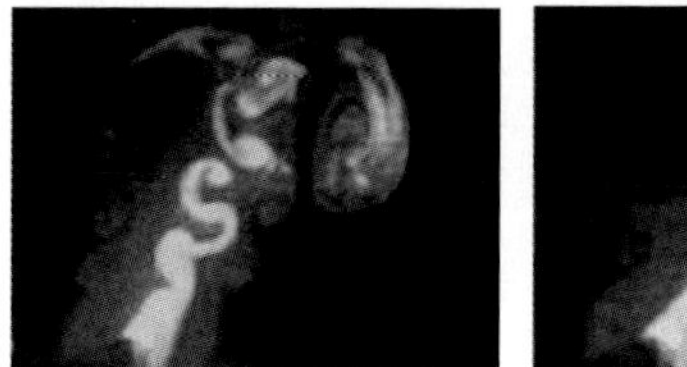

Fig. 3 Visualizations of the flow at $\Omega^* = 1.4$ (left) and $\Omega^* = 1.64$ (right).

identified around $\Omega_c^* = 1.5$, where the slope of the curve changes by more than a factor of two. This slope, which represents the jet-vectoring efficiency, is more than two times higher for supercritical values of Ω^*.

Given this result, we have investigated the flow in greater detail to see whether this critical behaviour could be associated with the disappearance of the von Karman global mode behind the cylinder when its rotating speed is increased. Flow visualizations given in Figure 3, obtained with a high speed CCD camera by illuminating an oil droplets seeded flow with a thin laser sheet, clearly illustrate this point. In the left view, the von Karman anti-symmetric vortex shedding is clearly seen. In the right view, it has disappeared, being replaced by a chaotic flow. For more details about both flow regimes, two film sequences may be viewed at the following internet addresses: http://www.youtube.com/v/0ar6iRGmzbc (subcritical regime) and http://www.youtube.com/v/7eseFoPMvsc (supercritical regime).

Spectral analyses of the hot-wire measurements of the longitudinal component of the instantaneous velocity in the jet flow behind the cylinder, located along the jet centreline at $z/h = 0.33$, are given in Figure 4. Below the critical speed, a unique sharp peak dominates the spectrum, a typical feature of a self-exited oscillator [6]. Above the critical speed, the spectrum is typical of a noise amplifier with broadband and lower peaks at a dominant frequency close to the one found in the free jet shear layers and identified as a jet cavity resonance.

The conjunction of theses observations about flow deviation angle, flow visualizations and velocity spectra indicates that the suppression of the global instability results in a more efficient jet vectoring for supercritical values of Ω^*.

3 Conclusion

In a rectangular jet of aspect ratio 4:1 controlled by a rotating cylinder, it is shown experimentally that the suppression of the global instability mode is correlated to an increase of the jet-vectoring efficiency. This first experimental evidence supports the theoretical scenario proposed by Hammond and Redekopp [1, 2] on wakes and Lim and Redekopp [4] on jets.

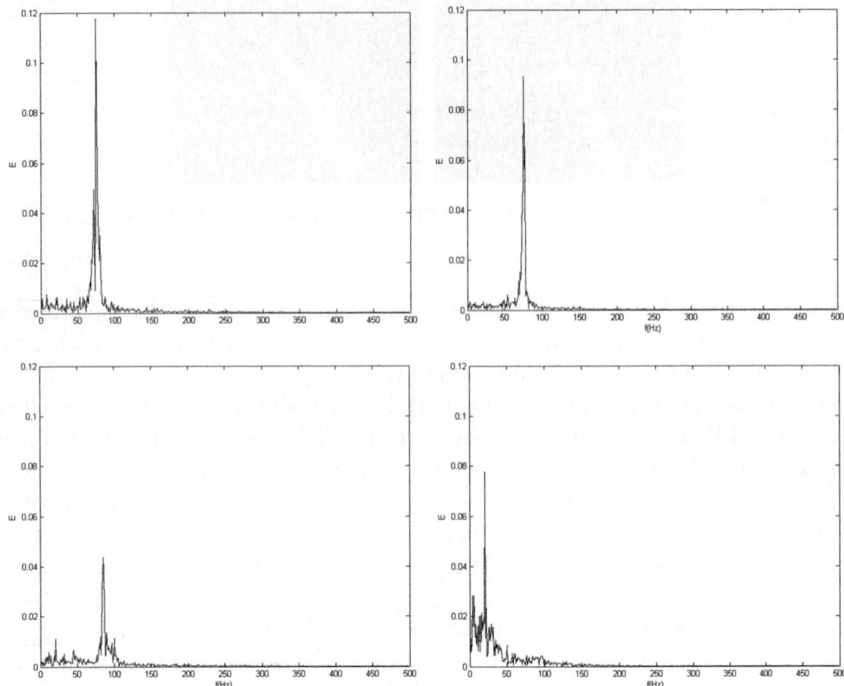

Fig. 4 Velocity spectra (a) $\Omega^* = 0$, (b) $\Omega^* = 0.14$, (c) $\Omega^* = 0.8$, (d) $\Omega^* = 3.1$.

Acknowledgment

The authors thank G. Toulouse for his fine actuator system design.

References

1. D.A. Hammond and L.G. Redekopp, Global dynamics and aerodynamic flow vectoring of wakes, *J. Fluid Mech.* **331**, 1997, 231-0260.
2. D.A. Hammond and L.G. Redekopp, Global dynamics and aerodynamic flow vectoring of wakes, *J. Fluid Mech.* **338**, 1997, 231–248.
3. P. Huerre and P.A. Monkewitz, Local and global instabilities in spatially developing flows, *Ann. Rev. Fluid Mech.* **22**, 1990, 473–537.
4. D.W. Lim and L.G. Redekopp, Aerodynamic flow-vectoring of a planar jet in a co-flowing stream, *J. Fluid Mech.* **450**, 2002, 343–375.
5. S. Mittal and B. Kumar, Flow past a rotating cylinder, *J. Fluid Mech.* **476**, 2003, 303.
6. P.A. Monkewitz and D.W. Bechert, Self-excited oscillations and mixing in a hot jet, *Phys. Fluids* **31**(9), September 1988.

PASSIVE CONTROL

Experimental Optimization of Bionic Dimpled Surfaces on Axisymmetric Bluff Bodies for Drag Reduction

Chengchun Zhang, Luquan Ren, Zhiwu Han and Qingping Liu
Key Laboratory for Terrain-Machine Bionics Engineering, Ministry of Education, Jilin University, Changchun 130022, China; E-mail: lqren@jlu.edu.cn

Abstract. The reduction of drag generated by axisymmetrica bluff bodies resulting from the control of the boundary layer with dimpled surfaces was investigated. A central composite design with 3 factors and 5 levels for each factor was used to optimize the parameters of the dimpled surfaces. Wind tunnel tests with a Mach number of 2.51 and a Reynolds number of 1.88×10^6 based on the maximum diameter of the model indicate that the dimples on the rearward configuration can reduce the viscous forebody drag by 4.98%, the base drag by 2.69%, and the total drag by 2.98%, respectively. By using the dimpled surface optimized by the quadratic regression equation, the total drag can be reduced by 3.81%.

Key words: Boundary layer, drag reduction, wind tunnel test, central composite design.

1 Introduction

The study of passive control methods of drag reduction has received much attention in recent years. The drag reduction of many types of riblet surfaces has been confirmed by Walsh [1], Bechert [2], and Choi [3]. Observing some biological surfaces, many investigators have realized that dimpled surfaces can also be used to control turbulent boundary layers and achieve a reduction of drag. Bearman and Harvey [4] found that, for Reynolds number $Re = 4 \times 10^4$ to 3×10^5, a dimpled circular cylinder has a lower drag coefficient than a smooth cylinder. It was measured that, at transonic speeds, dimples can reduce the total drag of bodies of revolution by 3% as compared to the smooth counterpart [5].

In our study, a central composite design was performed to optimize the dimpled surface configuration to reduce the drag exerted on axisymmetric bluff bodies. Through this analysis, the optimum dimpled surface was obtained.

J.F. Morrison et al. (eds.), IUTAM Symposium on Flow Control and MEMS, 399–403.
© 2008 *Springer. Printed in the Netherlands.*

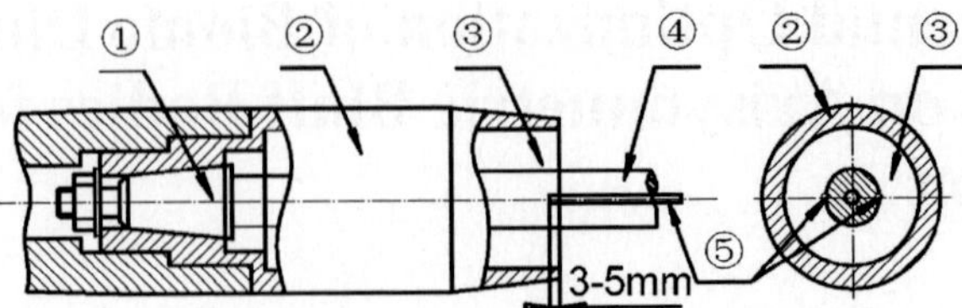

Fig. 1 Schematic of measurement system. ① is a strain-gauge balance; ② is the axisymmetric bluff body; ③ is a rear-facing cavity; ④ is the model support; ⑤ is a pressure tap.

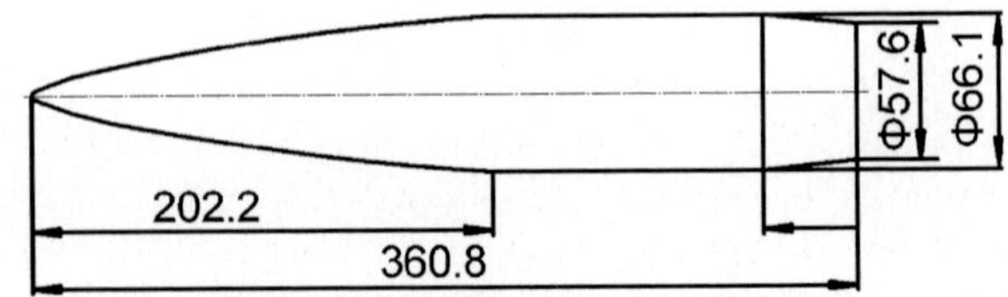

Fig. 2 Main dimensions of the bluff body.

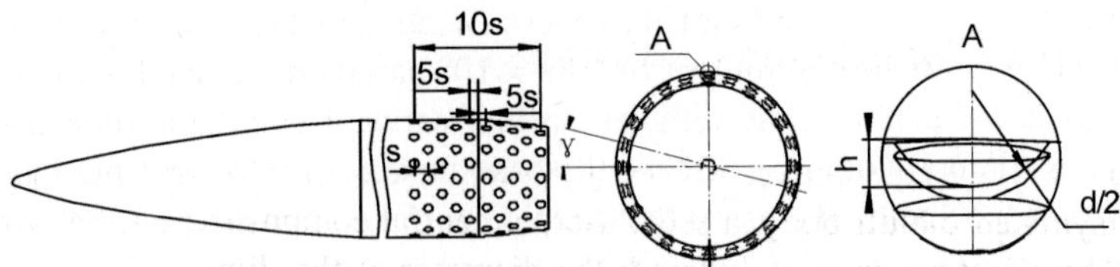

Fig. 3 Dimensions and parameters of the dimpled surface. d is the diameter of the dimples (the diameter of the spherical cutter); h is the depth of the dimples; s is the space between two adjacent dimples in the axial direction; γ, the central angle, is $15°$.

2 Experimental Apparatus and Testing Method

The tests were performed in a transonic, intermittent semi-return-type wind tunnel with a test section measuring $0.6 \times 0.6 \times 1.575$ m. The testing system, as shown in Figure 1, is controlled by a central computer. In Figure 1, the aerodynamic force coefficients and parameters of the flow are measured by a high accuracy strain-gauge balance built into the test model, and all tests were conducted using the same high-precision technique.

The maximum diameter and length of the model were limited by the wind tunnel blokackage ratio $\varepsilon_{\text{bloc}}$, the test Mach number, and the height of working section of the tunnel. The maximum diameter and length of the models are shown in Figure 2.

Figure 3 shows the positions and dimensions of the dimples. The models were made of steel and were produced using a numerical control machine. The grade of form and position tolerances was 7, and the values of the surface roughnesses were not more than 0.8 μm.

Table 1 Coded and natural values of the factors.

$X_j(z_j)$	$z_1(d)$/mm	$z_2(h)$/mm	$z_3(s)$/mm
$r(z_{2j})$	2	0.7	8
$1(z_{0j}+\Delta_j)$	1.72	0.62	7.2
$0(z_{0j})$	1.3	0.5	6
$-1(z_{0j}-\Delta_j)$	0.88	0.38	4.8
$-r(z_{1j})$	0.6	0.3	4
$\Delta_j = \dfrac{z_{2j}-z_{1j}}{2r}$	0.416	0.119	1.189

In the present tests, the aerodynamic coefficients were measured at angles of attack of $-2°$, $-1°$, $0°$, $1°$, $2°$, $3°$, $3.5°$, $4°$, $4.5°$ and $5°$, respectively. Results were corrected for the interference of balance system and deadweight of the model. The test Mach number was fixed at 2.51, and the freestream Reynolds number based on the maximum diameter of the model was $R_D = 1.88 \times 10^6$.

3 Design of Experiments and Results

To optimize of the parameters of the dimpled surface on the *rearward* section of the axisymmetric bluff body, a second-order central composite design (CCD) was used. Three factors were considered: the diameter of the dimples d, the depth of the dimples h and the space of the two adjacent dimples in axial direction s. Each variable assumed five coded levels $(-r,-1,0,1,r)$ to carry out the optimization, where $r = \sqrt[4]{n_c} = 1.682$, which can ensure the CCD is rotatable. The nature variables were standardized by the following formulas:

$$x_1 = \frac{z_1 - 1.3}{0.416}, \quad x_2 = \frac{z_2 - 0.5}{0.119}, \quad \text{and} \quad x_3 = \frac{z_3 - 6}{1.189}, \tag{1}$$

where x_1, x_2, x_3 and z_1, z_2, z_3 are the coded variables and the natural variables respectively. Table 1 shows the coded and natural values of the factors.

The number of trials N were based on the number of design factors $(k = 3)$ as follows:

$$N = n_c + n_r + n_0 = 18,$$

where $n_c = 2k$ is the number of cube points; $n_r = 2k$ is the number of star points; and $n_0 = 4$ is the number of center points in the design which are necessary to estimate within the experimental uncertainty.

The wind tunnel tests, carried out according to the scheme of CCD, indicate that the dimples on the rear section can reduce the viscous forebody drag and the base drag. At $\alpha = -2°$, the dimples ($d = 0.6$ mm, $h = 0.5$ mm, and $s = 6$ mm) reduce the viscous forebody drag and the total drag by 4.98% and 2.98% respectively, and at $\alpha = 3.5°$, the base drag is reduced by 2.69% with dimpled surface parameters $d = 1.72$ mm, $h = 0.62$ mm, and $s = 4.8$mm.

The program STATISTICA [6] was used to perform a regression analysis of the total drag reduction rate in order to estimate the coefficients of the regression equation. The effects of p-values higher than 0.05 are insignificant at the 95% confidence level and were discarded. Thus, a second order equation with the coefficient of determination $R^2 = 0.7836$, was obtained. Then, after discarding the insignificant effects at the significance level 0.05, the regression equation is given by

$$\hat{R}_{AT}(\%) = 0.9496 + 0.2388x_1 + 0.1636x_2 + 0.2258x_1^2 + 0.1657x_2^2 +$$

$$+ 0.1873x_3^2 - 0.1x_2 * x_3. \tag{2}$$

After substituting x_i by the natural variables z_i obtained from Equation (1), and restoring z_i to the parameters of the dimpled surface, a regression equation was obtained. Considering practical engineering situations and the applicability of the regression model, the optimizing dimpled surface must satisfy certain constrains, so the secondorder programming model can be written as

$$\text{Max} \, R_{AT}(\%) = 11.5339 - 2.8184d - 14.5654h - 1.9439s + 1.3048d^2 +$$

$$+ 11.7013h^2 + 0.1325s^2 + 0.7067h * s$$

$$\text{S.t. } g_1(d,h) = \frac{h}{d} \leq 0.5,$$

$$g_2(s,d) = s - d \geq 0, 0.4 \leq d \leq 2.1, 0.2 \leq h \leq 0.8, 4 \leq s \leq 8. \tag{3}$$

An optimization of Equation (3) yielded the best combination of parameters: $d = 2.1$ mm, $h = 0.8$ mm, and $s = 4$ mm, from which the peak value of $R_{AT} = 3.81\%$ as obtained. As the coefficient of determination $R^2 = 0.7836$, the variables cannot explain the data fairly well, and other variables should be considered in the further investigation.

4 Conclusions

The dimples on the rear section of a bluff body can reduce not only the viscous forebody drag but also the base drag effectively. Applying the model of second-order programming obtained by the CCD, the total drag reduction rate R_{AT} was maximized to 3.81% with the dimpled surface with $d = 2.1$ mm, $h = 0.8$ mm, and $s = 4$ mm.

Acknowledgments

The authors are grateful for the financial support provided by the National Key Grant Program of Basic (Grant No.2002CCA01200), the National High Technology Research and Development Program of China (863 Program) (Grant No.2003AA305080), the Natural Science Foundation of Jilin Province (No.20040703-1) and the Specialized Research Fund for the Doctoral Program of Higher Education (No. 20050183064). The authors are also grateful to the correlative technicians of the 701st Institute of China Aerospace Science & Technology Corporation for help in conducting experiments.

References

1. Walsh, M.J., Riblets as a viscous drag reduction technique, *AIAA J.* **21**, 1983, 485–486.
2. Bechert, D.W., Bartenwerfer, M., Hoppe, G. and Reif, W-E., Drag reduction mechanisms derived from shark skin, Paper presented at the 15th ICAS Congress, London, 86-1.8.3, distributed as *AIAA J.*, 1986, 1044–1068.
3. Choi, K.S., Near-wall structures of a turbulent boundary layer with riblets, *J. Fluid Mech.* **208**, 1989, 417–458.
4. Bearman, P.W. and Harvey. J.K., Control of circular cylinder flow by the use of dimples, *AIAA J.* **31**, 1993, 1753–1756.
5. Ren, L.Q., Zhang, C.C. and Tian, L.M., Experiment study on Drag Reduction for bodies of revolution using bionic non-smoothness, *J. Jilin Univ. (Eng. and Techn. Ed.)* **35**, 2005, 431–436.
6. STATISTICA (Data analysis software system) Version 6.0, StatSoft. Inc., Tulsa, 2001.

Flow Regularisation and Drag Reduction around Blunt Bodies Using Porous Devices

C.-H. Bruneau[1], I. Mortazavi[1] and P. Gilliéron[2]

[1]*CNRS UMR 5466, INRIA FUTURS Equipe MC2, Université Bordeaux 1, 351 Cours de la Libération, 33405 Talence, France;*
E-mail: bruneau@math.u-bordeau.fr
[2]*Technocentre Renault, Direction de la Recherche, Guyancourt, France*

Abstract. Porous layers are added on blunt bodies to change the shear forces and consequently to reduce the disorganisation of the flow or to reduce the drag coefficient. Numerical simulations of two dimensional flows around a riser pipe give a drastic increase in the regularity of the flow when a porous sheath is added. Also, setting some porous devices on a simplified ground vehicle geometry can reduce the pressure drag with an appropriate choice of the location. The pressure gradient in the near wake can be reduced by 67% and so a drag reduction of up to 45% can be achieved.

Key words: Flow regularisation, drag reduction, blunt bodies, porous devices.

1 Modelling and Numerical Simulation

Various active or passive control procedures have been used for decades to control the flows around blunt bodies [3], particularly in off-shore applications to reduce the vortex induced vibrations [6] and in the automotive industry to reduce the drag coefficient [4, 5]. Here we propose the addition of a porous layer between the fluid and the solid bodies in order to change the shear forces.

Using the penalised Navier–Stokes equations, which include an additional term U/K in the momentum equation, the flow around an obstacle and inside a porous medium is modeled (see [1, 2] for more details). The non-dimensional coefficient of permeability of the medium, K, is set equal to 10^{16} in the fluid, to 10^{-8} in the solid bodies and to 10^{-1} in the porous layers. It was shown by different approaches that solving these equations is equivalent to solving the Navier–Stokes equations in the fluid with a Fourier-like boundary condition instead of the no-slip boundary condition on the porous interfaces. Consequently, adding a porous layer between the solid body and the fluid induces a reduction of the shear effects in the boundary layer and thus allows the flow to be controlled. In addition, as the pressure is computed inside

J.F. Morrison et al. (eds.), IUTAM Symposium on Flow Control and MEMS, 405–408.
© 2008 *Springer. Printed in the Netherlands.*

Fig. 1 Vorticity field of the flow around the solid pipe (left) and around the pipe with a porous sheath (right) for the same time at $Re = 30000$.

the solid body, we can compute the drag and lift forces, F_D and F_L, by integrating the penalisation term on the volume of the body:

$$F_D = \int_{\text{body}} \frac{u}{K} dx, \quad F_L = \int_{\text{body}} \frac{v}{K} dx,$$

where the body includes the porous layers. To quantify the effect of the control we shall compare the static pressure coefficient, the drag coefficient, the root mean-square of the lift coeffcient CL_{rms} and the enstrophy.

Appropriate boundary conditions are applied to the unsteady equations. The time discretization is achieved by means of the second-order Gear scheme with explicit treatment of the convection term. The space discretization uses a second-order centred scheme for all the linear terms and a third-order upwind scheme for convection terms. The efficiency of the solution is improved by applying a multigrid procedure using a cell-by-cell relaxation smoother.

2 Results of the Passive Control

The first numerical test presented in this work is intended to establish the effects of the addition of a porous sheath upon the flow around a riser pipe. This point is crucial in the off-shore industry as the vortex induced vibrations can severely damage the pipes and reduce their life. The idea is to install a porous sheath 0.1 diameters thick around the pipe in order to modify the vortex shedding. The result, as shown in Figure 1, is a drastic increase in the regularity of the chaotic transitional flow, which becomes periodic as a Karman street is recovered. The CL_{rms} decreases from 0.293 for the flow without control to 0.081 whith the passive control.

The second numerical test concerns the flow around an Ahmed body on a road surface. To begin, we consider a square-backed, symmetric Ahmed body (without a rear window). We study the effects of applying porous elements at various locations under the surface of the body. In the previous test a porous sheath is added around the pipe; here, the porous layers are set in such a way that the surface profile of the Ahmed body is unaltered; however, there were discontinuities of the media at the ends of the elements. The flow around the body was highly sensitive to the location of the porous elements. For instance, putting porous elements only on the front of

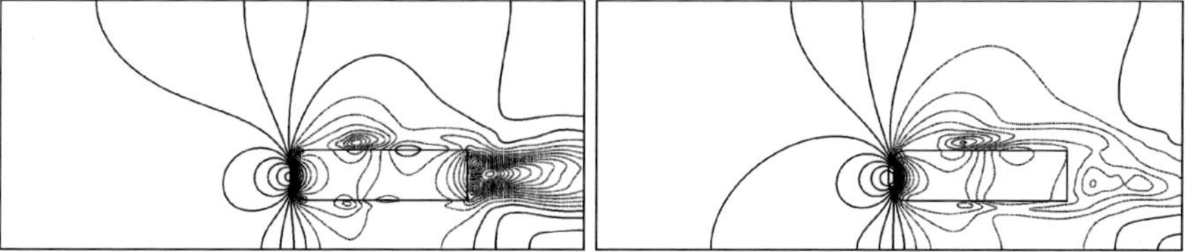

Fig. 2 Pressure isolines for the flow around the square-backed Ahmed body on top of a road at $Re = 30000$ without control (left) or with control (right).

Table 1 Asymptotic values of CL_{rms}, mean values of the enstrophy and the drag coefficient and minimum of the pressure p_{min} in the wake for the square-backed Ahmed body on a road surface at $Re = 30000$.

	Without control	With control
CL_{rms}	0.517	0.352 (–32%)
Enstrophy	827	533 (–36%)
Drag coefficient	0.526	0.354 (–33%)
Pressure p_{min}	–1.636	–0.510 (–69%)

the body can induce jets orthogonal to the body at the ends of the elements due to the presence of the solid body. In that case the aerodynamic performance was worsened as these jets increased the shedding and the transverse size of the wake. On the other hand, porous layers located on both lateral areas induce low horizontal jets on the rear-facing side of the body. The velocity decreased and the negative mean pressure p_{min} increased significantly as shown in Figure 2 and in Table 1; consequently the drag coefficient is drastically reduced. As the contribution of the front section is nearly identical with or without control, a 45% reduction of the total drag coefficient is achieved. There is also a strong reduction of the CL_{rms} and of the enstrophy, as seen in Table 1.

Now we extend the passive control to the 25-degree rear window Ahmed body on a road surface. If we use lateral porous elements as above with the one on top of the body ending at the rear window, the horizontal jet makes a $25°$ angle with the rear window and so the efficiency of the control is altered. Nevertheless, the pressure gradient in the wake is lower (as illustrated in Figure 3) and a 10% drag reduction is obtained. In addition, the CL_{rms} and the enstrophy are decreased in almost the same way as with the square-backed body.

3 Conclusions

An efficient passive control of the flow around a blunt body can be achieved using a porous interface between the uid and the body. The shear forces are reduced and consequently a strong decrease in the disorganisation of the flow is obtained.

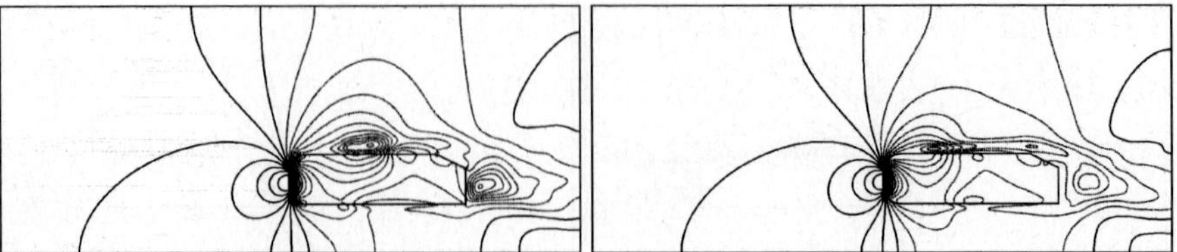

Fig. 3 Pressure isolines for the flow around the 25-degree rear window Ahmed body on a road surface at $Re = 30000$ without control (left) or with control (right).

In addition, by carefully selecting the location of the porous devices, a significant reduction of the drag coefficient of the Ahmed body can be obtained.

References

1. Ph. Angot, Ch.-H. Bruneau and P. Fabrie, A penalization method to take into account obstacles in incompressible viscous flows, *Numer. Math.* **81**, 1999.
2. C.-H. Bruneau and I. Mortazavi, Passive control of flow around a square cylinder using porous media, *Int. J. Num. Meth. Fluids* **46**, 2004.
3. H.E. Fiedler and H.H. Fernholz, On management and control of turbulent shear flows, *Program Aerospace Sci.* **27**, 1990.
4. P. Gilliéron and F. Chometon, Modelling of stationary three-dimensional detached airflows around an Ahmed Reference Body, *ESAIM Proc.* **7**, 1999.
5. P. Gilliéron, F. Chometon and J. Laurent, Analysis of hysteresis and phase shifting phenomena in unsteady three-dimensional wakes, *Exp. Fluids* **35**, 2003.
6. C.H.K. Williamson and R. Govardhan, Vortex-induced vibrations, *Annu. Rev. Fluid Mech.* **36**, 2004.

The Effects of Aspect Ratio and End Condition on the Control of Free Shear Layers Development and Force Coefficients for Flow Past Four Cylinders in the In-line Square Configuration

Kit Lam[1] and Lin Zou[2]

[1]*Department of Mechanical Engineering, The Hong Kong Polytechnic University, Hung Hom, Kowloon, Hong Kong; E-mail: mmklam@polyu.edu.hk*
[2]*School of Mechanical and Electronic Engineering, Wuhan University of Technology, Wuhan 430070, China; E-mail: mmlzou@whut.edu.cn*

Abstract. The effects of aspect ratio and end condition on the control of the development of free shear layers and the force coefficients of cylinders for flows around four cylinders at critical spacing ratio $L/D = 3.5$ in the in-line square configuration have been investigated numerically. The study demonstrated that the aspect ratio and end condition of the four cylinders produce a strong influence on the free shear layer development from the upstream cylinder and hence the pressure fields and force characteristics of the cylinders. The mean pressure and fluctuating pressure increase towards the mid-span of the cylinders. The development of free shear layers, the mean and fluctuating pressure as well as the force coefficients are critically affected by the cylinder aspect ratio in the range of H/D from 15 to 16.

Key words: Four cylinders, flow pattern, fluctuating pressure, three-dimensionality.

1 Introduction

Flows around circular cylinder arrays are frequently encountered in engineering applications. For flows around four cylinders in the in-line square configuration, some experimental and numerical studies have been carried out. For example, Farrant et al. [1] captured numerically the two-dimensional flow characteristics and associated interactive forces using a cell boundary element method. Sayers [2] conducted experiments in the spacing ratio range of 1.5 to 5. Lam et al. [3–5] carried out extensive experimental investigations for different spacing ratios and different orientations at different Reynolds numbers. In this paper, several three-dimensional numerical simulations for laminar flows around four circular cylinders in the in-line square configuration with aspect ratios $H/D = 6$, 8, 12, 14, 15 and 16 have been carried out using the finite volume method at Re $= 200$ in order to investigate the

J.F. Morrison et al. (eds.), IUTAM Symposium on Flow Control and MEMS, 409–413.
© 2008 *Springer. Printed in the Netherlands.*

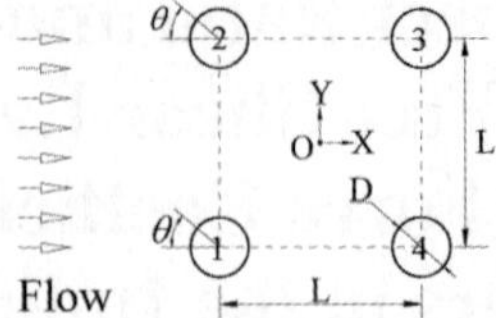

Fig. 1 Schematic diagram of the configuration for the four cylinders in in-line square configuration.

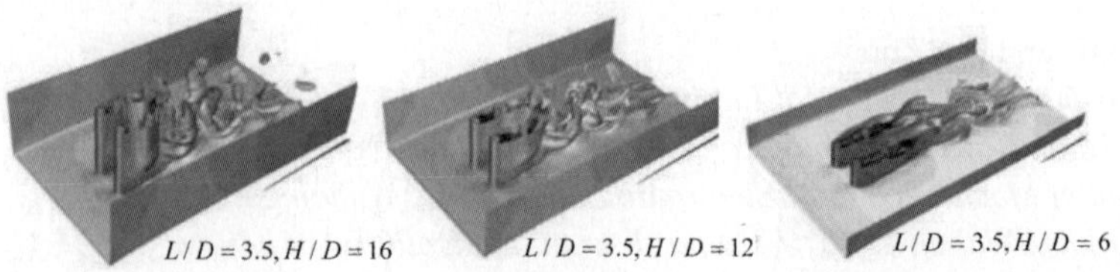

Fig. 2 The velocity fields over the iso-vorticity surfaces for four cylinders at different H/D.

effects of aspect ratio and end condition on the control of the development of free shear layers and the force coefficients of cylinders at the critical spacing ratio of $L/D = 3.5$.

2 Numerical Method and Models

The finite volume method with an unstructured hexahedral mesh is employed to solve the incompressible unsteady Navier–Stokes equations. As shown in Figure 1, the origin of the coordinate system is located at the centre point of the four cylinder arrangement. (X, Y, Z) denotes the coordinates along the streamwise direction, the transverse direction and the spanwise direction of the cylinder, respectively. Here, L is the centre-to-centre distance between cylinders. The aspect ratio H/D varies from 6 to 16 where H is the cylinder height. Only one half of the height of the cylinders is simulated in order to save the computational time.

3 Results and Discussion

In the present study, the laminar flows around four circular cylinders with different aspect ratios in the in-line square configuration are simulated at Re = 200. The detailed three-dimensional vortex structures are presented, together with the force and pressure coefficients.

As shown in Figure 2, for $H/D = 6$, the effects of the end wall will stabilize the free shear layers so that the downstream cylinders are completely covered by

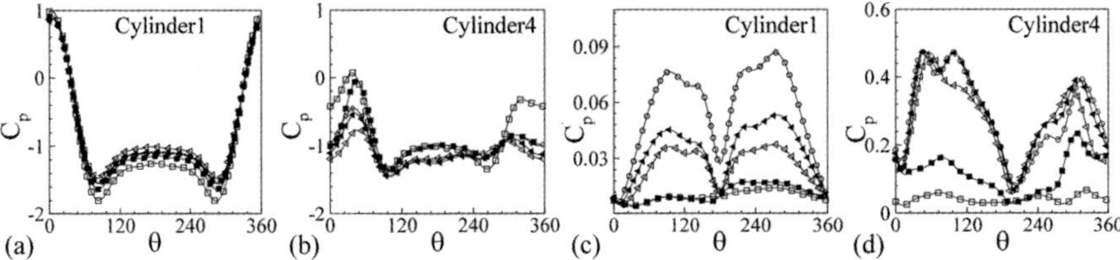

Fig. 3 The mean pressure (a, b) and fluctuating pressure (c, d) coefficient distributions at different spanwise positions at $H/D = 16$ and Re $= 200$. —□— $z/H = 0.0625$, —■— $z/H = 0.125$, —◄— $z/H = 0.25$, —◄— $z/H = 0.375$, —○— $z/H = 0.5$.

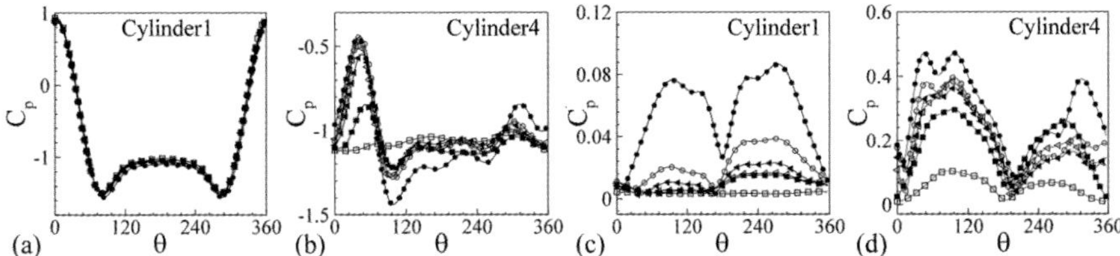

Fig. 4 The mean pressure (a, b) and fluctuating pressure(c, d) coefficient distributions at the mid-span with the different aspect ratios at $L/D = 3.5$ and Re $= 200$. —□— $H/D = 6$, —■— $H/D = 8$, —◄— $H/D = 12$, —◄— $H/D = 14$, —○— $H/D = 15$, —●— $H/D = 16$.

the free shear layers of the corresponding upstream cylinders and hence exhibit a shielding flow pattern over the entire span. The shear layers roll up into vortices far downstream. For $H/D > 8$, however, the effects of the end wall reduce significantly. From $z/H = 0.125$, distinct vortex shedding can be detected immediately behind the downstream cylinders. The coexistence of different flow patterns at different spanwise locations at the same aspect ratio was observed.

Figure 3 shows that the mean pressure coefficient distributions of the upstream cylinders in the range of $50° \leq \theta \leq 300°$ and the whole downstream cylinders are quite sensitive to variation in z/H. The fluctuating pressure of both the upstream and downstream cylinders also exhibited a strong spanwise dependency. The peak of the r.m.s. fluctuating pressure has an increasing trend towards the mid-span of the cylinders for all aspect ratios and displays a distinct jump between $z/H = 0.125$ and 0.25. This shows a flow pattern transformation due to the effect of the end condition.

Figure 4 shows that H/D has an important effect on the mean pressure coefficients of the downstream cylinders and the fluctuating pressure coefficients of both the upstream and downstream cylinders. The peaks of the fluctuating pressure coefficients increase with increasing aspect ratio and jump abruptly when H/D increases from 15 to 16. It suggests that the flow pattern transformation occurs with the variation of the aspect ratio. The development of free shear layers, the mean and fluctuating pressure as well as the force coefficients are critically affected by the cylinder aspect ratio in the range of H/D from 15 to 16. Figure 5 shows that the three-dimensional numerical simulations are in excellent agreement with the experimental results. The flow pattern between the four cylinders transforms from the

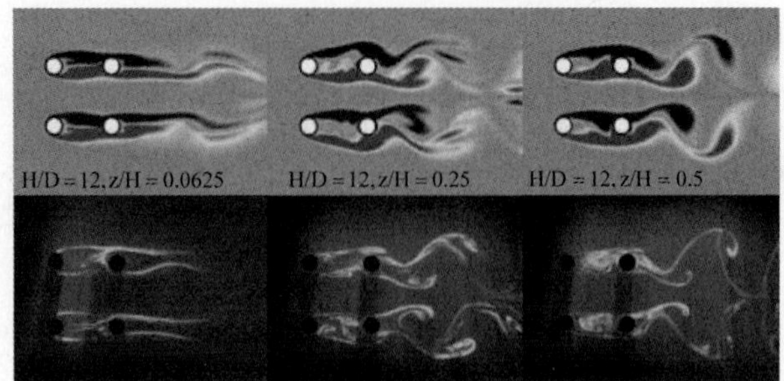

Fig. 5 Comparison of computation and the corresponding flow visualization results at three different spanwise positions for flows around the four cylinders in the in-line configuration.

shielding flow pattern to the reattached flow pattern along the spanwise position of the cylinders.

4 Conclusions

In this paper, the flows around four circular cylinders in the in-line square configuration with different aspect ratios have been simulated three-dimensionally. The results illustrate that the mean pressure and fluctuating pressure coefficient of the cylinders show a strong spanwise dependency on the cylinder aspect ratio H/D. The transformation of the flow pattern has a great effect on the mean pressure coefficients of the downstream cylinders, but a smaller effect on the upstream cylinders. At high aspect ratio, due to the highly three-dimensional nature of the flow resulting from the end effect of the wall, different flow patterns coexist at different spanwise stations on the cylinder at the critical spacing ratio of $L/D = 3.5$.

Acknowledgment

The authors wish to thank the Research Grants Council of the Hong Kong Special Administrative Region, China, for its support through Grant No. PolyU 5299/03E.

References

1. Farrant, T., Tan, M. and Price, W.G., A cell boundary element method applied to laminar vortex-shedding from arrays of cylinders in various arrangements, *Journal of Fluids and Structures* **14**, 2000, 375–402.
2. Sayers, A.T., Vortex shedding from groups of three and four equispaced cylinders situated in cross-flow, *Journal of Wind Engineering and Industrial Aerodynamics* **34**, 1990, 213–221.

3. Lam, K. and Lo, S.C., A visualization study of cross-flow around four cylinders in a square configuration, *Journal of Fluids and Structures* **6**, 1992, 109–131.
4. Lam, K. and Fang, X., The effect of interference of four equispaced cylinders in cross flow on pressure and force coefficients, *Journal of Fluids and Structures* **9**, 1995, 195–214.
5. Lam, K., Li, J.Y., Chan, K.T. and So, R.M.C., Flow pattern and velocity field distribution of cross-flow around four cylinders in a square configuration at low Reynolds number, *Journal of Fluids and Structures* **17**, 2003, 665–679.

Numerical Simulation on the Control of Drag Force and Vortex Formation by Different Wavy (Varicose) Cylinders

Kit Lam and Yufeng Lin

Department of Mechanical Engineering, The Hong Kong Polytechnic Univeristy, Hung Hom, Kowloon, Hong Kong; E-mail: {mmklam, mmyflin}@polyu.edu.hk

Abstract. Large eddy simulations of turbulent flow around wavy cylinders are performed at Re = 3000. The mean pressure distribution and mean streamwise velocity in the near-wake region are calculated and compared with those of a circular cylinder. The three-dimensional near-wake structures behind wavy cylinders were captured. It was found that due to the long vortex formation length resulting from the three-dimensional vortex sheet of the wavy cylinder, the mean drag coefficients of the wavy cylinders are less than that of a corresponding circular cylinder. A reduction of mean drag coefficient of up to 16% is obtained. Also, the fluctuating lift coefficients of the wavy cylinders are weakened significantly.

Key words: Wavy cylinders, drag reduction, turbulent flow, large eddy simulation.

1 Introduction

Flow induced vibration (FIV) around bluff bodies is a common problem in the general area of fluid-structure interaction. How to control the development of the free shear layers and hence control and even suppress the FIV of a cylinder is of great importance in engineering applications. Based on the concept of geometric disturbance, a circular cylinder with a cross section which varied along the spanwise direction (a "wavy" cylinder) has been studied. Earlier, Ahmed [1, 2] experimentally investigated the surface-pressure distributions of wavy cylinders. Recently, Lam [3, 4] has carried out extensive experimental investigations and found that the drag reduction of wavy cylinders can be up to 22% at Re = 10,000. Zhang [5] also investigated the three-dimensional vortex structures of a wavy cylinder by using technique of particle image velocimetry (PIV). The results show that the three-dimensional free shear layers developed from different wavy cylinders will give rise to complex three-dimensional vortex formations which can lead to the reduction of the cylinder drag coefficient and also the suppression of the fluctuating lift. In the present

J.F. Morrison et al. (eds.), IUTAM Symposium on Flow Control and MEMS, 415–419.
© 2008 *Springer. Printed in the Netherlands.*

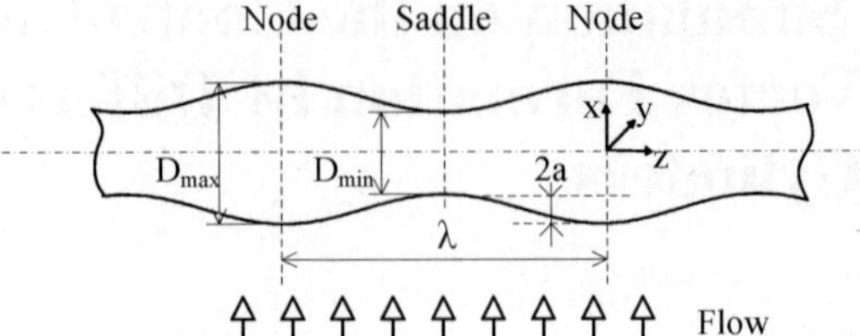

Fig. 1 Geometry of wavy cylinder models.

study, numerical investigations of the three-dimensional near-wake flow structures behind wavy cylinders have been carried out. The relationships between spanwise wavelength, wave amplitude and force coefficients are studied.

2 Numerical Method and Models

The turbulent model of large eddy simulation (LES) is adopted in the present simulation. The finite volume method (FVM) applied to unstructured grids was employed. The geometry of the wavy cylinders can be described by the equation

$$D = D_m + 2a\cos(2\pi z/\lambda).$$

Figure 1 shows the schematic diagram of the wavy cylinder, where D is the local diameter of the wavy cylinder along the spanwise direction, the mean diameter is defined by

$$D_m = (D_{max} + D_{min})/2,$$

a is the amplitude of the wavy surface, λ is the wavelength along the spanwise direction and z is the spanwise location as shown in Figure 1.

3 Results and Discussion

In the present study, three typical wavy cylinder models with different combinations of amplitude values ($a/D_m = 0.09$ and 0.15) and wavelength values ($\lambda/D_m = 1.52$ and 2.27) and a corresponding circular cylinder with diameter D_m were simulated at the same Reynolds number Re $= 3000$.

The mean pressure coefficient distributions around the cylinder and the mean streamwise velocity distributions near the wake of wavy cylinders are calculated. As shown in Figure 2, the base pressure coefficients of wavy cylinders are larger than those of the circular cylinder. The angle of maximum negative pressure coefficient of the saddle plane of wavy cylinders is smaller than that of the circular cylinder,

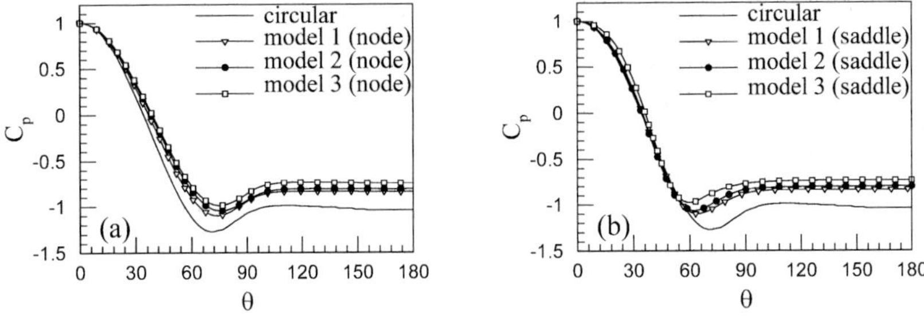

Fig. 2 Mean pressure coefficient distribution on wavy cylinder nodal (a) and saddle (b) planes compared with a circular cylinder.

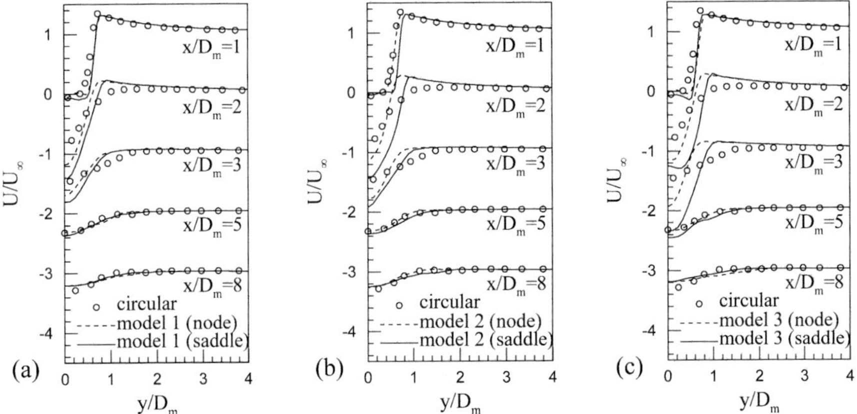

Fig. 3 Mean streamwise velocity distribution of different models at different locations in the wake of wavy cylinders compared with a circular cylinder at Re = 3000.

and is larger at the nodal plane. The pressure distribution indicates that the pressure drop is more rapid and the separation is earlier at the saddle plane than those at other spanwise locations of the wavy cylinders.

Furthermore, the wake vortex formation lengths of the wavy cylinders are longer than those of the circular cylinder (Figure 3). As a result, the mean drag coefficients of the wavy cylinders are less than that of circular cylinders (with reductions of up to 16% for model 3) and the fluctuating lift force is suppressed. Comparing with the circular cylinder, the mean streamwise velocity distribution of wavy cylinders at $x/D_m = 2$ and $x/D_m = 3$ is very different due to the three-dimensional nature of the vortex. The wake at the saddle plane is wider and longer than that at the nodal plane (Figures 3 and 4).

Due to a smaller separation angle, the wake width at the saddle plane is increased giving rise to a wide wake, while a larger separation angle at the nodal plane suppresses the shear layer development producing a narrower wake. Figure 5 shows that because of these complex three-dimensional effects, the three-dimensional vor-

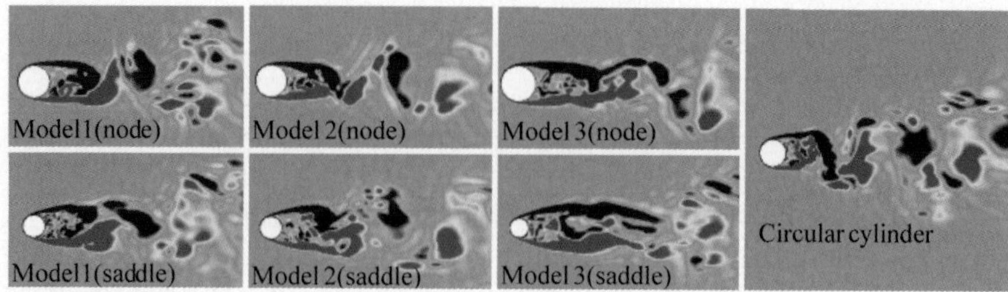

Fig. 4 The contours of instantaneous spanwise vorticity of wavy cylinders compared with a circular cylinder at Re = 3000.

Fig. 5 Velocity fields over isovorticity surfaces of wavy and circular cylinders. (a) $a/D_m = 0.09$, $\lambda/D_m = 2.27$; (b) $a/D_m = 0.09$, $\lambda/D_m = 1.52$; (c) $a/Dm = 0.15$, $\lambda/D_m = 2.27$; (d) $a/D_m = 0$, $\lambda/D_m = \infty$.

tex sheet of wavy cylinders rolls up into a mature vortex further downstream from the cylinder compared with the circular cylinder. All these effects will give rise to drag reduction and vibration suppression.

4 Conclusions

Three kinds of wavy cylinder surfaces of different a/D_m and λ/D_m were shown to generate peculiar three-dimensional vorticity structures which significantly modify the near wake structures and increase the vortex formation lengths behind the wavy cylinders. As a result, the vortex shedding behind the wavy cylinders is weakened and the base pressure of the cylinder increases. The simulations explain why cylinders with certain spanwise waviness have a significant effect on drag reduction and a corresponding suppression of vibration.

Acknowledgment

The authors wish to thank the Research Grants Council of the Hong Kong Special Administrative Region, China, for its support through Grant No. PolyU 5311/04E.

References

1. Ahmed, A. and Bays-Muchmore, B., Transverse flow over a wavy cylinder, *Physics of Fluids A* **4**(9), 1992, 1959–1967
2. Ahmed, A., Khan, M.J. and Bays-Muchmore, B., Experimental investigation of a three-dimensional bluff-body wake, *AIAA Journal* **31**, 1993, 559–563.
3. Lam, K., Wang, F.H. and So, R.M.C., Three-dimensional nature of vortices in the near wake of a wavy cylinder, *Journal of Fluids and Strucutres* **19**, 2004, 815–833.
4. Lam, K., Wang, F.H. and So, R.M.C., Experimental investigation of the mean and fluctuating forces of wavy (varicose) cylinders in a cross-flow, *Journal of Fluids and Strucutres* **19**, 2004, 321–334.
5. Zhang, W., Dai, C. and Lee, S.J., PIV measurements of the near-wake behind a sinusoidal cylinder, *Experiments in Fluids* **38**, 2005, 824–832.

Passive Multiscale Flow Control by Fractal Grids

R.E.E. Seoud and J.C. Vassilicos
Turbulence, Mixing and Flow Control Group, Department of Aeronautics, Imperial College London, London SW7 2AZ, U.K.; E-mail: richard.seoud@imperial.ac.uk

Abstract. Wind tunnel grid-generated turbulence based on a fractal square motif has been studied via hot-wire anemometry, at several free-stream velocities. The aim of the exercise was to see whether a fractal motif could enable multiscale flow control such that the 'objective functions' are the nature of the turbulence decay and its non-dimensional turbulence energy dissipation rate, C_ε. The outcome is that such a grid architecture does provide a unique homogeneous isotropic decaying turbulence field where the decay is exponential and the non-dimensional turbulence energy dissipation rate field evolves inversely with the Taylor–Reynolds number.

Key words: Fractal square grids, non-dimensional dissipation rate, turbulence decay.

1 Introduction

Multiscale flow control is a new concept whereby turbulent eddies are passively or actively forced into a flow by a distribution of objects or actuators of various sizes (as in the fractal objects and grids of [1,3,5] or where eddies of various sizes are fully controlled so as to maintain a flow that is laminar but with turbulent-like properties (as in the multiscale electromagnetic forcing of [4]) or where, in general, multiscale distributions of eddies are passively or actively forced up and down in intensity, position and/or time, in an overall laminar, transitional or turbulent state. The simultaneous manipulation of coexisting eddies of a broad range of sizes can dramatically alter interscale energy transfers as well as the dynamics, stretching and alignment of vorticity and strain rates [2], as well as pressure and acceleration fields (and thereby momentum transfers, in particular). Here we focus on planar fractal grids used in a wind tunnel to generate turbulence. These grids generate broad range turbulence energy spectra and unusual as well as controllable turbulence build-up and decay rates.

J.F. Morrison et al. (eds.), IUTAM Symposium on Flow Control and MEMS, 421–425.
© 2008 *Springer. Printed in the Netherlands.*

They also present unusual potential for many applications particularly because they allow the independent control of pressure drop and turbulence intensity.

2 Experimental Analysis

Detailed laboratory measurements and scaling studies carried out in our group [1] and independently by ourselves with a total of 21 planar fractal grids belonging to three different fractal families (fractal cross, fractal I and fractal square grids) in two different wind tunnels (open circuit and closed circuit) have shown that turbulence decay can be controlled by controlling specific parameters defining multiscale grids such as the fractal dimension D_f and the number of fractal iterations (which together determine an effective mesh size M_{eff}) and the ratio t_r of largest to smallest bar thicknesses on the grid. Specifically, in the case of fractal cross grids, the turbulence intensity u'/U scales and decays as

$$(u'/\langle U_{\text{inf}}\rangle)^2 = t_r^2 C_P f(x/M_{\text{eff}}), \tag{1}$$

where x is the streamwise distance from the grid and C_P is a normalised static pressure drop across the grid. In the case of fractal I grids,

$$(u'/\langle U_{\text{inf}}\rangle)^2 = t_r(T/L_{\text{max}})^2 C_P f(x/M_{\text{eff}}), \tag{2}$$

where T is the tunnel cross-sectional width and L_{max} is the maximal length on the grid). Note that turbulence intensity can be doubled without modifying the pressure drop by increasing the ratio of largest to smallest thicknesses on the fractal I-grid by 4. In the case of fractal square grids, the turbulence intensity first increases until a distance x_{peak},

$$x_{\text{peak}} = 75 t_{\text{min}} T/L_{\text{min}} \tag{3}$$

(where t_{min} and L_{min} are the minimal bar thicknesses and lengths on the grid) is reached downstream from the grid beyond which the turbulence decays fully within a distance l_{turb} from x_{peak} which is also controllable, for example by the mean speed $\langle U_{\text{inf}}\rangle$ as l_{turb} is directly proportional to $\langle U_{\text{inf}}\rangle$. Note that if the smallest thickness of the bars (millimetres) on the fractal grid is doubled, the distance downstream (metres) where the turbulence builds up is also doubled. The bulk of the present contribution is concerned with fractal square grids in a wind tunnel of size 0.46 m $\times$ 0.46 m $\times$ 4.5 m, with measurements made via hot wire anemometry. Free stream velocities ranged from 7 to 22 m/s, single wire and x-wire measurements were taken on the centre line as well as off the centre line in the decay region $x > x_{\text{peak}}$. Various velocity profiles have been obtained for the purpose of documenting the flow as completely as possible, in particular its small-scale and large-scale isotropy properties and absence/presence of turbulence production. Reynolds numbers Re_λ based on the Taylor micro-scale are unusually high for such a small wind tunnel and range between $0(100)$ and 1000 in the decay region. A particu-

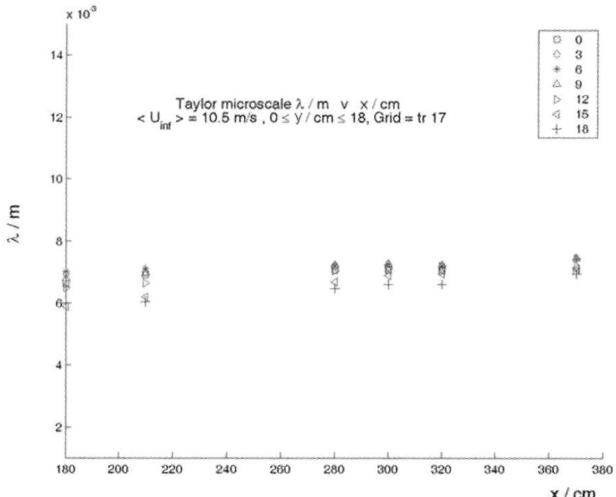

(a) Taylor microscale versus downstream distance at 10.5 m/s

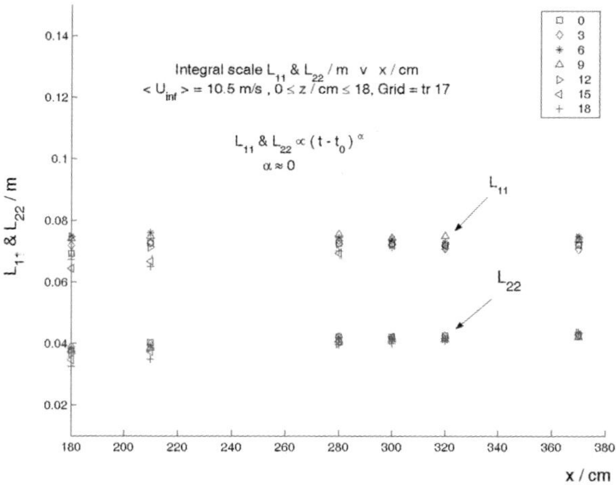

(b) Integral Scales, longitudinal,L_{11} and transverse, L_{22} versus downstream distance at 10.5 m/s

Fig. 1 Length scales.

larly striking observation is that fractal square grids can modify turbulence decay to the point that the integral and the Taylor time-scales remain about constant during decay (Figures 1a and 1b). This implies an exponential turbulence decay in homo-

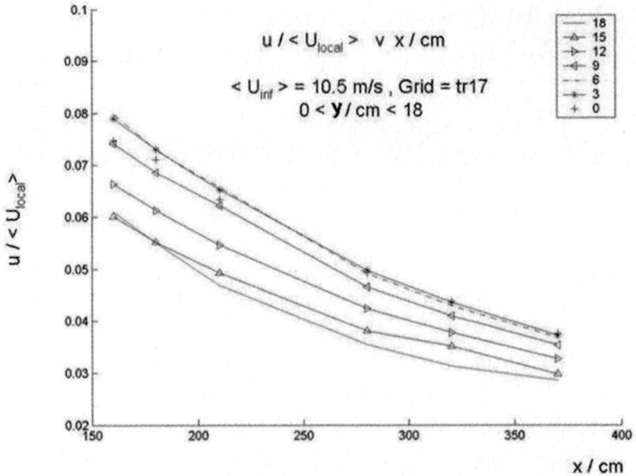

Fig. 2 Turbulence intensity $u/\langle U_{\text{local}}\rangle$ versus downstream distance, x/cm, at 10.5 m/s and at 7 stations across.

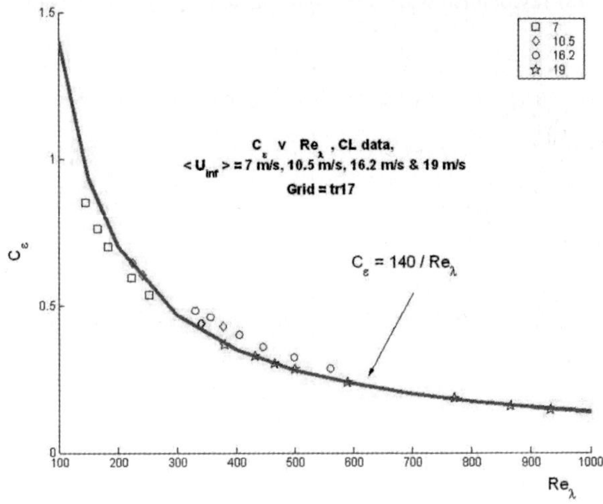

Fig. 3 Normalised dissipation energy versus Re_λ, at different free stream velocities.

geneous isotropic turbulence, and our decay results fit such an exponential very well (Figure 2).

Nevertheless, energy spectra have well-defined broad $-5/3$ power-law shapes over a broad range of times during decay. The only way in which these results can be made to be consistent with each other theoretically, assuming homogeneous isotropic turbulence decay, is for the spectra obtained at different stages of decay to collapse with the use of only one length-scale. This proves possible here because the integral and Taylor scales remain proportional to each other during decay; this is also observed in our measurements (Figures 1a and 1b). Finally, these observations and the hypothesis of homogeneous isotropic turbulence imply that the non-dimensionalised kinetic energy dissipation rate per unit mass is inversely proportional to Re_λ, something which also agrees very convincingly with our measurements (Figure 3). Hence, passive control by fractal grids causes the turbulence to become increasingly non-dissipative as the Reynolds number increases.

3 Conclusion

Passive multiscale flow control using fractal square grids has the capability to deliver a unique decaying homogeneous isotropic turbulence field. Two important properties of such a field are (i) that the turbulence intensity decays exponentially and (ii) that the non-dimensional turbulence energy dissipation rate, C_ε, evolves inversely with the Taylor Reynolds number, Re_λ.

References

1. D. Hurst and J.C. Vassilicos. Scalings and decay of fractal generated turbulence. *Phys. Fluids* **19**(3), 2007, 035103.
2. B. Mazzi and J.C. Vassilicos. Fractal generated turbulence. *J. Fluid Mech.* **502**, 2004, 65–87.
3. D. Queiros-Conde and J.C. Vassilicos. Turbulent wakes of 3D fractal grids. In J.C. Vassilicos (Ed.), *Intermittency in Turbulent Flows and Other Dynamical Systems*, Cambridge Press, 2001, pp. 136–167.
4. L. Rossi, J.C. Vassilicos, and Y. Hardalupas. Electromagnetically controlled multi-scale flow. *J. Fluid Mech.* **558**, 2006, 207–242.
5. A. Staicu, B. Mazzi, J.C. Vassilicos, and W. van de Water. Turbulent wakes of fractal objects. *Phys. Rev. E* **67**(6), 2003.

Hydraulic Model of the Skin Friction Reduction with Surface Grooves

Bettina Frohnapfel[1], Peter Lammers[2], Jovan Jovanović[1] and Antonio Delgado[1]
[1]*Institute of Fluid Mechanics, Friedrich-Alexander University
Erlangen-Nuremberg, Cauerstr. 4, 91058 Erlangen, Germany;
E-mail: bettina@lstm.uni-erlangen.de*
[2]*High Performance Computing Center Stuttgart, Nobelstr. 19, 70569 Stuttgart,
Germany*

Abstract. The reduction of skin friction in turbulent flows holds considerable promise for energy savings. The present work shows how and why skin friction and the dissipation are interrelated in turbulent channel flows. A hydraulic model formulation is presented for the skin friction reduction that can be obtained with a surface structure recently proposed for flow control. The model predictions are validated with results from direct numerical simulations.

Key words: Flow control, skin friction reduction, surface structures.

1 Skin Friction Reduction

It is well known that the skin friction coefficient

$$c_f = \frac{\tau_w}{\frac{\rho}{2}U_b^2} \tag{1}$$

(where τ_w is the wall shear stress and U_b the bulk flow velocity) suddenly increases when laminar to turbulent transition occurs. The transition delay or the reduction of the skin friction in turbulent flows therefore holds significant potential for energy savings. Employing the momentum and energy conservation for a turbulent channel flow of channel height H, it can be shown that c_f is related to the average total dissipation $\bar{\varepsilon}$ by:

$$\bar{\varepsilon} \simeq c_f \frac{U_b^3}{H}, \tag{2}$$

where $\bar{\varepsilon}$ is given as follows:

J.F. Morrison et al. (eds.), IUTAM Symposium on Flow Control and MEMS, 427–431.
© 2008 *Springer. Printed in the Netherlands.*

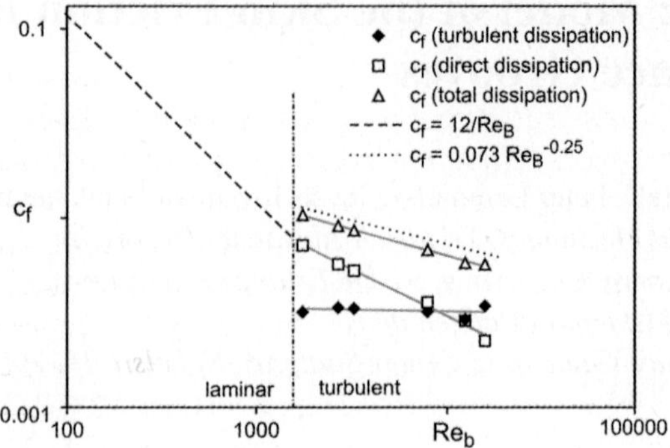

Fig. 1 Friction coefficient as a function of bulk Reynolds number in the laminar and turbulent flow regimes.

$$\overline{\varepsilon} = \frac{1}{V}\int_V \underbrace{v\left(\frac{\partial \overline{U}_1}{\partial x_2}\right)^2}_{\text{direct dissipation}} + \underbrace{v\overline{\frac{\partial u_j}{\partial x_i}\frac{\partial u_j}{\partial x_i}}}_{\text{turbulent dissipation}} dV. \tag{3}$$

Figure 1 shows a plot of c_f, based on $\overline{\varepsilon}$, versus Reynolds number, $Re_b = U_b H/v$ for turbulent channel flows at different Reynolds numbers [1,5,6,8,9] including the theoretical result for a laminar channel flow:

$$c_f = \frac{12}{Re_b} \tag{4}$$

and the typical correlation for a turbulent channel flow [2]:

$$c_f = 0.073 Re_b^{-0.25}. \tag{5}$$

It can clearly be seen that an increase of c_f at high Reynolds number is caused by turbulent dissipation. In order to obtain energy savings in the high Reynolds number regime, it is therefore necessary to reduce the turbulent dissipation. In [3] it is shown that the turbulent dissipation can be significantly reduced by forcing the turbulent fluctuations in the near-wall region to be predominantly one-component. In order to realize this state of turbulence at the wall a surface topology with grooves aligned in the mean flow direction as shown in Figure 2(left) is proposed. Inside the grooves, fluctuations in the spanwise direction are suppressed such that their intensity is almost identical to the one found in the wall-normal direction. The fluctuations in the streamwise direction are not restricted to grow by the surface topology and are therefore the most dominant ones inside the grooves.

1.1 Formulation of the Hydraulic Model

To estimate the performance of the proposed surface a hydraulic model is formulated. In [4] it is shown that a trend towards one-component turbulence is commonly found for different existing drag reduction techniques and that this trend resembles the one for decreasing Reynolds numbers in uncontrolled turbulent channel flows. Based on these results it can be concluded that we may expect relaminarization of a turbulent flow near the wall if two of the three velocity components are fully suppressed. The hydraulic model formulation is therefore based on the assumption that the flow inside the grooves will laminarize if these are of the same order as the sublayer thickness. The skin friction within the grooves is given by Equation (4), with the Reynolds number based on the characteristic length and velocity scales (which are on the order of the groove dimensions) and the wall shear velocity on the wall sections between the grooves where Equation (5) holds. Thus the main contribution to the skin friction arises from the wall sections which separate the grooves from each other. The resulting drag reduction DR for the grooved surface is therefore given by:

$$DR = 1 - \frac{\tau_{w,\text{grooved}} A_{\text{grooved}}}{\tau_{w,\text{smooth}} A_{\text{smooth}}}, \tag{6}$$

where τ_w is the wall shear stress of the grooved surface structure and the smooth surface of a regular turbulent channel flow, respectively, and A denotes the corresponding surface areas. Figure 2 (right) shows the model predictions for a surface structure in which the grooves are separated by a distance of three times their width ($b = 3a$). Drag reduction asymptotically tends to 25% for very high Reynolds numbers when the skin friction inside the grooves can be neglected:

$$DR_{\text{max}} \approx 1 - \frac{b}{a+b} = \frac{a}{a+b}. \tag{7}$$

1.2 Comparison with Computational Results

In order validate the model description, the obtained drag reduction is compared with results of direct numerical simulations of a channel flow with grooved surfaces. These simulations are based on the lattice-Boltzmann method. A detailed description of the employed simulation technique can be found in [7] where the applicability and accuracy of this method for channel flow simulations is discussed in detail. Simulations were carried out at two different Reynolds numbers; the details of the simulations are given in Table 1. Based on the obtained velocity profiles, the wall shear stress is calculated and the resulting drag reduction is obtained by comparing with a channel flow without grooves (Equation (6)). The results are included in Figure 2 (right) and show good agreement with the model predictions.

Table 1 Overview of the simulations carried out for the channel flow with grooves where Re_c is the Reynolds number based on centerline velocity and Re_τ the one based on the wall shear velocity.

Re_c	Re_τ	resolution $N1 \times N2 \times N3$	resolution in wall units	a^+	b^+	c^+
				in wall units		
6650	187	$4096 \times 264 \times 240$	1.4	5	14	5
5230	148	$4096 \times 364 \times 360$	0.8	4	15	4

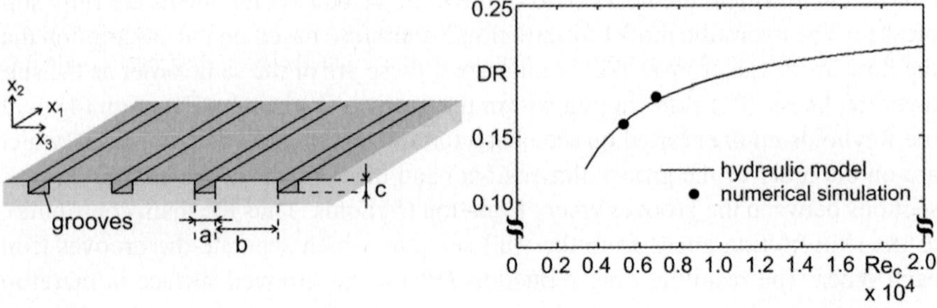

Fig. 2 Surface structure with grooves aligned in the flow direction (left). Drag reduction of the grooved surface based on the hydraulic model in comparison with numerical results (right).

2 Outlook and Conclusion

A hydraulic model is presented for the drag reducing performance of a grooved surface structure which was designed to minimize the turbulent dissipation in the flow. The obtained model predictions show very good agreement with the drag reduction obtained in direct numerical simulations. Future work will concentrate on experimental verification [10] of the obtained drag reduction and more refined numerical investigations.

References

1. R.A. Antonia, M. Teitel, J. Kim, L.W.B. Browne: Low-Reynolds-number effects in a fully developed turbulent channel flow. *J. Fluid Mech.* **236** (1992) 579–605.
2. R.B. Dean: Reynolds number dependence of skin friction and other bulk flow variables in two-dimensional rectangular duct flow. *J. Fluids Engrg.* **100** (1978) 215–223.
3. B. Frohnapfel, P. Lammers, J. Jovanović: The role of turbulent dissipation for flow control of near-wall turbulence. In *Notes on Numerical Fluid Mechanics and Multidisciplinary Design*, C. Tropea, S. Jarkilic, H.-J. Heinemann, R. Henke, H. Hönlinger (Eds.), Springer, Berlin (2007) in print.
4. B. Frohnapfel, P. Lammers, J. Jovanović, F. Durst: Interpretation of the mechanism associated with turbulent drag reduction in terms of anisotropy invariants. *J. Fluid Mech.* **577** (2007) 457–466.

5. J. Kim, P. Moin, R. Moser: Turbulence statistics in a fully developed channel flow at low Reynolds numbers. *J. Fluid Mech.* **177** (1987) 133–166.
6. A. Kuroda, N. Kasagi, M. Hirata: A direct numerical simulation of the fully developed turbulent channel flow. In *Proceedings of International Symposium on Computational Fluid Dynamics*, Nagoya, Japan (1989) pp. 1174–1179.
7. P. Lammers: Direct numerical simulations of wall-bounded flows at low Reynolds number with the lattice-Boltzmann method. Ph.D. Thesis, University of Erlangen-Nuremberg (2004) [in German].
8. R.D. Moser, J. Kim, N.N. Mansour: Direct numerical simulation of turbulent channel flow up to $Re_\tau = 590$. *Phys. Fluids* **11** (1999) 943–945.
9. R. Volkert: Determination of statistical turbulence quantities for a turbulent channel flow based on direct numerical simulations. Ph.D. Thesis, University of Erlangen-Nuremberg (2006) [in German].
10. B. Frohnapfel, J. Jovanović, A. Delgado: Experimental investigation of turbulent drag reduction by surface embedded grooves. *J. Fluid Mech.* (2007) in print.

Vortex Shedding behind a Tapered Cylinder and Its Control

O.N. Ramesh and R.S. Chopde
*Department of Aerospace Engineering, Indian Institute of Science, Bangalore,
India; E-mail: onr@aero.iisc.ernet.in*
2Defence Research & Development Organisation, Hyderabad, India

Abstract. The vortex shedding phenomenon behind a tapered cylinder is sought to
be controlled by means of a placement of a smaller cylinder placed outside the wake
of the main cylinder. The quenching of vortex shedding is evident from flow visualisation and hotwire anemometry studies and this is attributed to the suppression of
global instability modes.

Key words: Vortex shedding, wake, control, global mode.

1 Introduction

Strykowski and Sreenivasan [1] studied the low Reynolds number vortex shedding
phenomenon behind a circular cylinder and sought to control the wake vortex shedding by placing a small cylinder (control cylinder) placed in the wake of a bigger
main cylinder; both these cylinders were of uniform cross section all across the span
(2D). It was observed by them that shedding can be altered and even suppressed altogether over a limited range of Reynolds number, by proper placement of the control
cylinder outside the wake. The main import of their work was to associate an absolute instability mode (with its typical large temporal growth rates) with Karman
vortex shedding. It was argued by them that the control cylinder had the effect of altering the local stability of the flow by smearing and diffusing concentrated vorticity
in the shear layers behind the body.

We extend the same methodology of control to 3D flow behind cones in this
study. The premise behind our study is that if we somehow understand the control
methodology for 3D vortex shedding then we somehow understand more of the
3D vortex shedding itself and clarify even the physics of 2D vortex shedding also.
It must be noted that in the present study, we seek to control a very strongly 3D
vortex shedding phenomenon but still with a 2D control cylinder (in Strykowski
and Sreenivasan's experiments both the main cylinder and control cylinder are 2D

J.F. Morrison et al. (eds.), IUTAM Symposium on Flow Control and MEMS, 433–436.
© 2008 *Springer. Printed in the Netherlands.*

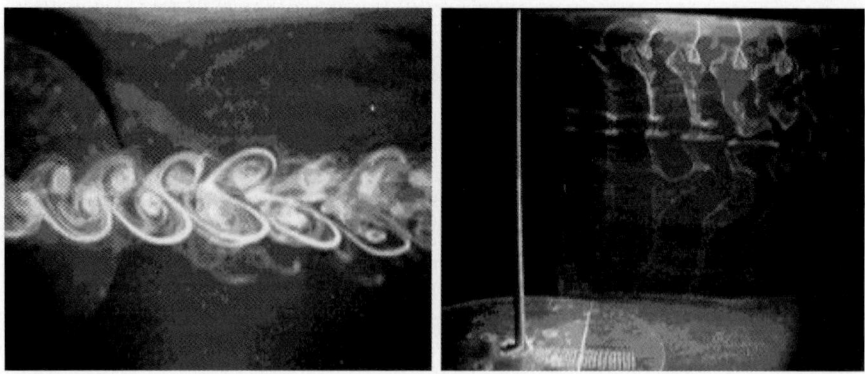

Fig. 1 Shedding – left: top view; right: spanwise view.

in geometry). In our present case, one could then expect the 2D control cylinder to alter the growth rates by different amounts along the span since the local Reynolds number is continually changing across the span of the cone. As a consequence one could perhaps expect partial quenching over part of the span whereas there is still shedding over the remaining part of the span due to the control rod. This expectation turns out to be over-conservative in that for certain values of Reynolds numbers (based on the mean diameter), vortex shedding over the whole span is quenched in a dramatic fashion.

2 Experimental Set-Up and Results

The dimensions of the cone are: spanwise length is 266 mm, taper ratio $T = 53$, a fairly steep cone. The Reynolds number of the flow (based on the mid span diameter) is about 72. When there is no control cylinder, vortex shedding is present as can be seen from the smoke flow visualisations below. Figure 1a shows the top view and Figure 1b shows the spanwise view of the phenomenon of low Reynolds number vortex shedding over the cone.

Now following the lead of Strykowski and Sreenivasan [1], we introduce a much smaller control cylinder outside the wake of the main cylinder. The control cylinder has a constant diameter ($d = 1.2$ mm) all across the span ($l = 266$ mm), i.e., it is 2D. The control cylinder is placed outside the wake such that its streamwise location $X = 0.8Dm$ and the normal distance $Y = 1.1Dm$, where Dm is the mean diameter of the cone.

Figures 2a and 2b respectively show the top and the spanwise view of the resulting flow pattern with the introduction of a control cylinder. It can be very clearly seen that there is complete quenching of the vortex shedding phenomenon all across the span. Hot wire measurements also corroborate these findings. This quenching is

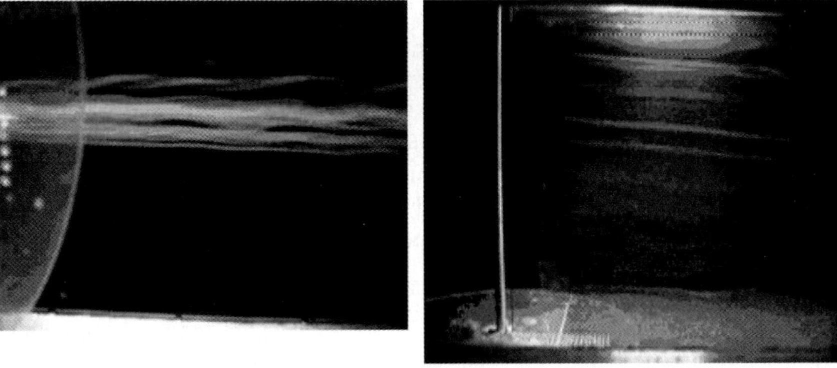

Fig. 2 Quenching; left: top view; right: spanwise-view.

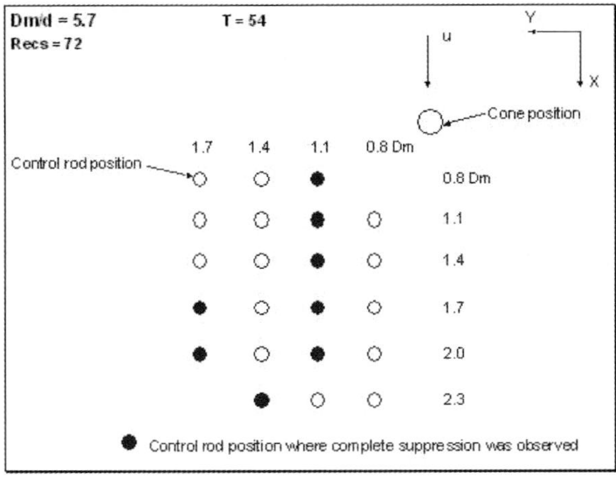

Fig. 3 Domain of control rod placement for suppression of vortex shedding.

seen to occur for a selective range of control rod placement in relation to the main conical cylinder position as shown in Figure 3.

It may at first sight appear surprising that a 2D control cylinder should be successful in suppressing a very complex 3D shedding phenomenon. This is because vorticity smearing due to the control rod (envisaged in the 2D scenario of Strykowski and Sreenivasan hypothesis) may be expected to be non-uniform all over the span in a 3D geometry as the gap between the main cylinder and the conrol cylinder varies with the span. Hence, it is clear that an explanation must be sought for this control methodology not on such local kinematic arguments but on global stability arguments. Moreover, since a naïve local Reynolds number argument such as above does not take the global nature of the instability modes into account, it leads to modest expectations from the control methodology.

The vortex shedding phenomenon in a 2D flow is known to be dominated by the so-called global modes. These are instability modes with a zero group velocity and the same frequency all over the wake. Theoretical studies, which use model equations (such as Ginzburg–Landau equation) to study the vortex shedding phenomenon with spanwise variations in geometry [3], indicate that a spanwise shear weakens the global mode structure. This means that, compared to the wake of a 2D cylinder, the wake of a cone is associated with a weaker global mode. Hence, the presence of a control rod in a 3D wake perhaps dampens the growth rate readily in an already weakened instability mode. This might perhaps explain the success of a 2D control cylinder in suppressing a 3D vortex shedding as shown above. In other words, the control rod works on the global instability modes and alters their growth rates all across the span even when the gap between the cone and the control cylinder varies along the span.

It is noteworthy that in [1] experiment suppression was possible only up Reynolds number of 80 but in the control experiments for tapered cylinder it was possible to suppress vortex shedding even when Reynolds number based on local diameter of tapered cylinder was 112 at the thicker end and 52 at the thinner end. This too suggests that an already weakened global mode was suppressed even at higher Reynolds numbers.

The present study has three major implications: (1) it demonstrates and explains the control methodology in cone wakes; (2) it confirms the existence of a global mode in the cone wake and indicates that spanwise shear does weaken this global mode structure; and (3) it clarifies the control methodology and the physics of the Strykowski and Sreenivasan [1] experiment.

References

1. Strykowski, P. and Sreenivasan, K.R., On the formation and suppression of vortex shedding at low Reynolds numbers. *J Fluid Mech.* **218**, 1989, 71–107.
2. Huerre, P. and Monkewitz, P.A., Local and global instabilities in spatially-developing flows. *Ann. Rev. Fluid. Mech.* **22**, 1990, 473.
3. Albarede, P. and Monkewitz, P., A model for the formation of oblique shedding patterns and "chevrons" in cylinder wakes. *Phys. Fluids A* **4**, 1992, 744–756.

Control of a Separated Flow over a Smoothly Contoured Ramp Using Vortex Generators

Thomas Duriez, Jean-Luc Aider and Jose Eduardo Wesfreid
*Laboratoire PMMH UMR CNRS 7636, ESPCI, 10 rue Vauquelin, 75231 Paris
Cedex 05, France; E-mail: aider@pmmh.espci.fr*

Abstract. In this communication, we study experimentally the modification of a
boundary layer by a line of four cylindrical vortex generators. We show how the
base flow modification and, specifically, the first non-linear mode (the zeroth mode)
can lead to new parameters characterizing the efficiency of the vortex generators.
We finally apply the vortex generators to a smoothly contoured ramp showing a
clear delay of the separation when the parameters are properly chosen.

Key words: Separated flow, contoured ramp, vortex generators.

1 Introduction

Vortex generators (VG) are well known as eficient tools for the control of flow separation [1, 2]. They are commonly used in various industrial applications (e.g. in
aeronautics, to enhance airplane lift force in near stall situations; or in chemical industry to increase the efficiency of static mixers). Nevertheless, the question of the
choice and design of the proper VG for a given application is still an open problem.
As a matter of fact, many different perturbations of the boundary layer can create
streamwise vortices, but the eficiency of the system will depend on many parameters. Our objective is to find some generic criteria to characterize the eficiency of a
set of vortex generators.

2 Experimental Set-Up

We use a low-speed water tunnel made of plexiglas to allow optical measurement
from any direction, including downstream. The flow is driven by gravity using a
water reservoir kept to a constant height. The rectangular cross section of the tunnel
is 140 mm high and 150 mm wide in the ramp configuration and 100 mm high and
150 mm wide in the flat plane configuration. The test-section is 800 mm long and

J.F. Morrison et al. (eds.), IUTAM Symposium on Flow Control and MEMS, 437–441.
© *2008 Springer. Printed in the Netherlands.*

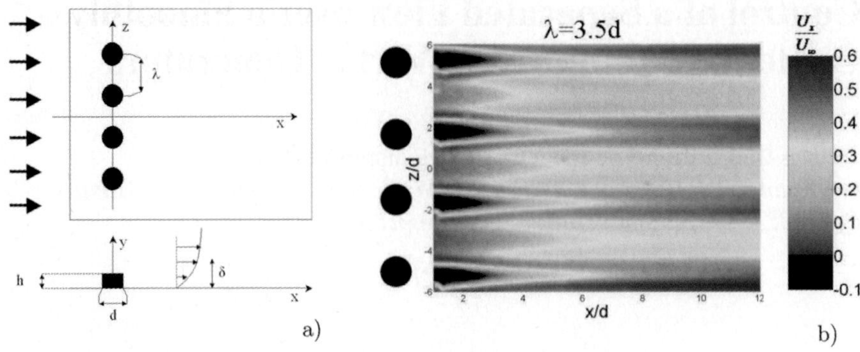

Fig. 1 (a) Definition of the axes and parameters describing the vortex generators line. (b) Contours of longitudinal velocity measured downstream of the CVG for $z = 3$ mm and $\lambda = 3d$.

can be changed from a flat plate to a smooth contoured ramp with a curved cross-section which has a constant radius $R = 100$ mm and a height $H = 40$ mm.

The mean freestream velocity is $U_\infty = 0.04$ ms^{-1}. The typical boundary layer thickness at the beginning of the test section is about 10 mm. The four vortex generators are located 80 mm upstream the curved ramp for the separated flow and 100 mm downstream from the beginning of the test section for the flat plate study. As shown in Figure 1a, the origin is taken at the beginning of the curved ramp for the separated flow and at the lower wall, at the symmetry point of the VG line, for the flat plane study. We use four small cylindrical vortex generators (CVG) defined by their diameter $d = 8$ mm and their height $h = 6$ mm. The spacing between the VG, or wavelength λ, is fixed and equal to $3d$.

3 Flat Plate Boundary Layer Modification

Preliminary studies have been made on a flat plate in order to investigate the modification of the boundary layer downstream of the four CVGs. We make a 3D reconstruction of the boundary layer using two-components PIV measurements in different planes parallel to the wall. We show in Figure 1b a typical longitudinal velocity contour downstream of the CVG for $z = 3$ mm. It clearly shows spanwise modulations with inflow (accelerated, in red) and outflow (decelerated, in light blue) regions which can be interpreted as non-linear perturbations. When the inflow regions are large, the result is a global decrease of the boundary layer thickness and then a global increase of the boundary layer velocity gradient. This should induce a modification of the base flow.

For this Reynolds number the flow is steady and can be separated in different linear and non-linear perturbations:

$$U(x,y,z) = U_{\text{base}}(x,y) + u_L(x,y,z) + u_{nL}(x,y,z), \tag{1}$$

where U_{base} is the base flow boundary layer, u_L is the first harmonic perturbation (wavelength λ), while u_{nL} is the non-linear perturbation. We can write the different perturbations as Fourier decompositions, averaged along the z direction on only two wavelength (between $z = -24$ mm and $z = 24$ mm) to avoid most boundary effects:

$$u_L(x,y,z) = u_1^*(x,y)\exp\left(i\frac{2\pi}{\lambda}z\right), \tag{2}$$

$$u_{nL}(x,y,z) = \sum_{k\geq 0, k\neq 1} u_i^*(x,y)\exp\left(ik\frac{2\pi}{\lambda}z\right), \tag{3}$$

$$\langle U(x,y)\rangle_z = U_{\text{base}}(x,y) + u_0^*(x,y), \tag{4}$$

where $U(x,y,z)$ is the measured flow-field with VG, while $U_{\text{base}}(x,y)$ is the measured base flow *without* VG. The mean flow modification (or zero-mode) $u_0^*(x,y)$ is then calculated after z-averaging the flowfield with VG on a multiple of the forcing wavelength and substracting the measured base flow. The vertical profiles $u_0^*(x,y)$ computed for every x position can be integrated over the vertical direction y to give a global estimate of the zeroth mode along the x direction:

$$E_0(x) = \int_0^\infty u_0^*(x,y)dy. \tag{5}$$

From the longitudinal evolution of E_0 we can evaluate different quantities that could be useful to choose some critical parameters, like the minimum longitudinal distance from the separation the VG should be placed. As a matter of fact, E_0 can be seen as a measure of the balance between inflow and outflow regions: if $E_0 > 0$, the flow is dominated by inflow regions, and respectively outflow regions for negative values. We call this value the "inversion length" L_{inv} and we show its evolution as a function of the spacing of the VG in Figure 2. We see that the case $\lambda = 3d$ leads clearly to the shorter L_{inv}. For this spacing, we know that the flow will be dominated by the inflow regions and should lead to a delay of the separation.

4 Application to a Smoothly Contoured Ramp

In this section, we briefly show the influence of the line of CVG over a separated flow over a smoothly contoured ramp. We choose the spacing $\lambda = 3d$ and put the CVG $10d$ upstream from the beginning of the curved ramp so that the flow will be clearly dominated by the inflow regions when it reaches the separation line. We show the result in Figure 3: the natural separation is delayed by 20%, i.e. 5.5 mm. This result seems to confirm that the inversion length is a good parameter to choose the location of the VG line.

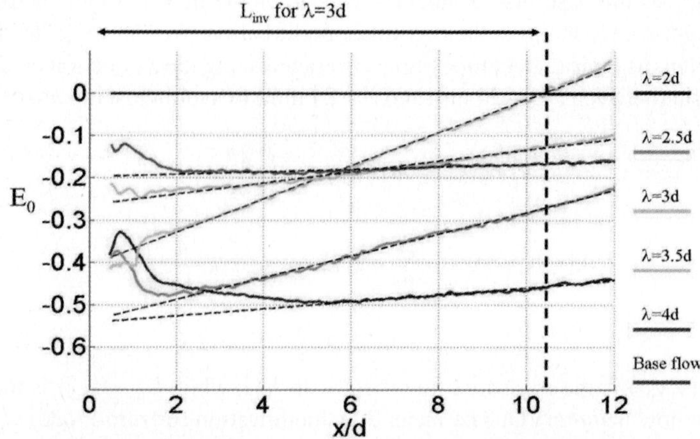

Fig. 2 Evaluation of the inversion length Linv from the evolution of $E_0(x)$.

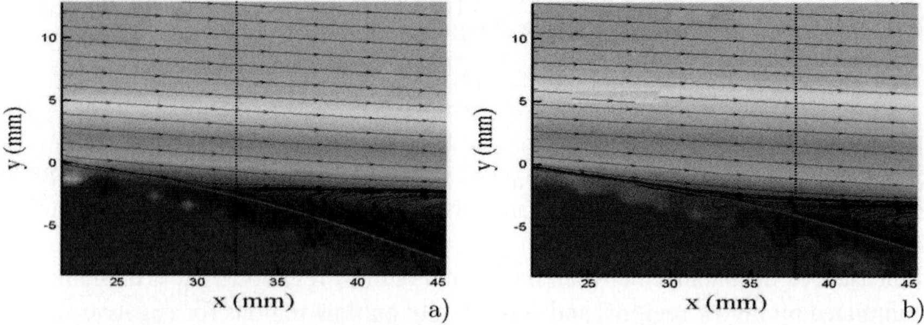

Fig. 3 Streamlines of time and space averaged flow over a smoothly contoured ramp without VGs (a) and with VGs (b). The average separation line is delayed further downstream. The position of the separation is marked with the dotted lines.

5 Conclusion

Using experiments over a flat plate, we were able to propose different mechanisms involved in the modification of the separation. The non-linear analysis shows the existence of a zeroth mode, which is associated to a mean flow modification of the boundary layer. With the proper choice of the parameters and, especially, the right spacing, the boundary layer can become dominated by the inflow regions, and then the mean separation over a contoured ramp could be delayed. We propose a new characteristic parameter, the inversion length, which appears to be useful to evaluate the right location of the line of vortex generators upstream of the separation line.

References

1. Angele, K.: Experimental studies of turbulent boundary layer separation and control. Ph.D. Thesis of the Royal Institut of Technology, Stockholm, Sweden (2003).
2. Aider, J.L., Beaudoin, J.F., Wesfreid J.E.: Drag Reduction of a 3D Bluff Body Using Vortex Generators. In *Proceedings of 4th Conference on Bluff Body Wakes and Vortex Induced Vibrations*, June 21–24, Santorini, Greece (2005).

Biomimetic Flight and Flow Control: Learning from the Birds

Ranjan Vepa
*Department of Engineering, Queen Mary, University of London, Miles End Road,
London E1 4NS, U.K.; E-mail: r.vepa@qmul.ac.uk*

Abstract. In this paper we consider the various methods employed by birds to generate lift and control it. We focus on three particular aspects, namely the method that a bird employs to compensate the transport lags, the method of rapid lift generation employed and growth of lift to a steady value and finally the angle of attack at which a bird flies to generate maximum lift. Based on the study of these methods we establish mathematical control models for compensating the transport lags, and establish a constraint for the aeroelastic tailoring of a wing to maintain a steady angle of attack even when flexible modes of vibration are present. Finally the unsteady aerodynamic modeling of vortex flows for active control applications is discussed.

Key words: Biomimetic flow control, transport lag, vortex lift, aeroelastic tailoring.

1 Features of Bird Flight: Control Mechanisms and Strategies and Techniques Employed by Birds

The notion of biomimetic control, i.e. techniques that mimic biological animals in the way they exercise control, rather than just humans, has led to the definition of a new class of biologically inspired robots that exhibit much greater robustness in performance in unstructured environments than the robots which are currently being built. One of the objectives of biomimetic control is to surpass and go beyond bio-mimicry. Therefore we consider the various strategies adopted by birds in different phases of their flight. Birds are superior flying machines with multi-element aerofoils capable of controlling the flow around them quite effortlessly and there are several associated flow mechanisms.

(a) *Flight with prescribed effective AOA or Constant Lift*: The first mechanism associated with large aspect-ratio wings is to configure the flight feathers or *remiges* as a variable-camber aerofoil such that the effective angle-of-attack is always a

J.F. Morrison et al. (eds.), IUTAM Symposium on Flow Control and MEMS, 443–447.
© 2008 *Springer. Printed in the Netherlands.*

constant. It follows that the aerofoil operates so the potential flow component of the lift coefficient is almost constant irrespective of the speed of flight. It is then possible to operate at an optimal angle-of-attack without suffering from the adverse effects of flow separation. A bird is aware by virtue of its associative memory of the angle of attack at which it should fly in order to generate a particular magnitude of lift. Thus it operates at a prescribed angle of attack and any additional lift it may require is generated by controlling the vortex component of the lift.

(b) *Flapping translation in three dimensions*: In one sense birds are different from man-made aircraft in that some birds, such as the sparrowhawk, are able to clap and fling. This is a beating motion about an axis that is pitched forward. Rapid beating results in a quick increase in the angle of attack which in turn generates lift about an axis normal to the axis of beating. Thus it is able to generate a lift component equal to the weight of the bird and a thrust component that keeps it in aerial equilibrium and hover.

Another wing motion that is present in larger birds but is technically equivalent to clap and fling is flapping translation. The wing motion of a bird will generally consist of four fundamental motions: (1) A flapping motion in a vertical plane, (2) a lead-lag motion, which denotes a posterior and anterior motion in the horizontal plane, (3) a feathering motion, which denotes a twisting motion of the wing pitch, and (4) a spanning motion, which denotes an alternatively extending and contracting motion of the wing span. Flapping-translation involves a combination of the first three modes. A birds wing is constantly changing its relative forward velocity as it flaps up and down, slowing down at the ends of the downstroke and upstroke, and then accelerating into the next half stroke. Furthermore, the wing base will always be moving slower than the wing tip, meaning that the wing velocity increases from base to tip.

(c) *Vortex-lift control and maximisation*: A third mechanism, suitable for low aspect-ratio wings, is based on stable operation at a maximum constant lift with the conventional lift being augmented by a leading-edge suction type vortex. Moreover, birds have the ability to align the vortex lift in any direction so it can either be employed to enhance the lift or downforce, or act as a propulsive or braking force. Thus birds are able to fly at angles of attack as high as $35°$ to $40°$.

The manner in which a dragonfly controls the generation of vortex lift is probably the best example of this mechanism. In normal counter-stroking flight there is a leading-edge vortex present over the forewing and attached flow on the hindwing. The leading-edge vortex lift, though present in thin aerofoil theory, does not contribute to the lift as it is directed forward. The vortex lift is formed during rapid increases in angle of attack due to the existence of the rounded leading-edge and the aft movement of the leading-edge vortex. In steady flight these rapid increases in angle of attack occur during wing rotation at the start of the downstroke. However, dragonflies can go from attached flow to leading-edge vortex flow at any stage of the wing-beat by rapidly increasing their angle of attack.

Falcons such as the Hobby are able to accelerate rapidly in pursuit of their prey. They have kinked trapezoidal wings with the inner wing nearer the body being swept forward and the outer wings being swept backwards. They have the ability to soar rapidly to great heights while scanning large areas on the ground for food and are also able to glide at low speeds. Their wings are characterised by "emarginations" or gaps in the wing tips which act like wing tip fences. This allows the wing to sustain a higher magnitude of vortex lift over every wing strip in the vicinity of the leading-edge and nearer the bird's body. Removal of this leading-edge vortex over a spanwise segment, without stalling, is achieved by spanwise pressure gradients that cause the vortex to travel down the wing to the wing tip fence, where they are accumulated into a system of tip vortices that are shed safely. Thus the far end of the wing is employed to maximise the vortex lift component in a clever and imaginative way.

At high speeds birds use a complex strategy to clap, fling and beat their wings in such a way that the gaps between them are able effectively inhibit the loss of lift. The gaps are also controlled in real time to act as 'slats' and most birds are able swing a 'canard' like surface relative to the main wing. These features significantly influence the controllability of the associated vortex flows.

(d) *Eliminating the transport lag effects*: Compensating for the Wagner-like effect eliminates the transport lags without altering the circulatory forces acting on the aerofoil. Compensation for this effect involves employing a higher initial value at the start of an impulsive change in the angle of attack to generate a steady lift force. Both the high frequency clap and fling mode and the lower frequency flapping translation mode are capable of generating a steady lift force thus effectively compensating for the transport delays.

The dragonfly, for example, is able to rapidly increase its angle of attack by a process of high frequency beating, which actually involves a gradual increase in the amplitude accompanied by a relatively rapid increase in the beat frequency. This is completely equivalent to a sudden unit step change in the angle of attack. Thus the Wagner problem concerning the growth of lift due to a sudden change in the angle of attack is of fundamental importance. In fact, what is important is the angle of attack time history that is essential to generate a step change in the lift distribution, a problem we refer to as the inverse Wagner problem. Birds employ a pair of clap and fling or flapping wings to execute modes such as flapping-translation which effectively maintain constant lift and compensate for the Wagner effect [1].

(e) *Inhibiting the separation bubble's upstream travel*: Birds have an effective way to control the movement of the separation bubble; during the landing approach or in gusty winds, the feathers on the upper rear surface of bird wings tend to pop up. Usually the separation bubble starts to develop on a wing in the vicinity of the trailing edge and following this event, locally reversed flow begins to occur in the separation regime. Under these locally reversed flow conditions, light feathers would pop up, acting like a brake on the upstream travel of the separation bubble towards the leading-edge. The effect has been simulated in a wind tunnel [2], by attaching self popping porous movable flaps with notched trailing edges, to obtain equal static pressure on both sides of the flap under attached flow conditions. The flaps are attached in the rear part of the airfoil and could pivot on their leading-edges. Unlike active flow control techniques this is a passive technique and requires no active control.

By examining the various strategies for bird flight, airplane designers have established a number of relatively new techniques for flow control. They are responsible for the development of a number active and passive systems, and generally fall into certain standard classes: (i) stripwise control of wing aerodynamics; (ii) active shape control: controlling aerofoil thickness, camber and flap mode-shape; (iii) control of leading-edge droop; (iv) on-off control by mini-tabs; (v) active control of wing-tip fences. These are briefly discussed in the next section.

2 Biomimetically Inspired Active and Passive Techniques of Flow Regulation

A number of biomimetically inspired active and passive techniques of flow regulation are evolving in the literature based on bird flight. We identify some of the primary techniques here.

1. Active flow control of multi-element aerofoils for constant lift flight including the flexi-flap: The primary idea here is to either employ multi-element aerofoils to maintain constant lift or to employ variable camber and variable thickness aerofoils to maintain a constant effective angle of attack leading to a constant lift. The prediction of unsteady forces and moments on such aerofoils is based on representative thickaerofoil theories.
2. Active suppression of wing flutter by transport lag elimination: In this case the inverse of the classic Wagner problem, which involves finding the angle of attack variation so the growth of lift is constant, is employed to compensate for the effects of lags in the growth of lift. Active control laws for the total elimination of the transport lags may also be synthesised.
3. Aeroelastic tailoring of wings for constant AOA wing vibration: This is a passive concept which involves the application of constraints that must be met, so that the normal-modes are such that the wing motion is with a constant AOA. This effectively allows for the passive realisation of flapping translation and to a limited extent, the clap and fling motion.

3 Modeling and Control of the Vortex Lift

By far the most significant outcome of the studies of bird flight is to develop new approaches to modeling and control of vortex lift. The influence of the leading-edge vortex on the unsteady forces and moments has been ignored in the past, and can now be included by consideration of the dynamics of vortex sheet evolution in terms of the total circulation. Asymptotic studies of the initial growth of lift as well as in quasisteady situations in thick and cambered aerofoils , which capture the leading-edge vortex effect, can be combined to obtain realistic models for the vortex lift.

4 Conclusions

In this paper not only have some features of the dynamics of bird flight been identified, but the features have also been applied to propose a method of compensating for the delay in the growth of lift. The concept of aeroelastic tailoring is employed so the flexibility effects can contribute positively so as to maintain steady lift on the wings. Finally, studies of bird flight have provided new insights into the importance of the leading-edge vortex, which is ignored in the classical unsteady aerodynamic theories.

References

1. Sane, S.P., Review: The aerodynamics of insect flight, *The Journal of Experimental Biology* **206**(23), 2003, 4191–4208.
2. Schatz, M., Knacke, T., Thiele, F., Meyer, R., Hage, W. and Bechert, D.W., Separation control by self-activated movable flaps, AIAA-2004-1243, 2004.

Author Index

IUTAM Bookseries

springer.com

IUTAM Bookseries

1. [illegible] Sediment Sources, 2007.
2. [illegible] IUTAM Symposium on Dynamics and Control of [illegible] 2007. ISBN 978-1-4020-[illegible]
3. [illegible] IUTAM Symposium on [illegible] in Complex Geometry, 2007. ISBN [illegible]
4. [illegible] IUTAM Symposium on [illegible] for [illegible] 2007. ISBN 978-1-4020-[illegible]
5. [illegible] IUTAM Symposium on [illegible] Segregation, 2008.
6. [illegible] ISBN 978-1-4020-[illegible]
7. [illegible] IUTAM Symposium on [illegible] 2008. ISBN 978-1-4020-[illegible]
8. [illegible] 2008. ISBN 978-1-4020-[illegible]